2

New Seeds and Poor People

TITLES OF RELATED INTEREST

New Seeds and Poor People

Michael Lipton *with Richard Longhurst*

London
UNWIN HYMAN
Boston Sydney Wellington

Published by the Academic Division of
Unwin Hyman Ltd
15/17 Broadwick Street, London W1V 1FP, UK

Unwin Hyman Inc.,
8 Winchester Place, Winchester, Mass. 01890, USA

Allen & Unwin (Australia) Ltd, 8 Napier Street,
North Sydney, NSW 2060, Australia

Allen & Unwin (New Zealand) Ltd in association with the Port
Nicholson Press Ltd,
Compusales Building, 75 Ghuznee Street, Wellington 1, New Zealand

First published in 1989

British Library Cataloguing in Publication Data
Lipton, Michael, *1936–*
    New seeds and poor people.
    1. Developing countries. Food production
    industries. Effects of foreign assistance
    for food supply by developed countries.
    I. Title   II. Longhurst, Richard
    338.1'9'1724

    ISBN 0-04-445326-4
    ISBN 0-04-445327-2 pbk

Library of Congress Cataloging-in-Publication Data

Lipton, Michael.
    New seeds and poor people.
    Bibliography: p.
    Includes index.
    1. Rural poor—developing countries.   2. Farmers—Developing
    countries.   3. Agriculture—Economic aspects—Developing
    countries.   4. Green revolution—Developing countries.
    I. Longhurst, Richard.   II. Title.
HC59.72.P6L56   1989   338.1'09172'4   88-34745
ISBN 0-04-445326-4 (alk. paper)
ISBN 0-04-445327-2 (pbk.)

Typeset in Baskerville by Saxon Ltd, Derby, and printed in Great
Britain by Billing & Sons, London and Worcester.

Contents

Foreword and Acknowledgements

This book originated as a consultancy to Michael Lipton, in the context of a 1984–5 study directed by Dr Jock Anderson, and funded by the countries supporting the fourteen International Agricultural Research Centres (IARCs)* comprising the Consultative Group on International Agricultural Research (CGIAR). This 'Impact Study' aimed to assess how far the work of the IARCs – which concentrate almost entirely on food production – had been appropriate, technically and socio-economically, to improving the position of farmers, workers, and consumers in less developed countries (LDCs).

It soon became clear that – although the Impact Study could usefully ask how well the IARCs had stimulated *national* research institutions (and indeed several other consultants reported on that issue) – the joint contribution of the two groups of researchers could not be evaluated separately. Indeed, Lipton's terms of reference asked him to assess the impact of modern varieties of cereals – whether from the IARCs or from national centres – on poverty and income.

Richard Longhurst joined him in preparing the principal draft of the chapter in the consultancy report dealing with nutrition. He also supplied many comments and ideas for other chapters. During the preparation of the report (published as CGIAR Study Paper No. 2, 'Modern Varieties,

*A glossary of acronyms used in the text can be found on pp. xii–xiv.

International Agricultural Research, and the Poor', 1985), the authors benefited greatly from support and comments from many people, including Jock Anderson, Barbara Harriss, Bob Herdt, Polly Hill, Shiv Nath, Kutlu Somel and Don Winkelman.

The authors are very grateful to Barbara Taylor, Ann Watson, Lynette Aspillera and Ding Dizon for their hard and precise work in preparing various stages of the manuscript. Sara Crowley contributed excellent research assistance in the early stages of the project. Paul Richards has provided considerable encouragement and support, playing a major role in bringing the original consultancy report to the stage of a book.

The authors are grateful to the CGIAR for agreeing that the study could be used as the basis of a book. It has been totally rewritten, and updated to April 1988. Large sections (including Chapter 7 and most of Chapter 2, both prepared by Michael Lipton) do not appear in the original Report. Full discussions have been added of the interactions between MVs, poverty groups, and (i) biotechnology, and (ii) population change.

During this process or rewriting, several further specialists commented very helpfully on a gradually expanding draft. Ed Clay, Lloyd Evans, John Lyman, and Tom Walker were especially helpful. For Chapter 2 in particular, Gerry Dempsey and Norman Simmonds not only provided detailed comments, but helped reshape our views on topics about which we were ignorant or out-of-date. Steen Joffe played a similar role in the writing of Chapter 7, sections e–f.

We have, as economists, felt compelled to invade several areas of natural science that were quite unfamiliar to us. In some parts of this book, we are reporting our learning processes. We think this may help other economists and social scientists, because the structures of the natural sciences around plant breeding partly shape, partly interact with, its effects on poverty and development. However, a learning process contains, inevitably, mistakes and oversimplifications. The former are accidents; the latter are partly deliberate (especially in the discussion of biotechnology). Both, however, lead us to underline the usual disclaimer: of our many helpers and advisers, none is implicated in our views or errors.

Glossary of Acronyms

ADB	Asian Development Bank
ADC	Agricultural Development Council (New York)
AER	*American Economic Review*
AJAE	*American Journal of Agricultural Economics*
AR	agricultural revolution
ARTI	Agrarian Research and Training Institute (Colombo)
BIDS	Bangladesh Institute of Development Studies (Dhaka)
BPH	brown planthopper
BT	biotechnology
CIAT	IARC for field beans, cassava, rice and tropical pastures (Cali, Colombia)
c.i.f.	inclusive of cost of carriage, insurance and freight
CIP	IARC for potatoes (Lima, Peru)
CGIAR	Consultative Group on International Agricultural Research
CV	coefficient of variation
CIMMYT	IARC for wheat and maize (Londres 40, Mexico)
DC	developed country
EDCC	*Economic Development and Cultural Change*
EEC	European Economic Community
EJ	*Economic Journal*

EPW	*Economic and Political Weekly* (Bombay)
FAO	Food and Agriculture Organization of the United Nations
FSR	farming systems research
GE	general equilibrium
GNP	gross national product
GSV	grassy stunt virus
HR	horizontal resistance
HYV	high-yielding variety
IARC	International Agricultural Research Centres
ICARDA	IARC for arid-zone research (Aleppo, Syria)
ICRISAT	IARC for semi-arid tropics (Patancheru, India)
IDS	Institute of Development Studies (Brighton, UK)
IFPRI	IARC for food policy research (Washington DC, USA)
IITA	IARC for root crops and farming systems (Ibadan, Nigeria)
IJAE	*Indian Journal of Agricultural Economics*
ILO	International Labour Office of the United Nations
ILR	*International Labour Review*(Geneva, Switzerland)
IR	inverse relationship
IMF	International Monetary Fund
INCAP	Institute of Nutrition of Central America and Panama
IRRI	IARC for rice (Los Banos, Philippines)
ISNAR	IARC for support of national agricultural research systems (Wageringen, Holland)
LDCs	less developed countries
MABs	monoclonal antibodies
MPC	marginal propensity to consume
MR	moderate resistance
MS	moisture stress
MVs	modern varieties
N	nitrogen
NPK	nitrogen, phosphorus, potassium
NR	near-immune resistance
NSS	National Sample Survey (India)
OECD	Organization for Economic Co-operation and Development

P	phosphorus
PARRD	Poverty-Alleviating Researcher's Regional Dilemma
PBRs	plant breeders' rights
PPS	photo-period sensitivity
rDNA	recombinant deoxyribonucleic acid
SADCC/	Southern African Development Co-ordinating
SACAR	Committee/Southern African Co-ordinating Committee for Agricultural Research
SSA	sub-Saharan Africa
TC	tissue culture
TFP	total factor productivity
TPB	tropical plant breeding
TVs	traditional varieties
VR	vertical resistance

New Seeds and Poor People

1. Modern Varieties and the Poor

(a) Greening without revolution

Has there been a 'green revolution' in tropical and sub-tropical food production? Certainly, since the early 1960s, plant breeders have brought a 'greening' to major cereal crops in many parts of many less developed countries (LDCs). Independence in tropical and sub-tropical colonies, and growing concern about hunger there, led breeders to apply to these countries' food crops two principles hitherto mostly confined to temperate crops: 'hybrid vigour' and dwarfing (Chapter 2, d and e). First came the maize hybrids, adapted in the 1950s from the USA and the colony of Rhodesia, and later spreading across large parts of Central America and East Africa. Next, since the mid-1960s, short-strawed, fertilizer-responsive varieties have spread throughout East Asia (rice) and Northern Mexico and the Indian and Pakistan Punjabs (wheat). Modern varieties (MVs) have also spread to many other parts of Asia and Latin America (Table 1, p. 2). In many areas with MVs, food production (per acre per season) has doubled or tripled in 20–30 years, outpacing population growth; short-duration MVs have permitted many farmers to take two crops a year; and more land has been put into cereals, because MVs made them more profitable or safer. History records no increase in food production that was remotely comparable in scale, speed, spread, and duration.

Table 1 *LDC Areas in Wheat, Rice, Maize, Mid-1980s*

Region	Wheat Area, 1982-3 ...mn. ha....			Rice Area, 1982-3 ...mn. ha....			Wheat and Rice Area, 1982-3 ...mn. ha....			Maize Area, 1983-6[f] ...mn. ha....		
	Total	MVs	% MVs	Total	MVs	% MVs	Total	MVs	% MVs	Total	MVs	% MVs
Asia (non-communist)[a]	32.1	25.4	79.2	81.1	36.4	44.9	113.2	61.8	54.6	44.1	15.7	35.5
Asia (communist)[b]	29.1	8.9	30.6	41.2	33.4	81.0	70.3	42.3	60.2c	27.0	19.2	71.1
Near East[d]	24.8	7.6	30.6	1.2	0.1	8.4	26.0	7.7	29.6	5.1	2.4	46.4
Africa[e]	1.0	0.5	50.6	4.3	0.2	4.7	5.3	0.7	13.3	29.0	14.9	51.3
Latin America	10.7	8.3	77.6	7.6	2.5	32.9	18.3	10.8	59.0	50.5	27.3	54.0
All LDCs	97.7	50.7	51.9	135.4	72.6	53.6	233.1	123.3	52.9	178.0	79.4	44.6

Source: Dalrymple, 1986, p. 85, and 1986a, p. 108, for wheat and rice areas under MVs; and 1986, p. 86, and 1985, p. 109, for proportions of these areas in totals under crops. CIMMYT, 1987, pp. 30-43, for maize.

Notes:

[a] Excludes Taiwan and West Asia.

[b] Excludes North Korea. Incomplete estimate for short varieties in China.

[c] Corrected from 58.0 [Dalrymple, pers. comm.].

[d] North Africa, West Asia and Afghanistan.

[e] Excluding the Republic of South Africa, and North Africa; including Sudan.

[f] 1985-6 area in MVs as proportion of 1983-5 average area harvested.

Wheat and rice MVs are almost all semi-dwarf (a few intermediate height) and derive, respectively, from CIMMYT and IRRI or CIAT, or from national developments of these or similar progeny. Maize MVs are commercially purchased – either hybrids, or else open-pollinated varieties released later than mid-1976.

Green, yes; revolution, no. The term 'green revolution', now much maligned as a journalistic exaggeration, did echo a real perception of scientists in the late 1960s: that, without political upheavals, MVs could produce 'revolutionary' improvements in the well-being of many of the world's poor. However, in most MV regions, the greening has not been revolutionary in this sense, either. Perhaps 40 per cent of rural populations in the developing world now cultivate mainly MVs. Yet, except in East Asia (including China), the poor in these 'MV regions' are neither much rarer nor much stronger, absolutely or relatively to the groups that held power before the MVs arrived.[1] In the 1960s, most socially aware tropical-plant scientists would have seen their main role as 'keeping food output growing faster than population'. In the 1980s, even where this has been achieved – and even though most of the extra food comprises cheap cereal MVs, grown by smallholders and/or worked by farm labourers – the achievement has not *sufficed* to improve poor people's food intakes much.

It is not the plant breeders' job to 'transform' the distribution of income and power. Yet their hopes and motivations, and those of the financial backers of aid to international research, have always centred on the belief that increased and more stable food production would mean less poverty and hunger. Can this be achieved without the 'transformation'?

The hopes and motivations remain realistic. Without modern plant science, poverty would have got far worse still. However, if the plant scientists are to achieve the hope of bringing about 'revolutionary' changes in poor people's well-being, their research design will need to go beyond the aims of growing more food at less risk and lower cost. These designs will need to take much more explicit account of *power*: both purchasing power and political power.

Remote farmers, unirrigated areas, rural people growing food for their own use: these are even more deprived of such power than were most farmers so far benefited by MVs. Therefore, it becomes more important for agricultural research designers to allow for the impact of power (in both senses) on their outreach to the poor, as scientific 'greening' reaches staple food crops and areas so far left behind. It is beginning to spread to cassava, yams, millets and sorghums;

and to uncertainly watered regions, including some in Africa. But such crops and regions contain the least 'powerful' of farmers; can they be helped to hang on to the gains from research?

The problem of 'poor power', and its impact on who gains from agricultural research, becomes even more important as the composition of the world's poor people changes. In the heroic age of the 'green revolution', 1963–70, they were mostly tropical and sub-tropical 'small farmers'. Increasingly, and not only in Asia, the poor are mainly 'landless and near-landless labourers'. It is a different task, and probably a harder one, to steer the benefits of agricultural research towards labourers than towards smallholders. Unorganized, dispersed rural labourers are usually the least powerful group of all.

In this new environment, how can plant scientists improve their impact on the poor?

Let us look at one outstanding example of how MVs have transformed the land, its plants and its productivity. In the crop year 1965–6 in the Indian Punjab, 1.55 m. hectares were planted to wheat and 0.29 m. hectares to rice, together comprising 38 per cent of the gross cropped area;[2] the average yield of wheat was 1.24 tonnes per hectare, and of rice 1.0 t./ha. In the crop year 1980–1, farmers found it worthwhile to plant 2.81 m.ha. to wheat and 1.18 m.ha. to rice, together comprising 59 per cent of the gross cropped area. This was mainly because MVs of wheat, and later of rice, had transformed yields: of wheat, to 2.73 t./ha., a rise of 120 per cent; and of rice, to 2.74 t./ha., a rise of 174 per cent. In this Indian State of 16.8 m. people in 1981 (of whom about 12.1 m. lived in rural areas), the MVs had meant grain output (i) increasing about twice as fast as population, (ii) produced with less severe year-to-year variations,[3] (iii) at somewhat lower prices, and (iv) with more employment per hectare.

Yet even in the Indian Punjab, which has consistently been at the cutting edge of technical change in LDC food production, the proportion of people unable to afford minimum safe diets[4] has fallen very slowly, and even that only recently. In 1965–75 – while the MVs were spreading over more than 70 per cent of the Indian Punjab's farmland (and more than doubling its food yield) – there may have been no improvement at all in human nutrition, in the proportion of poor people, or in the average severity of their poverty.

In several large areas of the impoverished tropics and sub-tropics, a similar tale can be told. There have been massive rises in yields of staple food crops eaten, grown, and worked mainly by poor people. There have been positive effects on employment and on the availability, cheapness and security of food. Yet there have been only delayed, scanty, and sometimes faltering and imperceptible improvements in the lot of the poor. In most developing countries, even those with major 'green revolution' areas and significant growth in food output per person at national level, the proportion of people who have moved out of poverty in the dynamic areas has been almost balanced by the proportion that has *become* poor, especially in rural areas which — because their crops or soil-water regimes *appeared* less amenable to research — have been little affected by MVs.

Yet many leading districts in many LDCs — not just in the Indian and Pakistan Punjabs but also, for example, in substantial parts of China, the Philippines, Indonesia, Bangladesh, North Mexico, Taiwan, and East Africa — after centuries of rather slow growth doubled the yields of major food crops between 1958 and 1978. Yet in 1988 most people who saw this technical triumph, and certainly most administrators and politicians who built plans on it, see in many LDCs a bitterly and mysteriously disappointing poverty impact. We believe this sense of mystery is right.

Many scientists would deny that any mystery exists. They would point out that few parts of the developing world, even with MVs, achieved such rapid growth as the leading districts, partly because of physical conditions, partly because agricultural investment or research was underfunded; that population grew quickly, eating up many of the benefits from MVs; and that income and power were distributed in ways that steered the remaining benefits to the better-off. Anyway, agricultural scientists often doubt that social enquiry can accurately identify — let alone assign to MVs or other causes — any but the most dramatic reductions in poverty. Yet such scientists did make big claims for the field performance of

Year	Rice		Wheat		Sorghum		Pearl Millet		Maize		Five Main Cereals	
	MV Area	% Total	MV Area	% Total	MV Area	% Total	MV Area	% Total	MV Area	% Total	MV Area	% Total
	mn. ha.		mn. ha.		mn. ha.		mn. ha.		mn. ha.		mn. ha.	
1965-6	0.007	...	0.003	...	...	...	...	...	...	...	...	...
1966-7	0.9	2.5	0.5	4.2	0.2	1.1	0.1	0.5	0.2	4.1	1.9	2.0
1967-8	1.8	4.9	2.9	19.6	0.6	3.3	0.4	3.3	0.3	5.1	6.0	6.1
1968-9	2.7	7.3	4.8	30.0	0.7	3.7	0.7	6.2	0.4	6.8	9.3	9.4
1969-70	4.3	11.5	4.9	29.5	0.6	3.0	1.2	9.2	0.5	7.7	11.4	11.2
1970-1	5.6	14.9	6.5	35.5	0.8	4.6	0.2	8.9	0.5	7.9	15.4	15.1
1971-2	7.4	19.6	7.9	41.1	0.7	4.1	1.8	15.1	0.4	7.7	18.1	18.1
1972-3	8.2	22.3	10.2	52.3	0.9	5.6	2.5	21.1	0.6	10.4	22.3	24.4
1973-4	10.0	26.1	11.0	59.3	1.2	6.9	3.0	21.6	0.9	14.5	26.0	25.3
1974-5	11.2	29.6	11.2	62.2	1.3	8.1	2.5	22.4	1.1	18.7	27.3	27.6
1975-6	12.4	31.5	13.5	65.8	2.0	12.2	2.9	25.0	1.1	18.8	31.9	30.7
1976-7	13.3	34.6	14.5	69.4	2.4	15.0	2.2	21.1	1.1	17.7	33.6	33.1
1977-8	16.1	40.0	15.8	73.7	3.1	19.2	2.6	23.7	1.2	21.7	38.9	37.4
1978-9	16.9	41.7	15.9	70.2	3.1	19.0	2.9	25.8	1.3	23.4	40.1	38.0
1979-80	16.0	40.6	15.0	67.8	3.1	18.3	2.9	28.0	1.4	23.6	38.4	37.3
1980-1	18.2	45.4	16.1	72.3	3.3	21.1	3.6	31.3	1.6	26.8	43.1	42.8
1981-2	19.7	49.2	16.8	75.6	3.9	24.9	4.5	38.8	1.6	26.9	46.5	44.2
1982-3	18.8	48.8	17.8	75.7	4.4	26.7	4.7	43.1	1.7	30.1	47.5	46.4
1983-4	21.7	52.7	19.4	78.6	5.3	32.3	5.4	45.8	1.9	32.6	53.7	49.9
1984-5	23.4	56.9	19.6	82.9	5.1	32.5	5.2	49.3	2.1	35.5	55.4	53.3

Sources: Gross cropped areas from FAI, 1986, pp. II 30-31, II 101, and Dalrymple, 1986, p. 37, and 1986a, p.40.

Table 2 *Share of MVs in Major Cereals, India, 1966-85*

MVs, and over large areas were right to do so. Also, later social enquiry, as this book will show, conclusively demonstrated that MVs were usually good for small farmers, as well as for big ones; for levels of employment, as well as for returns to landowners; for stability, as well as growth; and for food available to consumers, as well as farm incomes. So a politician, mystified by the smallness of MVs' poverty impact, should not be satisfied by some agricultural scientists' sudden modesty.

Nor is the sudden modesty of some economists about the poverty impact of extra food output any more convincing. Sen [1981, 1986] rightly argues that hungry poor people need, not (or at least not only) extra food output or availability, but extra *entitlements* to food – normally from land, employment income, or transfer payments – to improve their level of living. Some economists believe that this explains the failure of huge MV-based rises in food output to do much to alleviate poverty. But the original conception of MVs in the early 1960s – by researchers, funding agencies, politicians, agricultural scientists and economists alike – was that they must raise poor people's entitlements to food. Were not the poor mostly 'small farmers' and urban consumers? Would not MVs raise farm income for the former, while restraining food prices for the latter? To the extent that this has not happened, the apparently scanty effects of MVs in reducing poverty remain a mystery – especially as abundant research confirms that MVs do tend to reach 'small farmers', reduce risk, raise employment, and restrain food prices.

This book is written to help clear up the mystery, and in the process to guide agricultural researchers towards improving their impact on poverty. The organization of the chapters is, in part, chronological. The first stages of MV research focused on technical issues; Chapter 2 enquires whether the physical goals of plant structure, sought by agricultural research since 1960 or so, have been good for the poor. Around 1969–74, analysts became increasingly worried about MVs' impact on 'the poor', especially on 'small farmers'; Chapter 3 deals with that issue. Only in the last 10–15 years has it become increasingly clear that the poor, in more and more tropical LDCs, increasingly depended on labour to acquire food; the analysis of MVs' impacts on the poor thus turned to effects on labour, wages and employment (Chapter 4) and on consumption, food prices, and nutrition (Chapter 5).

However, this is a bitty approach, tacking on other effects to an analysis which still assumes that 'poor people' are affected by MVs mainly as small-farm households. If these can sow MVs and grow more food, they will eat better and get less poor. However, for various reasons to be discussed later, the poor have become less able to rely on owned land for livelihoods. These depend increasingly on rural labour (especially in Asia) and urban work (especially in Latin America). And we need a systemic analysis, as discussed in Chapter 6, to sort out the effects on them of MVs. For, *as the poor come to rely increasingly on labour income for food consumption, so their benefits from MVs become more vulnerable to dilution or diversion due to systemic effects.* We believe that this is 'the heart of the mystery'; but we ask readers to wait until Chapter 6 before deciding whether they agree, and Chapter 7 before deciding whether or how future agricultural research might improve upon past outcomes.

At this stage we add only two sentences in explanation. (i) The apparent gains to employees as *consumers*, when MVs raise food supply and thereby restrain prices, are largely passed on to employers – because (with increasing unemployment and a growing workforce) they can respond by restraining money wages. (ii) The apparent gains to employees as *workers*, when MVs raise employment requirements per crop year, are real but have been substantially eroded by labour-displacing innovations – weedicides, threshers, etc. – which are associated with MVs (through the financial and socio-political system, not because MVs make them much more profitable), and perhaps by insufficient linkage of MVs to non-farm employment.

(b) Seeds of frustration?

MV researchers bear little of the blame for the slow progress in reducing poverty. More money for MV research might have helped to spread its benefits to less-favoured regions and people. Apart from this, however, researchers could do more to ask how MVs could be better tailored – and designed – to avoid, or to improve, the impact on poor people's welfare of population growth and of power-structures. (We return in Chapters 6–7 to these issues.) Yet, without the extra employment income and food supplies created by the 30–60 m. tonnes of grain due to MVs, many of the world's poor would today be poorer still, and millions now alive would have died.[5]

Poverty cannot often be blamed directly on MVs, as was once argued (Chapter 6, j). Some areas with MVs have experienced labour-displacing mechanization, or the squeezing out of small farmers. But only seldom can such events reasonably be blamed on the MV technology (Chapter 4, f). Indeed, the MVs have to some extent failed the poor partly because they have not spread enough to offset mechanization or land-hunger, and especially because they have not spread to areas of insecure water supply, where most poor people's livelihoods still depend mainly on growing food.

But a huge and complex social, economical and technical system is a system still. Major technical transformations do not simply slot into old social realities, but are used by – and affect the power and options of – the groups of people who make those realities. Hence, even if MVs cannot usually be blamed for other major changes (such as mechanization or land polarization), neither can MVs be arbitrarily treated as independent of such changes.

For example, population growth (apart from leading to the cultivation of less and less promising land) has raised the supply of unskilled labour; and mechanization has lowered the demand for it. Both trends have restrained real wage-rates and employment-per-person. This has partly offset the effects of MVs (via cheaper food and more employment) that tend to raise labourer's real income.

However – just as we reject the techno-pessimism that *blames* the MVs for rural unemployment, poverty and hunger – so we reject the techno-optimism that would push ahead new MV research as though its outcomes were *independent* of these evils. Techno-optimists would, in effect, (i) maximize public and private food-crop research, (ii) leave it to others to improve social and demographic conditions, and (iii) expect the extra food output and the new technologies to improve poor people's lives at any constellation of social forces, and with minimal effect, good or bad, on that constellation. But systems hang together much more than such techno-optimism suggests. Applied agricultural researchers can and should predict the impact of adopting their recommendations in specific political and demographic circumstances, just as they already do in specific agroclimatic (and to some extent economic) circumstances.

Researchers fully recognize – even if they are able to seek the most poverty-reducing crop-mixes, varieties and techniques – that such aims require quite different methods and results as between (say) a humid area and a rainfed semi-arid area. What is less often understood is that research methods and results should also differ radically as between areas where most of the poor are (i) smallholders, (ii) near-landless rural employees of big farmers, or (iii) townspeople – and as between countries like Taiwan, where almost all farmland can be put into MVs, and countries like Bangladesh, where effects on big non-adopting areas must be allowed for.

Scientists plainly cannot accept the popular perception that, once MVs spread in a country, it is no longer in deep trouble from mass hunger. For example, in 1960–4, before the MVs spread, India had very small Government food stocks; in 1965–7 famine was only narrowly averted. Between 1960–1 and 1983–4, all-India yields of wheat rose from 8.5 to 18.5 quintals per hectare; of rice, from 10.1 to 14.6; and of sorghum, from 5.3 to 7.3. Leading MV areas did much better still [CSO (India), 1984, p. 55]. Nationally, food output has outpaced population. Since 1975 even such poor people's crops as sorghum and finger millet have greatly benefited from MVs. Yet – at least until 1977, and probably even by 1985 – *the incidence and severity of hunger hardly changed.*[6] With the extra output-per-person due to MVs, India first replaced food imports; then became a small net food exporter; and finally (and at vast cost) built Government stocks, averaging 25–30 million tons of foodgrains. For how could the hungry afford the extra food? Unless the MV strategy is to support a prolonged programme of food gifts or subsidies, or to be employment-generating or concentrated on poor people's crops and areas,[7] the strategy – during rapid growth of the workforce and labour-displacing mechanization, *and in the absence of concomitant redistributive social change* – is at risk of being strangled by the stagnation in poor people's purchasing-power. This has pushed ever-increasing proportions of those crops into deteriorating stocks (or exports) instead of into hungry people, not only in India but in most of South and South-east Asia and large parts of Latin America.

(c) Separately, all seems well

Taken separately, almost all effects of MVs seem to help the poor. This in essence is because their physical and chemical characteristics are selected to make more efficient and more stable use of sunlight, water and plant nutrients – even when farmers cannot afford to buy many inputs – and to use those nutrients to grow more, cheaper, safer food (Chapter 2). MVs now, in most environments (though not all: Simmonds, 1981, p. 360), outyield traditional varieties (TVs) even at low levels of inputs and management. New risks do exist: the narrow range of genetic materials in some groups of MVs increases the long-run risk that some variety of insect or fungus will 'like' and destroy many of them; 'soil mining' can be caused by high-output, low-input strategies. However, such risks have been overstated, and are often smaller than with TVs. Indeed, many MVs owe their good *average* performance precisely to greater avoidance of risk: to their better capacity to cope with disease attack and moisture stress. This should help the poor most, because they have fewest defences against risk, and so should gain most from safer seeds. However, for this to work properly, it is probably desirable that breeders increasingly select for 'horizontal resistance'; otherwise, the poor may come to rely on costly, and often ultimately self-defeating, pesticides (Chapter 2, j).

So far, the poor have voted with their pockets for MVs. Indeed, if poor people are small farmers (Chapter 3), they ultimately adopt MVs on at least as high a proportion of their land as big farmers, and achieve at least as high yield, cropping intensity, and overall farm efficiency. Seldom are they dis-possessed before they can gain from MVs – though early adopters, because they get the best prices, gain more; they tend not to be the smaller farmers. Also, many poor farmers live outside MV areas, and have thus enjoyed little increase in yield; those among them with grain surpluses to sell (not the very poorest) have lost as extra output from MV areas kept farm prices down. Nevertheless, many millions of poor 'small farmers' would be worse off today if MVs did not exist, and only a few would be better off.

However, Asia's poor today, and Africa's poor tomorrow, (pp. 000) live mainly and increasingly from farm labour, not as farmers (even 'small farmers') on owned or rented land. MVs increase the employment prospects of such labour (Chapter 4), but less so than in their early years. Around 1970, if MVs doubled yield for a particular cropped area, they raised employment by about 30–50 per cent. Today, the figure is nearer 10–30 per cent. Farm employers – having successfully lobbied for cheap credit, fuel, or machine-hire – increasingly handle labour scarcity, in the wake of MVs, by mechanization. They are then likelier to respond to MVs by hiring more tractors and threshers rather than more workers.

The initial thrust of MV research against poverty – prefigured by a celebrated enquiry into India's cereal needs [Ford Foundation, 1959] – sought to anticipate land shortage (as population grew) by raising foodgrain yields, initially in a few lead areas, and thus to benefit poor people as consumers. Soon afterwards, the emphasis increasingly moved towards small farmers – to the need to benefit poor people as food producers [ADB, 1977]. Now, the trick was (i) to let the extra food output restrain consumers' food prices, but (ii) to ensure that MV technology raised yields fast enough to offset any restraint in the prices paid to food producers.[8] This can work brilliantly if MVs spread rapidly where the rural poor are mostly small farmers, but what if they are mostly landless or near-landless workers? Almost no thought has yet been given to how – if at all – agricultural research, while raising and stabilizing yields, can or should seek cropping strategies that raise labour's share in the production process and its rewards. The MV researcher's goal, of helping the poor as consumers and producers, needs to be reinterpreted, to allow for the growing preponderance of labourers (instead of farmers) among poor rural producers.

As *consumers* (Chapter 5), the poor clearly gained from the more locally and reliably available, less inflation-prone food output that MVs assured. The poorest fifth of South Asians and Africans spend about 80 per cent of income on food,[9] well over half of it on cheap cereals and roots; for the richest fifth, both proportions are far lower. So, when food prices are restrained or stabilized, the poor gain proportionately much more than the rich. Since MVs usually produce 'coarse' rice or wheat varieties at a price discount, the consumption gains from such 'poor people's foods' are even more concentrated on the needy. Yet, as we have seen, extra MV output has often displaced imports – or built up stocks – because, despite mass hunger, effective demand for the extra food output was insufficient. Poor people's purchasing-power did not increase to buy the extra food, partly because of wage effects (Chapter 4); rich people were unwilling to spend extra income on coarse cereals. Of course, this 'adding-up frustration' is removed to the extent that MVs spread to *very* poor farm families, who consume their own extra output, and therefore do not rely on extra effective demand from others (Chapter 3, e); some of this has happened, but not enough.

The *nutritional* improvement most required of MVs by the poor – more calorie intake – has thus been partly frustrated (although without MVs the shortages would have been even worse). Nutrition-orientated MV researchers need to focus on increasing the access of the poorest – mainly labourers – to calories. Ironically, the 'nutrition research' into MVs rich in proteins, with good cooking quality, and so forth, is usually at best a diversion, and often actually bad for the poor. However, recent work in nutrition has identified, much more accurately than before, what sorts of people are at risk from calorie intakes deficient relative to their requirements, when they are at risk, where, and why. We shall suggest how agricultural researchers might improve the consumption impact of MVs by steering their work towards such people.

(d) Adding-up frustrations – and omissions

However, most of the major 'first-round' economic effects of MVs, analysed in Chapters 3–5, look good for the poor –

whether as small farmers, as rural employees, or as food consumers. Chapter 6 explores why these effects have not, so far, 'added up' to as good an outcome for the poor as the MV pioneers expected. Obvious countervailing factors (population; anti-rural power-structures; rising input costs; mechanization; major areas and crops enjoying few MVs) played their parts, but do not explain persistent mass poverty in parts of the 'green revolution' heartland (Chapter 1, a). Anyway, to the extent that a country's politics, institutions, and MV and other technologies form a total and interlocking system, it is rather lame to explain away 'poverty despite MVs' by pointing to other things that just happened, in dozens of countries, to be going on at the same time.

To some extent, standard economics can explore the impact of MVs on the poor by going beyond the 'one effect at a time' approach of Chapters 3–5 and instead calculating how the spread of MVs affects workers, farms, consumers, industry, etc, in big, national, interlocking systems. This approach (Chapter 6) can go a long way and can tackle vital issues. It can even be used to examine how extra Euro-American food output – due to research and to farm subsidies – can transform, via world food prices, the upshot of MVs for the Asian poor (see Chapter 7, k). But such 'general equilibrium' analysis, and indeed standard economics as a whole, avoids a policy issue – and faces a central logical problem – in handling big changes, like the spread of MVs. The policy issue is that 'long chains of deductive reasoning' in economics[10] can be brutally short-circuited by political responses to such big effects. The logical problem is that the effects and responses are so big that they alter not only the economic parameters, but the forms of the economic relationships themselves.

This is not because MVs are like some earlier 'agricultural revolutions' (ARs) – such as, say, the Neolithic Settlement[11] – in being linked, almost by necessity, to radical changes in the nature and structure of power. The spread of MVs has been much more like, say, the eighteenth-century AR in North-west Europe – compatible with many structures of rural ownership and power. But the speed, scale and spread of MVs (although not their impact on the internal balance of rural power) have far exceeded those of any earlier technical change in food farming. So, therefore, have the effects on land use, food output, the capacity of the land to sustain growing urban populations, and much else affecting the texture (though not the structure) of political life. The results of such large disturbances cannot be fully explored by adding up economic effects, even in 'general equilibrium'.

Chapter 6, therefore, also briefly explores another, historical, approach: the comparison of MVs with earlier ARs. These affected the poor mainly through interactions with socio-political and demographic structures, rather than just through the supply and demand for labour, land, and food. The message of earlier ARs is that, in default of policy change, even major innovations, that employ more labour and grow more food, often help the better-off more than the poor. Even if the bio-economics, as with MVs (Chapter 2), is pro-poor, this may not suffice to prevent an AR from deepening some forms of poverty.

(e) How researchers can fight back

What, then, can researchers do? MVs differ from earlier agricultural revolutions in being developed in large part by an international, formal, public-sector research system, which need not respond to purely commercial or career-scientific incentives. In the light of this, Chapter 7 considers how food-crop research strategies – having achieved major increases in output, and major reductions in the average cost of production – might adapt to meet three challenges. The first challenge is to remedy the past failure, despite major advances in production, to do enough – or to do the right things – to reduce significantly the incidence of poverty and hunger in most countries and regions with major MV spread. Although agricultural researchers cannot usefully try to manipulate socio-political (or demographic) systems, they can do much to improve the impact of their work on the poor by allowing for its interaction with such systems.

The second challenge lies in the areas and crops with sparse or no MVs, especially in Africa but also in ill-watered areas elsewhere. Should researchers concentrate on these? Sometimes an area or crop has been neglected because it has less potential than other research lines – but even its slight improvement through research may be vital to poor people with few alternatives. Sometimes, as with wheat, favourable environments are largely under MVs already and promise only slow improvement, so that 'the greatest potential . . . lies in the more marginal environments' [CIMMYT, 1984a, p.7]. If so, how can researchers into these environments obtain there the good output effects of their work in well-endowed areas, while improving on the inadequate poverty impact of such work?

The third challenge lies in the new prospects, methods, and risks of agricultural research: new prospects, mainly in nitrogen fixation; new methods, mainly due to biotechnology; new risks, mainly via depletion of the genetic bases of crops and of the regenerative capacities of soils. Can such 'futures' – and the (possibly concomitant) increase in the privately owned and marketed proportions of plant research activity – be 'socially engineered' to improve the impact on poor people?

In meeting such challenges researchers (and those who finance them) must consider, not only directions within MV research, but alternatives to it. Are there more promising ways to help the poor than MV research – or ways complementary to it? This might mean augmenting the productivity of tropical and sub-tropical food-growing land in other ways. Or it could involve quite different approaches. These might include: research and technical change that seek to use more land (e.g. by reducing the impact of more intensive crop mixes on depleting soils and shortening fallows in Zaire), instead of using land more intensively; research into non-food crops in developing countries (at least into those in relatively price-elastic world demand, such as cotton); 'non-research' policies for agriculture, concentrating on prices, or irrigation and other investment, or current farm inputs; or even 'non-farm' policies, aimed at absorbing the poor into off-farm activities, and using their output to buy from the USA's and EEC's burgeoning 'grain mountains'. Certainly, we need to put MV research into the context of alternative approaches to technical change, and to other possible ways forward for the poor.

(f) Technology, power and people

The issue of how MVs of main food staples affect the poor – and of what follows for agricultural research strategy – is big enough, yet may help us to understand an even deeper issue: the interplay of institutions with technology and human actions, as causes of social transformation and hence of changes in the lot of the poor. Almost certainly, without a sensible view on that issue, we shall not efficiently predict the impact of MVs on the poor.

This book therefore raises – explicitly in Chapter 6 – the question of the interplay among technical, socio-institutional and human changes. Is the intellectual thrust (and hence, to some extent, the impact on society and its poor) of the new science underlying great waves of technical progress – MVs or steam-engines or computers – determined largely by market demand, as many economic liberals believe; by class power, as many socialists argue; or by the internal logic of autonomous science, as many practising scientists claim? Are certain institutional structures or changes needed, especially in parts of Africa, before scientific findings can be efficiently adapted, developed or communicated? Or can technical change, if 'big' enough, steamroller through apparent obstacles of social structure or human under-education, transforming society and those very obstacles with it? Historically, such 'technical quick fixes', while neither frequent nor painless, have not always been as illusory as that persuasively dismissive phrase might suggest.

(g) Six central questions

The above wider issues, of political and scientific philosophy, are raised in the hope of clarifying six main questions, considered in turn in Chapters 2–7, about the impact on poor people of MVs. Do the *physical characteristics* of MVs lead to gains or losses for the poor? Do MVs help *poor farmers*, absolutely or relatively to rich farmers? Do *rural workers* gain or lose income, or shares in income, via employment or wage-rates? Do *poor consumers* gain or lose, nutritionally or other-wise? Does economics, political science, or history help us to predict the *interactive effects* of all these sequences – both on poor people directly and on the various social and institutional contexts that (with techniques, tastes, human skills, and prices) largely determine the ebb and flow of human poverty? What *responses from researchers* might, in future improve the effects on the poor – especially in view of new methods, prospects and risks in biological research into food crops, and of the spread of MVs to new crops and regions, above all in Africa?

These are complex questions. First, each question suggests ways for MVs to help or harm poor people (i) absolutely, (ii) relatively to the rich, (iii) both, or (iv) neither. For example, rural workers' income has often risen absolutely (even in the short term) in the wake of MVs, but fallen relative to landowners' income.

Second, in respect of the answers to all the six questions, MVs can affect poor people's welfare by changing not only its level per poor person, but also its distribution between the poor and the extremely poor, between regions, and between present and future, and its stability and predictability.

Third, no person is *only* a farmer, an employee or a consumer. Most poor people are in at least two groups, and many are in all three. It can be misleading to separate behaviours and responses in ways that artificially 'cut up' individuals or households – ignoring, for instance, the fact that a household may buy different *consumer* goods because MVs have altered its *farm* income.[12]

Fourth, the answers to our six questions vary according to 'who the poor are' in a given region. Most of them are urban consumers in Latin America, small farmers in Africa and increasingly rural labourers in South and South-East Asia. The answers also vary according to a country's political and institutional set-up.

(h) Swings of fashion

With many complex questions, and with over a hundred poor countries growing MVs, it is not surprising that the impact of MVs on the poor has spawned a vast literature. Nor is controversy surprising; but the sharp changes of direction, almost swings of fashion, perhaps are. First came the 'green revolution' euphoria of 1967–70. In the second phase, there were growing fears that the MVs enriched large farmers at the expense of small, and landowners at the expense of labourers. The later 1970s saw a third phase; several reassessments suggested that in MV-affected areas the poor gained absolutely, but lost relatively.[13] Small farmers adopted after large ones – but did adopt, and raised yields. Farmworkers found that the effect of MVs in boosting the demand for their labour seldom brought much higher wage rates – but employment rose. Above all, poor consumers gained, as extra cereals supplied by MVs restrained food prices. The big exception to this rather happier verdict on the MVs was that producers in non-MV areas, including many poor farmers, gained nothing from the new technology. Indeed, they often lost; the extra MV sales from the Indian Punjab (wheat), or in the Philippines from Central Luzon (rice), restrained farm-gate prices in impoverished Madhya Pradesh or Mindanao respectively.

In the 1980s, despite the African famines, the fourth phase, of extreme optimism about MVs, has begun. This rests only in part on somewhat science-fictional expectations from nitrogen fixation and biotechnology (Chapter 7, e–f). Also, as we shall see in Chapter 3, it is sometimes asserted that small farmers adopt MVs earlier and more intensively than big ones, that MVs raise the share of labour in income, and that poor consumers gain most of all. The known difficulties of the poor in borrowing money, in taking risks, in moving to new job opportunities, are de-emphasized. Only the problem of 'neglected regions' is still generally acknowledged. Except for some parts of Africa, the increasingly accepted view is that technically appropriate and profitable MVs, by being spread everywhere, will everywhere help the poor. Even for Africa, it is widely asserted that liberalization by the public sector, plus rather vaguely specified 'agricultural research', can produce transformation.[14]

In the 'second phase' the near-consensus among social scientists, that MVs threatened the poor, was absurdly gloomy. However, the revival in the 'fourth phase' of the early euphoria about MVs needs critical review. In order to provide one, we look back (at opportunities seized and lost) and ahead (to suggest how research can adapt to rapidly changing composi- tions of poverty). Also, we examine how agricultural research findings, national and international, are inserted into political systems. These systems, at least as much as 'production functions' and other aspects of pure economics, determine who gains and who loses from MVs. Researchers can and should allow for such effects in setting priorities for work likely to help the poor.

Why have there been these huge swings in the pendulum of opinion about the effect of MVs on the absolute and relative position of the poor? One reason is plain optimism (or pessimism). This is often based on a couple of years of good (or bad) harvests; or on falling (or rising) oil prices.

A second reason is intellectual fashion. This is intensified by the temptation to pigeonhole the pessimists as 'Marxists' and the optimists as 'neo-classicals' or economic liberals.

A more important reason why fashions change is that MV research, and its results, also change – partly in response to earlier criticisms. For example, the first 'rice revolution' cross- bred semi-dwarf varieties with traditional rices used in much of South and South-East Asia. The resulting semi-dwarf varieties, notably IR-8 and TN-1, were much more fertilizer- responsive, but also had higher requirements for inputs and management. This package raised yields, but was hard on poorer farmers. The second and third rice revolutions com- bined dwarfing with, respectively, improved disease-resistance and shorter duration to avoid moisture stress. These two approaches – largely in response to criticisms of TN-1 and IR-8 – developed more robust, 'poor-friendly' varieties. Sim- ilarly, the disadvantages of maize hybrids for the poor have been largely removed by synthetics; hybrid seeds must be distributed each year to maintain yields, but synthetics enable farmers to retain seed yet keep a 'mix' of vigorous plants in the same field for several years (Chapter 2, d), helping remote or poor farmers who cannot rely on buying the right hybrid on time each year.[15]

Finally, fashions change because – even given the available MVs – the rural scene features different groups, learning at different rates. Frequently, for example, big farmers adopt first, so that early observers are gloomy about the gains of small farmers from MVs. However, the small farmers usually catch up later, once they have seen that the risks are not too great; then the next 'generation' of observers becomes very hopeful about the spread of MVs to 'poor' farmers. However, later still, it turns out that late adopters gain much less than early ones, because prices have been reduced by early successes.[16]

(i) Research on research

There may be good reasons – or reasons of pure chance – for 'swings of the pendulum'. However, if they happen often, something is probably wrong with the machinery. Have economists, sociologists, administrators, and natural scientists designed appropriate methods for judging, and if necessary redirecting, agricultural research? We return to these issues in Chapter 7. But it needs to be made clear now that several things are badly wrong.

First, economists and other social scientists seldom take part in agricultural research design. Usually, they complain afterwards about the results, or the social outcomes of those results (for example, of MVs). They often know little or nothing about agriculture; few development economists know the difference between inbred and outbred crops, yet it is crucial to the design of research and its impact on the poor (Chapter 2, c). Conversely, many good scientists are quite ignorant of economic effects of their work, yet take regular decisions dependent upon crude, even wrong, mental pictures or models of those effects.

Second, if social and natural scientists do work together in agricultural research design and implementation, it is almost entirely at the top levels of international research. (At best, such co-operation can extend down to apex institutions in big regions, or in huge and diverse countries such as India or the USA).[17] Yet it is in local and adaptive research – all the way from incorporation of farmers' own, usually neglected, research findings [Richards, 1985] to field trial design and the treatment of farmers' feedback – that social science and natural science perspectives interlock most tightly. The interlocking involves issues much larger than is suggested by current emphases – however justified – on 'farming systems research'.[18] Just as international agricultural research usually produces poor returns unless there is good national adaptive research,[19] so interaction between biology and economics in international centres is of much more value if it is matched by research at the local level.

Third – a related issue – national research capacities in most poor countries, especially in Africa, have lagged far behind international research. Fourth, the two remedies currently most popular are to increase funding for national agricultural research and to devise and impose blueprints for its organization;[20] both remedies will fail unless supplemented by much greater and more systematic attention to research *content*. Fifth, the analyses of the farmers' own role in research are polarized: either crude 'topdownery' or naive populism.[21]

In this environment, 'research on research' has not got far beyond establishing, to some people's satisfaction, that agricultural research has high rates of return, except (for unknown reasons) in Africa.[22] As regards the impact of MVs on the poor, 'research on research' has been inadequate in two main ways. It has asked the wrong questions; and it has looked only at 'first-round' effects. One example of each error must suffice here.

'In adopting MVs or supporting inputs, or in getting high yields from them, do small farmers lag behind big ones?' is a question asked by almost all commentators. However, it is the wrong question, if we are interested in what MVs do to the poor (Chapter 3, a). The poverty or affluence of a farm family is affected not only by its land area, but also by the quality of its lands, its sources of non-farm income, and the number of family members. Yet of the hundreds of studies of adoption of, and returns to, MVs, almost all ask whether 'small' farmers (in terms of land area) lag behind. The right question, instead, is whether families with a low initial endowment of farm and non-farm income sources per member do so.[23]

As for the limited relevance of 'first-round' effects, much of the benefit of MVs for poor people arises because the higher yield of MVs makes more food available domestically, so that the rate of consumer price inflation is reduced. On this observation have been based several analyses of the amount of consumer benefit – and of its distribution to, and nutritional impact on, the poor.[24] When food price inflation is cut back, the automatic and simultaneous 'first-round effect' is that real value of consumers' wages rises. Since unskilled labour is in ample supply, it is highly responsive to such 'first-round' rises in its real wage. So much more work is offered at the new, higher real wage than previously. Employers find that labour supply far exceeds demand. So, on the 'second round', employers are able to slow down the increase in the *money*-wage, after food price increases have slowed down. That leaves the *real* wage increased little, if at all, as Indian data show [de Janvry and Subbarao, 1987]. Thus, when MVs moderate food prices, the gain to employees – in poor countries where unskilled workers are in excess supply already – is much less than is suggested by the first-round effect. On the second round, most of the gain is passed on to their employers.

The approach of MV research, in seeking to help the poor, has hitherto been to supply seeds (and linked techniques) most beneficial to 'small farmers', poor consumers, landless labourers, and, where possible, disadvantaged regions. This approach – partly because of the 'wrong question' and 'first-round' problems caused by the inadequate state of 'research on research' – is, while desirable, insufficient. Poor people will be helped by an MV to the extent that it improves their well-being in their total context. They are members of families and localities, not just (and certainly not always) 'small farmers'. They are employees, tenants, borrowers, etc., affected by outcomes of MVs after many 'rounds' of linkages via consuming, investing, employing, etc., not just by the immediate effects. General-equilibrium economics helps;[25] but other social sciences, and disequilibrium considerations, matter also, as do the 'lessons of history' about what rapid agricultural change does to poor people. We return to these issues in Chapter 6, k.

First of all, however, we need to link up the 'scientific'nature of MVs with their socio-economic consequences for the poor.

Notes and references

1 Griffin and Khan, 1977, on South and East Asia; Ghai and Radwan, 1983, on sub-Saharan Africa; Ahluwalia, 1985, on the recent Indian evidence (see below, footnote 6).

2 Bhalla and Chadha, 1983; Chadha, 1983.

3 On the thesis of increasing instability, see below, Chapter 3, h; this thesis is correct on its own definitions. However, 'worst-case' output and food consumption – both at farm level and nationally – have been increased as a result of MVs, both absolutely and as a proportion of average levels, and for poor people as well as overall.

4 For definitions, see Lipton, 1983. For evidence of stagnant poverty in the Punjab in 1963–75, see Rajaraman, 1975; Bardhan, 1984; and, on nutrition, Berg, 1978, p.3. On improvements since the mid-1970s see Chadha, 1983; Sheila Bhalla, 1979.

5 Of every 1000 new-borns in low-income countries, about 180 died before their first birthday in 1960, and about 140 in 1981. At least one-third of these deaths were associated with the synergism of infection with inadequate dietary energy. Hence the reduced death-rate must be partly due to the effect of MVs in, at least, preventing a decline in poor people's calorie intake (while health care improved).

6 Ahluwalia, 1985; Subbarao, 1987; Minhas *et al.*, 1987. The 1983 round of the National Sample Survey showed some improvement from 1977, so that the proportion of people below the poverty line fell to the 1961-2 level; but 1983 was an exceptionally good crop year. On the proper weighting of 'incidence' and 'severity', see A. Sen, 1981.

7 There is a conflict between concentration on areas of high and promising yield (to create maximum extra output and reduce prices for poor *consumers*), and on backward and less promising areas (to maximize gains to poor producers). See Chapter 3, i, and Brass, 1984.

8 (a) The 'compromise', by which the State pays high prices to producers but charges lower prices to consumers, tends to be very costly to the budget if pursued on any significant scale. (b) It is not quite correct to define the 'trick' as being to allow producer prices to fall more slowly than yields rise, because there are also input costs, which are pulled up as MV farmers demand more input so as to increase yield. Indeed, in view of diminishing marginal returns, input quantities often rise faster than yields.

9 See Lipton, 1983. This includes, in both income and food consumption, the value of on-farm production consumed by its producers.

10 These were warned against by the great economist Alfred Marshall [1890, p. 637].

11 The shift from hunting and gathering to settled agriculture, which began in China before 5000 BC., and in Europe around 2000 B.C. In a few areas of Africa, it is still under way today. See below, Chapter 7, d.

12 For the ways in which farm households behave differently as farmers because they *are* households – and differently, therefore, from large commercial farms – see Barnum and Squire, 1979; Low, 1986. This farmer–consumer integration can completely reverse farmers' response to price incentives [Singh, Squire and Strauss, 1986] – and perhaps also to some changes in incentives due to MVs.

13 For the first phase, see Brown, 1970. For the second, see Frankel, 1971; Borgstrom 1974; Griffin, 1975. For the third, see Ruttan, 1977; Lipton, 1978, 1979; and – despite its undeserved reputation as a vehicle of techno-pessimism – the balanced account in Dasgupta, 1977.

14 Hayami, 1984; Barker and Herdt, 1984; Berg, 1980. For words of caution, see Prahladachar, 1983.

15 On successive changes in rice production, see Herdt and Capule, 1983. On problems of poor farmers with maize hybrids, see Malaos, 1975. Note that the trend to ease these problems is not uniform; hybrids of open-pollinated crops like wheat and rice have been developed, requiring poor farmers, for the first time, to buy new seeds each season (Chapter 2, d).

16 Binswanger and Ryan, 1977; Dalrymple, 1979, pp. 720–1.

17 Research posts for economists in joint high-level teams, for example, are taken very seriously by such organizations as the International Rice Research Institute and the All-India Co-ordinated Rice Improvement Programme; but in most regional or local research institutes (even in India) such posts are non-existent, unfilled or of a 'dogsbody' nature, subordinated to natural scientists.

18 See below, Chapter 6, i, and Simmonds, 1985.

19 Evenson and Kislev, 1976.

20 Lipton, 1985.

21 An almost unique exception is Richards, 1985, which documents how some farmers' genuine experimentalism (e.g., among the Mende of Sierra Leone, with self-selected new rice varieties at different altitudes) and formal AR can both gain from mutual learning.

22 An excellent outline of this work (and of its limitations) is Pinstrup-Andersen, 1982. See also Lipton, 1985.

23 Strictly, sources of income per consumer-unit [Lipton, 1983]. Examples of undue emphasis on the 'wrong question' are Lipton, 1978, 1979.

24 On the price effect, see Evenson and Flores, 1978; on consequent consumer benefit, see Scobie and Posada, 1978, 1984, and Flores, Evenson and Hayami, 1978; on impact on the poor, see Pinstrup-Andersen, 1977. Both the above arguments, and the objection about money-wage response when labour supply is elastic, apply equally well if the extra food supply due to MVs – instead of causing food prices to fall – merely causes them to increase less than would have been the case without MVs. Then, on the second round, employers are able (given ample labour supply) to raise money wage-rates less than would have been necessary without MVs.

25 On linkages, see Bell, Hazell and Slade, 1982, and Hazell and Roell, 1984 (for a powerful critique, see B. Harriss, 1987). On general equilibrium, see Binswanger, 1980; Binswanger and Ryan, 1977

2. Physical Features of Modern Varieties: Impact on the Poor

(a) Plants, breeders, and the poor

This chapter presents the main morphological and physiological features of modern varieties (MVs), as sought by plant scientists, and enquires how these features – and the process of plant science that generates them – affect poor people. First, however, we consider two questions. What are plant scientists trying to do with food crops? What limits and options are set by these crops themselves? Only then can we explore how the likely results of plant science, as constrained or advanced by the food crops' own structures and environments, relate to poor people's requirements.

We approach the questions through Simmonds's outstanding guide to the logic of plant breeding, seen as 'the continuation of crop evolution', i.e. of natural selection combined with farmers' choices, 'by other means' [Simmonds, 1981, page v]. It might seem odd to pick a book on the *breeding* of *plants* in order to understand the goals and options in the *improvement* of *food crops*. Better varieties of crops are generated to specifications requiring many disciplines other than plant breeding. Plant pathologists define the biochemistry of diseases attacking the plant, and of its resistance to them; plant physiologists assess the prospects of developing different dimensions in roots, leaves and stalks, and the results for plant performance; agronomists examine the impact of alternative farming practices, timings, and systems on plant growth and yield; and so on. Moreover, food crops – while highly disparate – differ systematically from other crops in ways critically relevant to the poor, especially because only food crops are, in many places, grown mainly for the use of their growers rather than for sale.

Nevertheless, plant breeding (with its underlying 'pure science', genetics) has two special claims on our attention. First, it is the integrating discipline around which other plant sciences (including agricultural economics) cohere. For example, plant pathologists and physiologists seek breeders' advice on the mechanisms, partly genetic, determining the features of a variety that affect its resistance to a disease, and in turn advise breeders on the physiological and disease consequences of alternative breeding tactics and aims. Second, it is plant breeders, acting on this exchange of advice, who have developed the maize hybrids, and later the semi-dwarf rice and wheat varieties, that have brought yield breakthroughs in parts of many poor countries. True, plant breeding is not specific to food crops; but neither genetics nor interaction with environments differs all that much between food crops and others – and most human food crops have other uses too.[1]

(b) The goals of plant science, and of poor farmers

How do the overt goals for changing the physical or chemical nature of a food crop, as they are seen by the plant breeder – the standard-bearer and integrator of the 'green revolution' – affect the prospects of poor people, as smallholders or workers or consumers, for meeting their requirements from food crops? Breeders see quantity (as indicated by yield) – and, some way behind, quality – as overwhelmingly their main goals, plus a few 'oddments'[2] [*ibid.*, pp. 40, 64]. An economist might complain that net value added, not 'quantity', is what farmers and consumers want; and that yield-per-acre is not the main determinant of quantity – and hardly at all of net value added - if land is plentiful, as in a few parts of Africa such as Eastern Zambia and central Zaire. These objections, though important (and sometimes ignored), can be easily accommodated, by appropriate adjustments in breeders' goals and later in extension advice.

A poor farmer would still wonder about three missing items. They are stability, sustainability, and cross-crop effects. Even Simmonds's 'oddments' do not include these goals. This is certainly not because Simmonds slights them.[3] He (and most breeders) may well see them as long-term components of yield. Indeed, a lower risk of downward fluctuation in crop output, or of its long-run decline, is ultimately a form of increased yield.

However, poor farmers also value stability as such independently of yield, and even at its expense. They cannot afford to take big risks; they also find storage in advance of a possible bad harvest – or borrowing if it arrives – costly. So a poor farmer often prefers 'a safe two tons an acre' to an unstable average of three tons, made up of, say, one ton in half the seasons, five tons in the other half, and no way to predict which season will be which. So the poor farmer usually values stability *more* than breeders do. Often, on the other hand, he or she is so hard-pressed to survive now – and finds borrowing so expensive – as to be sometimes tempted to value sustainable yields, say ten years hence, *less* than most breeders or economists do.

Poor farmers would also question the breeder's normal goals of quantity and quality for single crops, grown in pure line stands for single seasons. Having little land per family worker, poorer farmers find that labour is more readily available, so that they are likelier to grow several crops a year than richer farmers with more land [Berry and Cline, 1979]. Also, it is a reasonable assumption that poorer farmers (being less able to take risks or afford fertilizers) are likelier to mix crops in a field. This can spread risks – for example, if one crop does better in wet years and the other in dry years. It can also partly compensate for lack of fertilizers, if one crop fixes or restores nutrients taken out by another (as beans restore nitrogen, used up by intercropped maize). And it can use family labour to provide at least some end-product, even when there is no work to be had on the main crop. Moreover – if there is enough family labour per acre, i.e. again on the less well-endowed farms – crop mixtures can also be a more profitable strategy than sole cropping, apart from reducing risk [Norman, 1974].

Plant scientists rightly simplify the farmer's complex problems in their research, so as to make it manageable. For that reason, they usually aim at high yields per season of major crops in pure stands.[4] But poor farmers also want stable yields, and often grow mixed crops. Hence researchers' very *definition* of their aim means that they should constantly refer back to farmers' practices and needs; an aim can imply a recommendation that needs adjustment in the light of farmers' own, somewhat different, aims. Also, varieties or practices, recommended as meeting the researcher's necessarily simplified model of the farmer's goals (and often best selected through mechanized trials on uniform land, i.e. in conditions very different from smallholders': Simmonds, 1981, pp. 211–12), need testing in the circumstances of real-life smallholders. For example, breeders find that an early-maturing variety almost always yields less [*ibid.*, p. 181]. But this is often offset, especially for poor farmers, because it (i) is probably less risky, since exposure to diseases, droughts and floods is briefer, (ii) is likelier to allow a second crop to be taken, even with slow and labour-using farm methods, i.e. to raise yield per acre *per year* (though lowering it in the season while the early maturer is growing) without incurring the costs of combine harvesting or tractorized ploughing, and (iii) provides food or income late in the slack season, when stores may well be exhausted and food purchases expensive.[5]

Possible conflicts between the breeder's goals of 'quantity and quality' and the concerns of the poor do not arise only for poor *farmers*. Breeders' criteria, e.g. in respect of weed management and hence breeding for herbicide tolerance [Simmonds, p. 61], can obtain 'quantity' and uniformity of crops in ways that destroy jobs for poor *labourers* (e.g. in weeding) – and thereby not only increase hunger but also reduce demand for the extra food grown. As for the poor as *food consumers*, they often eat mainly what they grow, or what they are paid in kind; they are generally not best served by the breeder's natural view that the quality goal is best indicated by the product's marketability [ibid., pp. 51–3].[6]

When the breeder's general goals of 'quantity and quality' are unpacked, further possible conflicts with poor people's interests arise in respect of food crops. Attempts to increase quantity are separated into attempts (i) to raise *biomass*, and (ii) to improve its *partitioning*, by 'enhancement of the yield of desired product at the expense of unwanted plant parts' [*ibid.*, p. 45]. But biomass has to be raised per unit of and scarce resources, natural (light, soil nutrients, rain) and added (labour, land, fertilizers, irrigation). Breeders, policymakers, rich and poor people, and markets do not all signal the same scarcities. On the input side, richer people tend to be short of labour, poorer people of land and capital; and different uses of the various resources are made by different crops or MV, or the same crop or MV in different places. On the output side – as with plant partitioning – which parts are 'unwanted' also depends on who one is. Commercial farmers and better-off consumers of food crops want high-level, low-cost outturn of high-grade whole cereal grain, preferably convertible into modern bread products. Such people are less concerned than the poor to obtain cheap broken grains, gleanage, straw for animal feed or thatching (e.g. sorghum stalks for housing in Northern Nigeria), leafy parts as vegetable supplements or grazing, etc. Similarly, high-grade, processible roots and tubers are prized, and other parts are 'unwanted', for the better-off. But deficit farmers with cash constraints (who find commercialization risky or costly), or poor consumers, may want precisely the 'plant parts' whose supply is squeezed by the normal breeding priorities.[7]

Common sense, however, counts. More food is likely to mean more saved lives than is more straw. Breeders, in view of the long time-lag between initiating research and releasing its end-product, well know that they must 'take a view of where economic advantage will lie 10–20 years hence [and] of likely environments on the same time scale [and the impact on management and] disease patterns'. But should we conclude that 'in determining objectives [breeders] will no doubt listen. . . but [their] own judgement. . . is probably better than that of most others' [*ibid.*, pp. 63–4]? Despite the big addition to poor people's welfare made by 'breeder-led' goals for food crops, these goals can and do produce outcomes in serious and non-obvious conflict with poor people's interests. Plant breeding must be left to plant breeders; priority setting must not be.

Priorities *are* set by *someone*, irrespective of the state of knowledge at the time. The assessment of research requirements for a crop can, in principle, be carried out by using research resources in a way that maximizes the expected ratio of benefits to costs. For alternative candidates for this 'best way', the ratio has to be calculated to allow for delay and risk [Lipton, 1985] at three stages between initiation of research and successful implementation on the farm. How long is it expected to take before there is a research result ready for release to farmers – and what is the probability of slower, or faster, results? Between release and adoption by farmers? Between adoption and attainment of higher net farm output? There are trade-offs between different sorts of cost, delay and uncertainty; between the three stages; between quick results and certain results; and between speed, certainty and cheapness. The three expectations, risks and trade-offs help, in each particular case, to decide whether to concentrate research on raising output in safer areas; on raising output of safer crops in risky areas; or on reducing risk to a given crop and area.

Research directors will want some helpfully general formulation of the problem: 'Should research resources be diverted from crop A to crop B?' To simplify, we assume that research success is achieved at a particular time from the start of work (say three years) or not at all; that each farmer adopts exactly two years after research success, or does not adopt at all; and that extra net output appears exactly one year after adoption, or not at all. (These assumptions are easily made more realistic, but they simplify the argument). Then, a unit of research expenditure should be diverted from Crop A to Crop B (for a given agro-climatic zone) if, and only if, we thereby increase the amount that results from multiplying three expected values:[8] that of *success* after three years; that of *adopting* area after two more years, given success; and that of discounted present worth of net *value added*, after a further year, due to the innovation (e.g. of an MV), per unit of adopting area. If desired, we can incorporate weights into the last two factors, to allow for the greater desirability of benefits if they accrue to poverty groups. We can also allow for any special drawbacks, e.g. planners' risk-aversion, over and above discounting, to delay and riskiness of research; for interactions between the expected values of success, of adopting area given success, and of value-added per acre given adoption; and for more realistic assumptions about the likely time-distribution of research success, of adoption, and of output and sales.

There will be limited information available to do these sums (estimating probabilities and delays for each crop). However, in all cases implicit weightings are made by research station directors in determining what crops and locations to emphasise. Therefore the sums are always done implicitly. The 'hidden agenda' of planners and research directors – e.g. the importance they attach in practice to sure-thing research, to poverty-focus, etc. – would be much clearer, if these processes were made explicit.

(c) Food crop types: elementary biology, breeding, and poverty

On the options and limits, set by plant structures and types upon breeders' goals in food crop development for poor people, the messages of natural scientists are clear and not really controversial. Yet, because they are 'technical' in somebody else's discipline, the planner or economist listens to these objective messages far less than to the much more controversial judgements about goals. We apologise, as economists, for expounding elementary biology – and, despite patient help from several scientists, for probably getting some of it wrong; but we need to explain just how the basic plant science, as we (partly) understand it, affects the options for poor people.

A few distinctions are crucial. They are (i) between plant *populations* heterogeneous and homogeneous in a field; (ii) between inbreeding and outbreeding *plant types*; and (iii) between features of a particular single plant's *characteristics* (such as height): (1) to what extent the characteristic is caused by heredity, environment, or their interaction; (2) in so far as hereditary, whether it is due to one gene-pair (major-gene) or several (polygene); (3) in each of the characteristic's causative gene-pairs, whether the two genes (one inherited from each parent, or from male and female parts of a single inbreeding parent) are the same, so that the pair is 'homozygous', or different, making it 'heterozygous'; and (4) if the latter, which (if either) effect of the pair of genes, e.g. tall or dwarf height, is 'dominant' and which 'recessive'.

We need to understand these distinctions because they interact with each other (and with breeders' decisions) to produce a paradox. Decisions on breeding strategy are motivated in large part – at least for public-sector and international research centres – by incentives, financial or idealistic, to produce varieties and methods especially helpful to poor consumers and farmers. Researchers have achieved major successes in this. Yet these very strategies – and their impact on other plant scientists and on farmers – can threaten plants' diversity, in ways potentially damaging to the poor. Pressures are towards (i) more homogeneous plant populations; (ii) inbreeding with homozygosity – or else hybridity with hetero-zygosity that can attain the results the farmers want only if the seeds are produced each year, in a uniform way, by a few research centres (instead of in many different ways by millions of small farmers); (iii) plant characteristics genetically control-led so as to be uniform across many environments, and often linked to a major gene.

The pressures are not all towards uniformity (Chapter 7, i). Where they are, a major reason has been to create and stabilize high-yield and low-risk sources of cheap food for poor people. But the pressures do generally favour the search for these 'poor-friendly' outcomes by methods that reduce diversity *in the field*: of crops; of varieties of a crop; of breeding strategies and control systems; and of the genetic origins of plant characteristics. Chapter 2, k, shows that – unless compensating diversity is available in, and readily dispersed to farmers from, research stations – the very strategies that have so helped the poor by reducing the short-run risk of plant diseases can well increase it in the long run. First, however, we must look at the basic biology of the above distinctions in a little more detail.

(i) If a plant population is homogeneous, all plants in a field – if it were possible for them to have the same 'environment' (farmers' treatments, light, water, temperature, terrain, nutrients, soil bacteria, worms, weeds, diseases, etc.) – would achieve the same characteristics, e.g. maturity to harvest, at exactly the same time. This condition is not strictly possible even in logic (some plants must be at the edges of fields). However, a field of homogeneous plants normally varies much less, is much more homogeneous than other typical fields, e.g. sown to a land race of wheat, or to mixed crops. Such fields of heterogeneous plants, while usually selected for uniformity in some desired respects such as harvest date, behave differently in other respects, e.g. resist different pathotypes, even if the 'environment' is the same.

The historical process, by which farmers have increasingly abandoned the responsibility for seed selection to professional researchers, has tended to move fields from mixed crops towards single crops; from mixed to single varieties; and even, to a considerable extent, from heterogeneous to homogeneous single varieties (for, although breeders' varieties, notably of rice, often include genetic materials from many different land races and even nations, the aim is to release to farmers at a much higher level of 'purity' than is normal in farmers' own retained seed). Researchers prefer tasks that can be simplified enough to be tackled systematically, and that produce stable and widely applicable outcomes. Complex mixtures of crops, and even heterogeneous varieties of a single crop, are – precisely because adapted to local conditions – normally too varied to be researched on a standard scientific approach.[9] Such an approach often achieves much faster improvements in yield, and even in short-run resistance to pests or droughts, than farmers can do – hence their readiness to delegate seed selection to researchers and to adopt, over large areas, their results. Good varieties drive out less-good – and, while raising yields, also raise long-run risks.

However, the relative security against drought or flood or disease, achieved by poor farmers through heterogeneity, has to be replaced in some way if risk levels are to be tolerable for poor people who farm – or eat – the much more homogeneous plant populations generated by researchers. Resistance breeding and germ-plasm collections (Chapter 2, j–k) are part of the answer, but such genetically near-uniform populations as the MV wheats of North India, Pakistan, and Bangladesh remain more prone to viral or fungal attack in the long run – probably due to the very process of homogenizing research that has made them more resistant (and higher yielding) in the short run – than the heterogeneous plant populations (including TV wheats) that preceded them.

(ii) Another key distinction is between inbreeders and outbreeders. All inbreeders and outbreeders have male and female organs on each plant. Outbreeding species have evolved a crossing mechanism; normally, opening flowers; normally, some physical factors inhibiting self-pollination [Ford, 1965, p. 45]; and (almost) random mating. Inbreeders have usually evolved closed flowers, and produce all but a few per cent of offspring by 'selfing', i.e. self-pollinating from the male to the female part of the same plant. Most plants are outbreeders, including major food staples such as maize, rye, and bulrush millet [Simmonds, 1979/81, pp. 23, 81], and each such plant inherits a set of genes from each of two parent plants. A few seed-propagated annuals, including some major cereals (wheat, rice, barley, oats), are inbreeders, normally inheriting both sets of genes from the same parent. Tubers such as yam and cassava are outbreeders if propagated by seeds, but of course retain the exact genome of the parent plant if propagated by clonal means. Though the distinction between inbreeders and outbreeders is not cast-iron,[10] it crucially determines the strategies by which plant breeders seek manageably uniform plant populations, because elementary genetics links inbreeding to homozygosity and homogeneity, and outbreeding to heterozygosity and heterogeneity; and because crops could survive under evolutionary pressure only if adapted to those links.

A breeder of an inbred crop such as wheat or rice, therefore, has normally to go for a homogeneous, homozygous plant population with the desired characteristics. If the variety is fairly robust, it will be stable once released – the farmer can keep the rice or wheat seed, and almost all offspring from the seed will show the same features as their parent.[11]

A breeder of an outbred crop, such as maize or millet, has a harder task to obtain stability and control. Because the crop is normally outbreeding, heterozygous and heterogeneous, farmers who – having planted an improved set of seeds – then retain mature plant seeds will, next season, experience severe deterioration in plant quality. Where plants reproduce by seeding, breeders can prevent this deterioration only by population breeding or hybridization (see Chapter 2, d). These preserve, respectively, either (more or less) stably varied or homogeneous plant populations. Hybrids need reissuing yearly, but yield somewhat better than the alternatives.

(iii) How are poor people's gains or losses from MV research related to our third group of distinctions: to the alternative pairs of features of a single plant's *characteristics*, as analysed above on pp. 33–4 under (iii)? Let us take the characteristic 'plant height'. We start by temporarily assuming that height is determined entirely by a single 'major gene' (which does not determine any other characteristics), and does not interact with environment in any particular plant. Hence a plant, once genetically programmed to grow to a particular height, will in any particular environment either mature at that height exactly, or die well before it can pollinate, but not adjust its mature height. (These crude assumptions are not needed, but simplify the discussion.)

Let us assume that the environment, and all of the genome except for one 'major gene' affecting height, are fixed. How do the other 'alternatives' that determine a characteristic of a *single plant*, such as plant height – (3) the homozygosity or heterozygosity, and (4) the dominance or recessiveness, of the (assumed) non-environmental major-gene effect – relate to the characteristics of *crop* (inbreeder/outbreeder) and of *plant population* (homogeneous/heterogeneous) already discussed? Why does this interaction induce uniformity of breeding outcomes, reinforcing the paradox that the uniformity harms the poor yet the outcomes help them,[12] and raising the question of whether we can retain the outcomes without the uniformity?

Suppose a plant inherits different major-gene effects, say tallness from the male parent and dwarfism from the female parent (or from male and female parts of the same inbreeding parent). If the plant becomes dwarf, that effect is called 'dominant', and the other effect (tallness) is called 'recessive'. If a different effect is inherited from each parent (or parts of the same parent) in this way, the gene-pair is called 'heterozygous'; if the two effects are identical (so that for the individual offspring dominance or recessiveness does not matter), the gene-pair is called 'homozygous'.

Let us write H for the assumed dominant effect of the height major-gene, say dwarfism, in a particular group of rice varieties; and h for the assumed recessive effect, say tallness. We describe a plant's height genes by writing, in order, the effect inherited from male and female outbreeding parents (or inbreeding parts). HH is obviously dwarf, but (since dwarfism is assumed dominant) so is Hh or hH; but a tall plant must be hh.

Consider a population of 4000 plants – 1000 each HH, Hh, hH and hh – each of them a pure inbreeder, and each producing the same number of viable offspring, say four. Now, all 4,000 offspring of an HH parent inherit H from both female and male parts, and are therefore HH. Similarly, all 4,000 offspring of an hh plant are hh. The 1000 Hh inbreeders should produce 1000 each of HH, hh, Hh and hH offspring, since there is an equal chance that any offspring inherits either H or h from the male side of its parent (and similarly from the female side). Exactly the same applies to the 1,000 hH inbreeder parents. Therefore – because a heterozygous inbred parent produces half heterozygotes and half homozygotes, but a homozygote inbreeder parent produces only homozygotes – the proportion of homozygotes rises from half (assumed; 2000 in 4000) in generation 1 to three-quarters (12000 in 16000) in generation 2, and further to seven-eighths in generation 3, fifteen-sixteenths in generation 4, and so on. In a pure inbreeder, homozygosity is approached very quickly. This trend does not depend on a major-gene origin of a characteristic (i.e. on (iii) (2) on p. 33, para 4); it applies to inbred inheritance of any gene pair. In practice there are limits, preserving some heterozygosity;[13] but the pressures towards homozygosity in inbreeders are so strong that inbreeding, homozygosity, and survival in particular environments have to be closely linked. A breeder or farmer, selecting rice plants that are short and early-maturing, will find that successive generations of survivors carry increasingly few recessives for tallness or late maturity. Indeed, in a physical mixture of varieties of an inbreeder, even natural selection alone causes a field 'rather quickly [to] come to be dominated by one (occasionally two) components' [Simmonds, p. 117].

The position for outbreeders is simpler [*ibid.*, pp. 73–4]. Start with two maize plants homozygous for height, one tall (*hh*), one short (*HH*). Crossing them produces heterozygotes (*hH* or *Hh*). Crossing these heterozygotes produces a population half *hH* or *Hh*, and one-quarter each *HH* and *hh* – and the proportions, given undisturbed pure outbreeding and no selection, stay that way. The mating system of outbreeders protects recessives such as *h*; even though tall *hh* plants die out if they are unsuited to the environment, this process is usually too slow to eliminate the recessive tallness *h*-gene. Thus outbreeders, because the mating system preserves heterozygosity, carry a load of recessives that (i) produces some unfit offspring,[14] but (ii) renders the crop better able than an inbreeder to adapt to new or changed environments; if these are suitable only for tall plants, since *h* is preserved in outbreeders, new tall *hh* plants can emerge and thrive.

How does all this affect breeding strategies?

(d) From plant types to breeding strategies

Simmonds [1981, pp. 124–35, 147–55, 163–4, 199–200] explains how the four 'fundamental populations' of a crop – themselves defined largely by the facts set out in section c above – each lead to a distinct 'breeding strategy' for plant scientists. We summarize Simmonds's account of these populations and strategies. We then show how the strategies offer distinct scope for actions, by planners or socio-economists, to alter the impact of MV research on the poor.

Some generalizations about outcomes are nevertheless feasible across the strategies. In all four – although breeders draw genetic materials from many places, creating plants more diverse in origin than most that farmers select for themselves – the best bred variety is selected by farmers and reissued (and kept pure) by scientists; thus 'successful plant breeding tends to narrow the genetic base of a crop' [*ibid.*, p. 323]. Such success (if the breeder's criteria were well selected) helps the poor, but the associated genetic uniformity often harms them. Sections e-h of this chapter trace the impact on poor people of successful breeding interventions – affecting plant response to

nutrients, light, water, and pests and diseases respectively -
across all four strategies. However, we shall not understand
what a government or a socio-economist can reasonably ask of
plant scientists, unless we discriminate among the four plant
populations with which they deal.[15]

The first 'fundamental population' is the *inbred pure line* -
almost always of a seed-propagated annual. Such lines are IR-8
or IR-56 rice, or 'Mexipak' (Sharbati Sonora) wheat. Pure lines
have existed only for about 150 years, and have been selected,
largely by scientists, from 'land races'; some of these land races,
especially inbred sorghums and West African red rices, still
exist. Inbred line – however diverse in its origins - increases the
tendency of inbreeder plant populations (e.g. wheat or rice)
towards homozygosity.[16] They are also highly homogeneous.
Glasshouse generations, descending from single seeds selected
from each plant in a large sample, can greatly accelerate the
breeding of inbred varieties; bio-technological methods can
already take this process a step further (Chapter 7, f). When
research methods diverge so greatly from farming conditions,
it is even more than usually important to conduct pre-release
trials of a recommended variety's safety and profitability in the
environmental, economic and managerial circumstances of
smallholders.

Open-pollinated populations are 'ancient', seed-propagated,
and – unlike the other three populations – heterogeneous:
'uniformity. . . is impossible and trueness-to-type is a statistical
feature of the population as a whole, not a characteristic of
individual plants'. Breeders improve these populations by
increasing the frequency of desired gene combinations. Populations are stabilized *either* by mixing plants so that the
frequency of gene combinations (and thus a reasonable degree
of uniformity) can be maintained by random mating and
farmers' seed retention, *or* by steady annual resupply of a plant
population reconstituted from parents by breeders. Plainly
the latter is less 'poor-friendly' unless resupply is unusually
competitive, timely, inexpensive, and reliable. The former
method, to some extent, combines two 'poor-friendly' features: farmers' control over seed supply, and adaptable
(heterogeneous and heterozygous) plant populations.

Hybrids, invented in the US in the early 1900s (though not developed until the 1930s), exploit the depression of yield due to forced inbreeding, especially in naturally outbred crops like maize, millets, and many sorghums. Inbred plants, in such heterozygous populations, often inherit harmful recessives from both sides of the same parent (i.e. become homozygous in a locus where the effect is harmful), and die before harvest, or produce low yield per plant. Crossing two inbred plants, with distinct gene structures, reverses the process, restores 'hetero-zygote advantage', and therefore normally produces offspring with heterosis or 'hybrid vigour' - yields above those of either parent. However, a breeders' learning process, leading to careful selection of inbred parents for 'combining ability', is essential; seed cannot be retained by farmers without major yield reduction in future generations, so they must rely, on hybrid seed researchers and distributors. Only hybrids, there-fore, offer obvious attractions to private sellers – even in the absence of 'plant breeders' rights'.[17] If private breeders of hybrid seed are fairly competitive, as in the USA – or if public seed distributors are fairly reliable, even for smaller and remoter farmers, as in parts of India and Kenya – then 'hybrid vigour' can produce massive benefits, even for poorer growers of maize or sorghum. More typically, certainly in most of Africa, neither public timeliness nor private competitiveness can be relied upon by poorer farmers; this justifies CIMMYT's preference for maize strategies based on open-pollinated population management as more appropriate – even if at somewhat lower yield potential – and casts doubt on IITA's preferred strategy of hybrids. Certainly, deliberate develop-ment of hybrids of natural inbreeders such as wheat and rice – e.g. via cytoplasmic male sterility (pp. 369–70) – whatever its gains via heterosis, carries serious risks for poor farmers who face monopolistic seed suppliers.

Clonal breeding is the vegetative propagation (from budd-ings, cuttings, etc.) of perennials – sometimes, as with cassava, farmed as quasi-annuals – to produce identical genetic copies of an original heterozygous plant. Like hybrids, clones depend on fixing a particular set of attributes, thus producing homog-eneity from a naturally heterogeneous population; unlike hybrids, clones can be propagated by any farmer who obtains them, and need not be reconstituted (and distributed again in each generation) from carefully-selected parents.

*   *   *

The above account relies on Simmonds [1981], whose summary (p. 127) of the 'homogenizing' changes, due to 'plant breeding. . . in a specific social-agricultural context' is especially useful. Variable land races have given way to uniform inbred lines 'nearly universally'; open-pollinated populations have 'recourse to fewer but more thoroughly selected parents', and give way to hybrids and where feasible to clones; and 'mixed clonal plantings [are replaced by] uniform cultivation of few clones over large areas'. Such pressures to uniformity and off-farm control arise because farmers want the short-term results; the new plant populations are more fruitful or more stable than the old. However, the gainers need not always include the poorer farmers, workers, or consumers. Such groups carry the greatest long-term risks if the stability breaks down and there is too little diversity in reserve.

(e) Nutrient response

Short varieties are designed to benefit from much higher fertilizer inputs than traditional tall varieties. This is because, even when extra nitrogen, phosphorus and potassium are turned into heavy grain heads, the short and stiff stalk holds the plant upright. Other plant characteristics, sought by breeders of short varieties, tall hybrids, and improved clones, also aim to improve the plant's response to scarce nutrients. However, though *some* fertilizer is *usually* needed for *substantial* benefit, it is mistaken to conclude [Borgstrom, 1974, p. 14] that 'fertilizers are a *sine qua non*' for farmers to benefit from MVs.[18] Early MVs, such as IR-8 rice, sometimes did worse unfertilized in poor soils than TVs [Barker, 1978, p. 49]; but recent MVs seldom do worse than TVs, even if farmers are too poor to purchase (or risk) fertilizers.

This is partly because plants do not live off nutrient sources (e.g. fertilizers) but off nutrients.[19] Usually, most nutrients come from non-fertilizer sources: often from irrigation-water, but mostly from the soil itself, enriched by bacteria, fungi, stubble, manure (often from several years ago), nitrogen fixation by legumes (from current or past seasons), or worm casts [IRRI, *Ann. Rep. 1972*, p. xxvi; Olson *et al.*, 1941, p. 188]. MVs are designed to make better use of nutrients irrespective of source, and hence normally turn them into more grain than do TVs, even if no such nutrients come from fertilizer [Swaminathan, 1974, pp. 11–13].

This greater all-round nutrient-using efficiency of MVs, far from damaging poorer farmers who cannot obtain fertilizers, can sometimes be especially helpful to such farmers. Thus CIAT – the international crop research centre responsible for beans, cassava and (in tropical America) rice – explicitly seeks 'agriculture that is less dependent on fertilizers', and stresses that 'genetic manipulation [can improve plants'] capacity to more efficiently use nutrients in the soil'. This 'minimum-input agriculture. . . underlies all [CIAT's] research strategies' and seeks to keep agricultural production cheap, especially for 'subsistence' farmers [CIAT, *Annual Report 1985*, pp. 1–2, 6]. Such 'extractive efficiency' (p. 47) has dangers, however; it is too early to assess the scope for this approach.

MVs do often render a low-input approach feasible in principle. Many MVs persistently outyield TVs at zero fertilizer input. But for a substantial advantage there are preconditions – and, unless they are met, it is risky to seek 'low-input high-output agriculture'. Before exploring them, we look at the evidence that the goal *is* sometimes feasible.

Even in the early years, many MVs outperformed TV predecessors without added fertilizers. While fertilizers were usually needed to maximize expected profit, its ratio to expected cost – an attractive safety indicator for the very poor farmer – was highest when none was added, for typical MVs of all five main cereals [Lowdermilk, 1972, p. 243; Kahlon, 1974, p. 5; *IRRI Reporter*, 3, 1973, p. 4; Pal, 1972, p. 95; IRRI, 1975, pp. 19–21; Ryan and Subrahmanyam, 1985, pp. 11–13]. Unfertilized, on the same plots, older rice MVs greatly outperformed TVs for over twenty-five seasons running – but on very fertile IRRI plots [IRRI, *Ann. Rep. 1973*, p. 100, and *1984*, p. 316], probably with water supply adequate for the natural occurrence of nitrogen-fixing organisms in the soil [Swaminathan, 1974a, p. 69].

More interestingly, recent MVs usually outyield TVs with zero inorganic fertilizer (although their yield advantage is proportionally greater at higher inputs) *even under quite unfavourable conditions.*

- Indicating the progress of breeding for efficient nutrient use at very low fertility levels, 'experiments. . . on land that grew unfertilized maize during this and several previous summers, and. . . of low fertility' showed recent releases 'significantly higher-yielding than the old variety at all N application rates', and apparently so at zero fertilization, though at low significance levels [CIMMYT, 1982, p. 111]. Maize MVs, selected to be suitable for maize to be picked green, now clearly outyield TVs even at zero fertilizer [IITA, 1986, p. 35].
- As for wheat, an ever-increasing emphasis on wide-spectrum pest and disease resistance enables most MVs to outperform TVs, if the moisture regime suits MVs at all, even with zero fertilizer input [Byerlee and Harrington, 1982, pp. 1–2; Parikh and Mosley, 1983, for wheat in Haryana].
- This advantage over TVs without fertilizers has also appeared since the late 1970s for many rice MVs [IRRI, 1978, pp.176–80] – for IR-36 even under moisture stress [Barlow *et al.*, 1983, p. 86].
- Most encouragingly, it has also appeared for sorghum, the classic 'poor person's crop', under farmers' own management and in rainfed conditions, with hybrid MVs selected to resist *Striga* (witchweed) – after drought and birds the main threat to sorghum in Africa and much of Asia [Rao, in ICRISAT, 1982, vol. 1, pp. 49–50; ICRISAT, *Ann. Rep. 1984*, pp. 53–4].
- For ragi millet *Eleusine corcorana*, also grown and eaten mainly by the very poor, the Indore series of MVs greatly raised monsoon-season output on marginal and drought-prone lands in Karnataka State, India, without fertilizer [Rajpurohit, 1983]; these improvements may be transferable to parts of Zimbabwe.

However, there are probably not so many cases in which big, safe, *sustainable* gains can be expected from MVs in poor soils with no inorganic fertilizer. Indeed, there are exceptions where MVs do worse than TVs in such conditions [Simmonds, 1981, p. 361]. Actions to improve phosphorus availability are often necessary, e.g. before MVs of millet and sorghum have much to offer in the Sahel, or of wheat in Chile [CIMMYT, 1983, pp. 136–7], or of barley in Syria. Some MVs can also require extra outlays for zinc before they outyield TVs [Narvaez, 1973, p. 269].

More generally, MVs with high yield potential at low or zero fertilizer may achieve it in part by allocating mass to grain at the expense of roots or leaves. If so, the MV may need better water provision (to allow for reduced root systems), or better weed control (to allow for smaller leaves and less shading), if they are to realize much of that potential.

In the great majority of cases, however, MVs – if grown at all – initially do at least as well as TVs even with no extra fertilizer.[20] Even then, however, a farmer living in an area suitable for MVs, but too poor to afford the risks, loans, or transport involved in acquiring fertilizers, may well not gain by adopting the right MVs. It is not that simple.

*   *   *

There is no free lunch. Extra grain, produced via a MV without extra fertilizer, must get the extra required nutrients from somewhere. A variety's extra efficiency in using nutrients seldom comes entirely from greater conversion efficiency (p. 47), and/or from shifting dry matter, due to nutrients, into grains instead of into stalks, leaves or roots.

Suppose a farmer replaces a TV by an unfertilized MV, while maintaining established crop-mixes, rotations, manuring, stubbling, etc. Such practices may well have proved adequate to maintain the soil's nutrient status over many decades, while only TVs were used; for TVs were not very good at turning nutrients, especially soil nitrogen, into dry matter. If the practices continue after the switch to the MV – which is 'better' at extracting soil nutrients and turning them into dry matter – then in the long run those nutrients will normally be depleted.[21]

The risk is increased because the switch to MVs provides incentives to abandon some of the old restorative practices. In particular, because MVs offer higher yields, farmers are tempted to abandon attempts to restore soil nitrogen by means of rotations or mixes (involving legumes or pulses, formerly alternated with TVs). It becomes very attractive, instead, to grow continuous, unbroken stands of MVs of the most successful crop, often twice a year. Poorer farmers, because poverty compels them to value immediate income highly over future income, are especially liable to temptation – above all if doses of nitrogenous fertilizers can be reduced substantially below recommended levels with little loss of profit and much risk-reduction *this* season [Mandac and Flinn, 1983], at the cost of possible sharp falls in soil fertility later, perhaps much later. Shorthold or insecure tenants – heavily represented among the poorest farmers of Java, Bangladesh and Central Luzon – are especially likely to be tempted by such 'soil mining', since such a tenant is unlikely to be farming the same land when its fertility falls off.[22]

These facts do not justify naive ecologism – the view that even economically justified and scientific modernisations of farming methods are almost certain to threaten sustainable environments. There are, however, clear implications, for national and international crop research seeking to help poor people, from the characteristics of MVs as 'efficient' nutrient extractors and from the incentives created by these characteristics. First, it remains right – under appropriate constraints, reviewed below – to seek MVs that outperform TVs at low or zero levels of fertilizer, because poor farmers often face special difficulties in obtaining or affording appropriate or timely fertilizer.

Second, however, to ensure sustainable farming[23] where such MVs are adopted, researchers need to compare alternative MVs and techniques (in assorted field conditions and farming systems) not just for improved 'performance without fertilizers', but also for how it is achieved:

- If by the MV's higher *partition efficiency* – a higher ratio of grain to roots, leaves, or stalk – can plants (and soils and people) sustainably tolerate the resulting effects, respectively, on water conversion, weed growth, or fodder availability: even through bad years, or in the long term?

- If by higher *extraction efficiency* – a plant system better at removing NPK, zinc, etc. from unimproved soils – can that MV be kept away from lands that are not rich in the nutrient(s) which it extracts? If not, poor people on such lands, who often cannot compensate by increased use of appropriate fertilizers, may be at great risk.
- An unfertilized MV's improvements in yield are 'safe' only to the extent that they are achieved by better *conversion efficiency*, i.e. by turning nutrients into more grain without depleting the soil, other parts of the plant, or other crops with which the MV is mixed or rotated. Soil depletion, in the long run, can render soils inadequate for safe growth, especially for non-grain plant parts (selected against, in order to raise partitioning efficiency) essential for water or nutrient management.

The difference between a plant's efficiency in nutrient conversion, extraction, and partition is researched by plant scientists.[24] However, the choice of which MVs or recommendations to research or release appears to be little, if at all, guided by the need for high conversion efficiency – and by the dangers of high extraction efficiency[25] – at the low, or zero, levels of fertilizer use typical of very poor farmers and regions. Thus CIAT's advocacy of 'genetic manipulation [to improve a plant's] capacity to . . . use nutrients in the soil' [CIAT, 1984, pp. 1–2] does not specify conversion efficiency, as opposed to extraction efficiency, as the goal; neither do its reported experiments usually appear to make the distinction. More worryingly, IITA's advocacy of, and successful search for, cassava cultivars that outyield TVs unfertilized, especially when combined with the already severe extractive effects of shortening fallows (due to population growth) in much of West Africa, may involve grave long-term risks. The risks are gravest for the poorer farmers, labourers, and areas, because these are relatively: tempted to adopt short-run yield-enhancing varieties and methods even at long-run cost; unlikely to overcome the problems (cost, risk, transport, access) of organic fertilizer; and unlikely to find new income sources, or to rehabilitate their lands, if exhaustion does occur.

Researchers, then, should consider making experiments to select MVs and methods that raise conversion efficiency of locally scarce nutrients into grain when no or few fertilizers are added – and to distinguish this from, and to test for long-run risks of, higher extraction efficiency and partition efficiency. Are there other strategies to improve the sustainable impact of MVs' nutrient characteristics on farms too poor to use, or soils too poor to justify, much inorganic fertilizer?

(i) Researchers should go beyond single-season 'response curves' showing yields at various fertilizer levels. These curves appear to select 'optimal' combinations of an MV and a fertilizer input, but to ignore later seasons. The effect via soil nutrients on farm profits, risks, and employment over several seasons, especially at low or zero levels of inorganic fertilizer use, needs more research. A good but rare example, in the wake of the new sorghum hybrids for poor soils (Chapter 3, section i), is ICRISAT's finding that continued sorghum cropping is bad for long-run nitrogen balance and hence long-run yields, and with some improved sorghum cultivars can be improved by introducing rotations with millet (though the best millet varieties are not yet specified: ICRISAT, *Ann. Rep. 1984*, p. 38).

(ii) Before choosing which MV to release, and which practices to recommend, researchers should enquire how increased organic fertilizer use can be made more profitable (or reduced levels deterred) and should review long-term benefits of such methods for different farming systems. For example, some MVs are better suited than others to reaper-binders and combines; these normally reduce organic nutrient restoration via stubble.

(iii) Rapid MV progress, even in initially unproductive areas and especially if combined with better water control, can make adoption of inorganic fertilization attractive, even to small farmers. This is a research priority, since such areas will not solve their central food problems without fertilizers sooner or later. More competitive provision of transport and credit, often requiring State involvement at least initially, speeds this process up, and thus normally[26] reduces the long-run dangers of depleting scarce soil nutrients.

(iv) Several new methods of nutrient enrichment in tropical and sub-tropical soils, mostly for nitrogen, are at various stages of research or diffusion. The methods range from slow-release fertilizers, through enhancement of nitrogen-fixing organisms (bacteria, mycorrhizal fungi, blue-green algae, azolla), to improved nitrogen-fixing capacity in the plants themselves. Perhaps surprisingly, some of this 'frontier technology' is especially promising for poor farmers – and labourers – in unproductive areas.[27] We return to these options in Chapter 7.

*　　　*　　　*

For poor people, then, MVs' impact upon nutrient efficiency is generally good, because MVs, in general, make better use of nutrients from all sources than TVs. Increasingly, this is also true when little, or no, inorganic fertilizer is added. Moreover, MVs increase the incentive to add some, or more, inorganic fertilizer, because they turn more of it into grain.

There is an important caution. It applies to a side-effect of the MVs' welcome recent spread into some marginal and rainfed lands in Asia, and of the much-needed thrust to increase this spread in Africa, and to apply it to lower-value crops such as cassava. This raises the likelihood that MVs will be grown in places too unproductive, even at the margin and with the higher yields produced by MVs, to make it worth the farmer's while to add much (or any) organic or inorganic fertilizer – especially for a poor farmer with short time-perspectives, and for a low-value crop. In such circumstances, the ecological risks may be very serious, and require much more systematic research.

Moreover, without fertilizers, it is seldom that really big yield increases – above, say, 30 per cent – can be achieved *merely* by switching from a TV to a MV. If yield advantages are that small, three factors may deter poorer farmers from adopting MVs.

(i) The grain produced by MVs in most cases has quality characteristics that lead to a 10–20% price discount (p. 214),[28] though this is less important for poor farmers, who want inexpensive calories for home use rather than sale.

(ii) Although MVs usually have more plants per acre, their straw is sometimes worth less per acre (as animal food or for cash) than for TVs, which grow longer, less stiff, and more digestible straw [Johnson, 1970, p. 188; Lowdermilk, 1972, p. 488; von Oppen and Rao, in ICRISAT, 1982, vol. 2]. Across twenty-four sorghum cultivars, a 10 per cent rise in grain yield was associated with a 3 per cent fall in digestible straw yield [McIntire *et al.*, 1985; result significant at 5 per cent].

(iii) MV seeds – especially hybrid seeds, and above all in remote areas or in the wake of bad harvests – are occasionally costly, and quite often difficult to obtain on time.

So MVs often require fertilizers, to be really attractive to farmers. Yet the nutrient requirements of many MVs are more complex to manage and subjectively riskier than the TVs they displace. The MVs' nutrient needs also often require farmers to obtain new information and new purchases: often extra nitrogen is needed to *optimize* with the MV but not the TV; sometimes, zinc constraints are revealed when more nitrogen is used; and so on. All this may disfavour small farmers via 'diseconomies of small scale even though [MVs *per se* do] not' [Burke, 1979, pp. 148–9]. To get urea or zinc or information, for a poor farmer considering putting half an acre under a new variety, is much more than ten times as costly (per unit of expected income) as for a rich farmer who can assign five acres.

Yet the proof of the variety is in the adopting. Tens of millions of tiny deficit farmers have adopted not just one MV but several improved MVs successively, not just in reliably irrigated areas such as the Punjab but in unreliably rainfed – and lightly, if at all, fertilized – areas of SW India, Bangladesh and Central Kenya. This spread to non-favoured areas has brought much higher output, some risks, and a few near-disasters (Chapter 3, h), but not, so far, due to failure of, or soil exhaustion by, MVs with low or zero fertilizer. Nor are any such risks *necessarily* greater for small, low-fertilizing farmers; sometimes, indeed, the small farmer's greater endowment of labour-per-acre means that eventually he or she uses *more* fertilizer, per acre of MVs planted, than the large farmer [Asaduzzaman, 1980; Herdt and Capule, 1983].

As the spread of MVs to more marginal areas and crops continues, however, risks from their 'extractive efficiency' will rise; and it will seldom pay poor farmers to offset such risks by fertilizing low-value crops such as cassava. So researchers need to seek MVs efficient at converting a given nutrient intake into human food, rather than at extracting nutrients from the soil – especially under moisture stress,[29] and in mixed-cropping systems. An MV that was especially suitable for such systems would not only have major equity benefit (because they are most often found on small farms, and are employment-intensive: Jodha, 1980), but would also share soils with crops that restored some of the nutrients the MV used up.

Finally, in assessing research priorities for nutrient use in MVs, long-run considerations may be unduly affected and obscured by the 'dance of the dollar', and of prices of oil and natural gas as feedstocks for nitrogenous fertilizer. It usually takes several years for rises (or falls) in the world value of feedstock and dollars to work their way into fertilizer prices.[30] Such price changes, in turn, provide incentives to farmers and governments, in respect of nutrient use and management, that are often irrelevant to the *long-run* price of fertilizers. In the early 1980s, expensive oil (and dollars) sharpened the urge of governments to finance research that, by the early 1990s, should increase the conversion efficiency of N into grain; this is good, because the long-run price of fossil-fuel-based feedstock has to be rising. But it is absurd if in the late 1980s such research is curtailed because cheaper oil (and dollars) now misinform governments that, since feedstock and hence fertilizers look cheaper, soil exhaustion and N-depleting MVs are perfectly all right. The short-term movements of fertilizer prices have almost nothing to do with their ineluctable long-term increase, which is based on the rising marginal cost of mining N-feedstock (mainly oil and natural gas) and rock phosphate. Some feedstock economies are feasible and desirable [Herdt, 1981] but, by the year 2000, inorganic N and P will be much more expensive than today. If the poor are then to gain much from MVs, researchers need to explore ways to improve MVs' nutrient-grain conversion efficiency, notably by reducing N losses from the plant [Craswell and Vliek, 1979], but also by better placing, timing, and selection of fertilizers.

(f) Light response

Plants get nutrients from many sources, natural (soil, water, air) and human (fertilizer, manure, compost. . .). There are thus many complex interactions and options to consider in developing MVs and techniques that help poor people make the best use of plant nutrients, especially as many interacting nutrients are required. Sunlight for photosynthesis, however, is available from only one source; its amount and timing cannot be changed or altered, or readily replaced. Therefore, amendments to plant types and farm practices are the only ways in which sunlight (unlike nutrients) can be better used to make food. Since light is always free, and since poor people are under no temptation to use it up, a poor person's plant gains as much as a rich person's, if scientists improve the efficiency of plants in turning light into food.

Can this be done? Probably, but not directly. 'Erect leaves to prevent mutual shading' [Peiris, 1973, p. 4], and also other means towards directly improved sunlight-to-grain conversions, were once seen as a major benefit to be gained from MV research. But these hopes have largely disappeared. Empirical work suggests that biomass responds little, if at all, to breeders' efforts to increase a plant's photosynthetic efficiency – i.e. its capacity to convert sunlight into biomass [Simmonds, 1981, p. 49] – although more needs to be understood about the subject. (It is true that MVs which bring an increased ratio of food to dry matter, or reduce the proportion of plants lost to diseases and pests, thereby increase the ratio of food output to biomass, and therefore to light – as well as to water, land and labour.)

However, another aspect of response to light – photo-period sensitivity (PPS) – is a more important variable for plant breeders than photosynthetic efficiency. With a high-PPS variety of any crop, while total sunlight still determines total biomass (given other conditions), the timing of the dark periods – arrival of night, of a season, or of cloud cover – strongly influences the time of flowering, and hence of maturity. Low-PPS plants, the early aim of MV breeders, can thrive despite variable day-length and cloud cover. This permits a crop to be planted whenever temperature, water, nutrients, and light (however timed) suffice to bring it to harvest. This, in turn, enables seasonal smoothing of food output, work availability and food prices, and conversion of one-season into two-season food agricultures, in many areas [Bolton and Zandstra, 1981; Dalrymple, 1979]. Since poor people have the greatest problems in carrying stocks, saving or borrowing – and are thus most damaged by seasonal fluctuation – they gain most from such PPS bio-engineering. Also, non-PPS plants often do well because they flourish even if late rains require that planting is delayed to a time that – with a PPS variety – would mean exposure to unusual day-lengths or cloud cover [Sen, 1974, p. 37; Lipton and Longhurst, 1975, p. 67]. Finally, these advantages tend to spread; low-PPS varieties are especially likely to be transferable between areas, as well as seasons, of differing light–dark patterns [Dalrymple, 1985, p. 31].

However, agricultural research should seek, not to breed non-PPS plants as such, but to develop varieties and practices that suit local priorities and risks. In many circumstances, PPS is desirable. First, some areas feature (i) scarce planting labour and (ii) uncertain but generally low rainfall; day-length-sensitive varieties of millet and sorghum can spread the peak of planting labour (i.e. have flexibility of planting date) because, even if planted on slightly different dates, they come to flower, and hence to harvest, at the same time. Second, retaining some PPS may be especially valuable in sorghum because of its vulnerability to bird damage, particularly in Africa; if the 'appropriate' day-length does not signal each stage of development to the plant, it suffers because anything that matures even a few days differently from the main planting will, in some regions, be devastated by birds. (More generally, the timing of pest attacks can require a fixed sowing date to avoid them; PPS plants then gain by reaching each stage of their development only when the day-length signals that the season is ready [Swaminathan, 1974, p. 40].) Third, extra inputs, especially fertilizer, can also hasten physiological maturity, thereby enhancing the risk of bird damage, unless a plant is PPS. Fourth, PPS plants are often wanted because crops must 'mature towards the end of the rainy season, when favourable weather for sun-drying occurs' [Beachell *et al.*, 1972, p. 91; compare Frankel, 1971, pp. 52–3]. Fifth, low-PPS varieties can readily induce post-harvest innovation that displaces labour [Duff, 1978, p. 148].

The poor, the weak and the 'tail-enders' are especially vulnerable to unexpected delivery failures constraining timely operations. They want plants that are PPS in some respects but not others: low-PPS plants that allow some choice of sowing date; but also plants that can respond to subsequent administrative delays (in, say, water releases or fertilizer arrivals) by adapting the growing cycle, in particular the light responses, without great yield loss. Is this feasible, and if so at what costs (in tons of yield foregone)?

More generally, agricultural research centres, if they use socio-economists *early* in research design, could usefully ask: 'What light-response characteristics will generate most, safest real net income for poor growers, workers, and consumers of the crops we are mandated to improve, in the major (especially the poorer) areas where these crops are important? What are the probabilities of achieving key PPS-related characteristics by research?' At present the second question is asked first. The chosen characteristics appear (at least to scientific laypersons such as ourselves) to be brilliantly researched. But the first question, about what the poor want and will benefit from, is asked, if at all, too late: after the research results have been extended to farmers and in a spirit of 'what went wrong', not during research design and in a spirit of 'what is right'.

If poor farmers need MVs that are non-PPS early in the growing cycle but PPS later – as much of the above suggests – then breeding priorities will change. One complication is the interaction of day-length and temperature. For example, in field beans: (i) a longer day delays maturing to flower; (ii) warmer temperatures speed it up; but (iii) warmer temperatures also increase the impact of longer days in delaying maturity, and (iv) of course, longer days go alongside greater chances of warmer temperatures; also (v) sensitivity to moisture stress is greatly increased after flowering, as is the case with millet, but exactly opposite to maize (both often intercropped with field beans) [CIAT, *Ann. Rep. 1985*, pp. 12–13; ICRISAT, *Ann. Rep. 1984*, pp. 85–6; Fischer *et al.*, 1983, p. 2]. The strength of all five effects itself depends on the group of varieties from which breeders develop releases. This sort of complexity, which exists for many crops, requires careful research planning, not a crash programme against PPS.

(g) Water efficiency and moisture stress in MVs

Many social scientists have argued that typically an MV performs worse than a TV when the water environment is unsatisfactory.[31] They have often at least implied that researchers stress yield potential very highly, but at the expense of robustness under moisture stress (MS). Certainly, such a research strategy would be geared to the needs of better-off farmers in irrigated or reliably rainfed areas; poorer farmers are less able to bear risks, and to acquire most forms of irrigation [Narain and Roy, 1980]. Such a strategy might possibly be justifiable even on grounds of poverty alleviation, if irrigated areas offered the best prospects of increasing avail- ability of inexpensive food and/or of employing poor labourers.

However, to seek 'yield first, MS resistance last' usually makes little sense. We doubt whether this approach was ever tried by any major competent research team. Certainly since 1970, scientists have not been going all-out for yield and neglecting MS. The dichotomy is misconceived: better resist- ance to MS and disease is probably the main route to higher yields for most major tropical food crops. However, 'MS' and 'drought resistance' are much more complex concepts – and, even if measured in a grossly simplified way, are less easy to improve by research – than either the critics or the defenders of MVs have recognized.

The African rainfall deficiencies of 1982–5 included not only very severe droughts (against which no MV or TV can fight), but also – as in Machakos, Kenya – quite modest disruptions in rainfall timing and amounts that nevertheless produced great hardship. This should remind us that MS- tolerant food crops are at the very centre of poor people's requirements, even survival. So we must try to unravel the complexities of MS: to explain its meanings, and the mechanics of research to tackle it in a pro-poor way. First, however, we dispose of the myths that MVs are especially drought-prone, that researchers do nothing about it, and that poor farmers consequently suffer. After unpacking the concept of 'MS' – as it affects researchers, and poor people as farmers, workers and consumers – we next examine some research approaches (and some technical and political problems in various agroclimatic conditions) that affect the development of MVs and associated farming systems to reduce vulnerability. We conclude that, increasingly, most MVs use water more efficiently than TVs; that this should mean that MVs perform much better than TVs under MS; but that the gains, especially to the poor, are much reduced by the politics and technology of water management, including its frequent and arbitrary separation from crop research. The section closes by suggesting how MV researchers might improve matters.

*   *   *

First, then, the myths. It is a myth that breeders neglect MS.

- Well before 1972, IRRI's main goal was *rice* MVs with improved tolerance of MS [IRRI, *Ann. Rep. 1972*, p. 85]. Since then, in view of the widespread belief that short shoots accompanied short roots and inefficient moisture search, IRRI has deliberately sought somewhat less short varieties, so that plant roots can better seek out water during MS – at some cost to yield, both because slightly increased lodging reduces fertilizer response, and because partitioning (grain/stalk ratio) deteriorates [Johnston and Clark, 1982, pp. 90–1].
- By the early 1970s, *millets and sorghum* breeding aimed mainly 'to withstand MS' [Swaminathan, 1974, p. 29]. This goal remains a leading priority for bread *wheats* [Rajaram *et al.*, 1984, pp. 14, 18, 20].
- *New crop* development through wide crosses, while not yet of major importance to very poor people, reveals the same priorities; triticale was developed to breed rye's drought-tolerance into a wheat-like plant, and may prove especially useful for hill farmers facing MS in semi-arid sandy soils [Biggs, 1982].

Breeders increasingly seek resistance to MS. That is not only because poor people suffer most from it. It is mainly because

MS resistance is not seen as an alternative, in which researchers try to obtain immunity in bad years instead of seeking better yields, but as often the most cost-effective research route to better yields in average years.

For maize, deficits of water for one or two days during tasselling or pollination may cause 22 per cent reduction in yield. . . Drought may account for an average loss of 15 per cent of production in the tropics, even when total rainfall is reasonably high, [and in addition damages] utilization of fertilizer and other inputs [Fischer *et al.*, 1983, p. 1].

Moreover, MV research so far has produced its most striking results in reliably watered areas, and it may be time to assume rapidly diminishing returns (rather than to go for 'more of the same'); thus the previous director of CIMMYT saw the 'greatest potential for increasing wheat production in the developing world . . . in the more marginal environments' [CIMMYT, 1984a, p. 7] – inevitably implying more emphasis on yield stability, because in such areas MS is a typical, not only a drought, phenomenon.

Increased enthusiasm about MS reduction has come from questioning of the old belief that dwarfing of shoots, because of its effects on root structure, means less-efficient moisture search [IRRI, 1981]. For maize (while shortness has yet to be combined with really high yields) short stature usually goes with early maturity and, probably for that reason, with drought avoidance or resistance [CIMMYT, *Review 1981*, p. 32].

But perhaps breeders have failed in these efforts to develop MVs good at handling MS? That is also a myth. Even many older MVs usually yielded more in *absolute* terms than MVs under moisture stress [IRRI, *Ann. Rep. 1975*, p. 156]. They generally matured earlier, and thus avoided MS by being 'not so dependent on the late rains' [IRRI, *Ann. Rep. 1968*, p. 22]. Given total water available, MVs are usually less sensitive to its timing than competing TVs [Palmer, 1972, p. 51]. Even the older MVs, by making it pay for the farmer to apply more nitrogen, saved water, since fertilization normally reduces water use (per unit of dietary energy produced) by over 35 per cent for rice [Swaminathan, 1974, p. 40] and wheat [Borlaug, 1972, p. 586]; such extra nitrogen does *somewhat* increase yield

loss during MS (see footnote 29 to this chapter), but from a
much higher base yield with appropriate MVs than with TVs.

Recently, the MS-motivations of breeders have produced
even more striking successes. By the late 1970s, CSH-1 hybrid
sorghum was achieving 'spectacular' yields at farm level in
some drought-prone parts of India, and in rather dry years
[Rao, in ICRISAT, 1982, vol. 1, pp. 49–50]. Work at IITA
questions even the accepted view that carefully selected hybrid
maize does worse than traditional open-pollinated populations
under MS [IITA, *Ann. Rep. 1985*, p. 71].[32]

Even if breeders try to develop MS-resistant MVs, and often
succeed, are these varieties sufficiently profitable, safe, and
accessible to poor farmers? That they are not is yet another
myth. Farmers in many unirrigated places have adopted MVs
mainly for improved resistance to moisture, and other, stress in
the Philippines, Pakistan and Tunisia [Barker, 1971, p. 121,
compare Herdt and Capule, 1983, p. 15; Rochin, 1973, p. 140;
Palmer, 1972, pp. 54–7]. To some extent, this applied also to
IR-20 *aman* rice in Bangladesh.

*   *   *

It is a myth that breeders (i) place lower priority on MS
tolerance than on yield (or view them as alternatives), (ii) have
failed to generate MVs with better MS tolerance, or (iii) have
not reached many poor farmers with such MVs. Yet it is a fact
that vast areas of unreliably rainfed rice, flood rice, and semi-
arid crops – mainly grown, worked, and eaten by poorer rural
people - remain under TVs, in large part because water
conditions are unsuitable for MVs. If MS research is so clever,
why are these people so poor?

The answer depends on sorting out types of MS problems –
and in identifying those best handled by alternative strategies
from plant breeders, by other plant scientists, by changed
economic options to farmers (whether via incentives or via
irrigation), or otherwise. The nature of the MS problem(s)
renders such 'sorting out' essential. A plant's behaviour under
MS is controlled by several genes, because it depends on many
aspects of plant physiology (root length and mass, leaf size and
transpiration, osmosis, etc.) and plant chemistry. Most of these
aspects – notably root characteristics – vary with micro-

environment as well as heredity. With outbreeders such as maize and millet, heterozygosity creates further problems in stabilizing even those features of a plant's behaviour under MS that are largely hereditary. Given the great complexity of the breeder's tasks in attacking MS, therefore, it is vital for scarce breeding skills to be concentrated on problems defined much more precisely than is conveyed by the bald term 'moisture stress'.

Here are some key questions.

- What is the probability that a given water shortage has to be handled, for how long, at what period of a growing cycle, in what terrain and temperature? This is crucial because 'the pattern of moisture availability [varies across] semi-arid environments; hence the drought tolerance of a variety must be . . . specific' to an appropriate, though possibly large, set of locations [Rajaram *et al.*, 1985, p. 10].
- Given the crop, is MS best tackled by seeking an MV with more resistance, better recovery, or higher yield given imperfect resistance and recovery?
- Are such targets best achieved by breeding for any of the above; or for varieties with timings that keep critical (i.e. very moisture-sensitive) periods in the crop's life-cycle away from times of year when the risk of stress is highest; or by changes in crop-mixes or farming methods, and not by varietal breeding at all?

To answer such questions, and to assign priorities in MV research, breeders need to know what proportions of a researched crop – and of the poor people who depend upon it – are to be found in various *types* of MS conditions. Is the farmer's main problem a 25 per cent risk of a 20–30 per cent water shortfall in the fourth week after planting? Or is it a 30–40 per cent risk of a 10–20 per cent shortfall eight weeks after planting? For some such combinations of risk and shortfall, breeding strategies against MS could be promising; for others, hopeless, so that different timings of planting (or different crops, or irrigation) should be sought. At least, we need to know roughly how many poor people depend – as growers, workers or consumers – on crops vulnerable to the various types of MS conditions. We now lack such knowledge, though it is discoverable from a few person-years of research in most of South and East Asia. Without this knowledge, it will be impossible to decide what sort of 'drought-resistant varieties' or techniques researchers should be looking for. Breeders should complain loudly about the inadequacies of agricultural statistics on MS!

For different crops, vulnerability to MS reaches a peak at different times. Thus greater security demands different – sometimes opposite – responses. Like beans [CIAT, *Ann. Rep. 1985*, pp. 12–13], both sorghum and millet are relatively little harmed by MS before flowering. These cereals can suspend inflorescence when water is scarce, renewing it later (provided the wait is not too long) – and can increase tillers and panicles in moderate MS, compensating for reduced grain yield per panicle. Under similar stress, maize – which is especially vulnerable to pre-flowering MS – tends to reduce both ear size and ear number, with greater (and less reversible) damage to yield [ICRISAT, *Ann. Rep. 1984*, p. 88; Fischer *et al.*, 1983, pp. 2–3].

Plant scientists' models, as well as inter-crop differences, can lead to distinct priorities for the reduction of damage from MS. Tolerant bean varieties, according to CIAT [*Ann. Rep. 1985*, pp. 14–15], have *high* leaf transpiration during MS, indicating that the plant has still plenty of moisture to transpire, i.e. has a more efficient root system. Yet cassava scientists, at the same outstanding institution, observe that MS tolerance in this crop is shown by resisting MS through initially retaining water and *reducing* transpiration [CIAT, 1984, p. 30]. Would bean breeders develop varieties with high leaf transpiration in MS, and cassava breeders low-transpiring varieties?

Crop characteristics, and breeders' models, largely determine approaches to MS. Fischer *et al.* [1983, pp. 3–7] see sorghum as offering better scope than maize to breeding for drought resistance, because sorghum has more (i) root osmotic potential, (ii) leaf capacity to continue photosynthesis even at low water potentials, and (iii) 'developmental plasticity', or capacity to postpone a stage of growth if water is scarce at a crucial moment. This latency, especially important in cassava [CIAT, 1984, p. 30], has, in the case of maize, proved difficult to combine with high yield. A major difficulty in modelling (and therefore combating) vulnerability to MS is that many genes affect it, as does the environment; the relationship of MS resistance to the gene structure is not understood in detail yet, though biotechnology may help in the long run.

Adaptation of maize – or other 'difficult' crops in this respect – to MS is thus seen by Fischer *et al.* as relying less on breeding than on three types of farmer adjustment.

- Improved *agronomic practices* – better weed control, to prevent unwanted plants from sharing scarce water; more mulching – are standard advice, not always easy (or economic) for hard-worked poor farmers.
- Escaping periods of rainfall risk, by *changing the sowing date* (or selecting a quick-maturing MV) is limited – in maize but in few food crops – by extreme sensitivity to MS early in the growth cycle, combined with ignorance of the precise days of high rainfall risk.
- That leaves *changing the crop-mix*. This is the poor, risk-averse farmer's classic route to spreading the risk from MS – via staggered planting, intercropping with robust legumes, or planting crops (or varieties) with different maturities.

However, the impact of MVs on the crop-mix, because specific crops and varieties have been so very successful, has usually been to reduce diversity, and often to encourage shifts towards MS-prone crops. Thus the success in producing high-yielding, somewhat MS-resistant, varieties of rainfed *rice* has induced many Indonesian farmers to shift out of *cassava*, which (although offering lower value-added per acre) is usually affected by MS much less than even the more resistant rice varieties [Roche, 1984, pp. 9–10, 36]. Similarly, the shift to high-yielding *maize* hybrids in parts of Kenya has been at the expense, not mainly of maize TVs (which may *not* be more drought-resistant: p. 58) but of much more drought-resistant *millet and sorghum*. Again, the Mexipak-based MVs of *wheat* that have spread throughout North-west Bangladesh, North India and Pakistan are probably better at handling MS than wheat TVs [Rochin, 1973; Lowdermilk, 1972] – but have induced many poor farmers to adopt them at the cost of more robust but lower-yielding non-wheat crops.[33]

Of course, these choices are not forced; poor people too should be allowed to reveal preferences for income over security by adopting innovations! But should they – or their vulnerable children, or their even poorer employees – bear the full consequences, if moisture stress proves severer than expected? An old Spanish proverb echoes the spirit of such free-market vindictiveness: 'Take what you want, said the Lord God; take it, and pay for it'. Poverty and equity aside, one wonders whether this is the way to encourage poor people to innovate next time. In any case, it is not easy to argue that an amended crop-mix, in the wake of MVs, has proved an obviously pro-poor means of reducing risk from MS.

This brings us back to the direct breeding option in handling MS. Consider a crop where this looks promising: sorghum. What do breeders and their supporting plant scientists look for? What might a socio-economist say to them on behalf of poor people? Most of the time, breeders classify a variety for MS simply by looking at its yield reduction during drought. Sometimes, they are less retrospective. For example, in 1983–4, 700 selected sorghum varieties were screened for three characteristics [ICRISAT, *Ann. Rep. 1984*, pp. 17–18]. *Desiccation tolerance* or avoidance, given the MS, was measured by the lowness of the proportion of leaves scorched. *Recovery resistance* is 'ability of a line to continue to produce new leaves once the rains begin'. *Agronomic score* is 'ability of a stressed line to produce grain when the rains come'.

From a farmer's viewpoint, only the last is a clear gain. Desiccation tolerance and recovery resistance help the farmer, i.e. raise output,[34] only if a plant's yield, or its yield stability under MS, is inhibited by leaf area (presumably a surrogate for photosynthetic capacity and/or for transpiration) – rather than, say, by root mass, by water uptake per unit of root mass, or by N absorption capacity. Moreover, the farmer is interested in output per acre, or per person-hour of work, not per plant; a plant that half recovers from drought – only to take nutrients and water from otherwise better-adapted neighbours that had resisted drought damage altogether – may *reduce*, not raise, yield per acre or per worker. Agronomic criteria for an MV's behaviour under MS, then, fail to reflect poor people's needs for crop cover that, under MS, maintains its contribution to poor farmers' income, to employment, or to the supply of low-cost dietary energy.

*   *   *

Four general problems affect the impact of MVs on farmers' security against MS, whether those farmers are poor or not. First, MVs are damaged, even in irrigated areas, by the politics of MS. The improvement of MV over TV rice is less for tail-enders than for users near the irrigation source, and the yield gap for MVs is greatest for them [Herdt and Wickham, 1978, pp. 5, 22]. Uncertain water deliveries are to blame. They limited the acceptance of MV rice in some parts of Bangladesh to better-off farmers, who owned water sources or could buy priority for their use [van Schendel, 1981, p. 150]. In part, technical improvements in irrigation design could enable water to be controlled more accurately, and thus could help poorer farmers to overcome MS [Wade, 1988; Jairath, 1985]. A more central issue, however, is the severe – but remediable – segregation of irrigation specialists from practitioners (and researchers) in MVs, and in general from agricultural scientists.[35]

Second, some research stations are badly located to analyse MS [Biggs and Clay, 1981, p. 332]. 'IRRI is poorly situated for rainfed rice research because of the high [and] protracted rainfall'; but it is not clear whether it follows that IRRI should be confined to fundamental research and generation of germplasm [O'Toole *et al.*, in IRRI, 1982, p. 217]. Instead, 'perhaps more effort should be made . . . for farmers [and] scientists to meet' [Vergara and Dikshit, 1982, p. 199]. Natural scientists would then be working alongside the village-level research of the socio-economists, and listening to farmers, not lecturing them, about water requirements and choices. ICRISAT irrigates 20–30 per cent of planted area in the research station during the main monsoon season, and much more at other times; this may well be justified by the need for quick results, but it points up the need for field trials, and for close contact with hundreds of individual small farmers in various stressed soil-water regimes. ICRISAT, notably through its village studies and its (rainfed) farming systems experiments, recognizes this requirement for realistic MS research; not all research stations, especially in some national systems in Africa, can say the same.

Third, a related issue is that 'farmer-like' research stations can help overcome MS via the deliberate development of intermediate varieties. In Sri Lanka, H-4, a fairly tall (but stiff and fertilizer-responsive) medium-yield rice, spread fast and far because it was developed in research stations with badly drained soils. This enabled researchers to anticipate *field* problems of water control [H. Weeraratne, personal communication].

Fourth, major problem areas (e.g. semi-arid winter crops,[36] upland and deepwater rice) remain where, despite major spending, IARCs have not yet achieved major improvements in water use efficiency. We point out in Chapter 3, i, that such 'neglected areas' are the core poverty problem for farmers – not necessarily for consumers or labourers – and that such farmers have been almost unassisted by MVs (unless their households are net buyers of food at prices restrained by MVs, or supply migrants to MV areas). Indeed, many people outside MV areas have been harmed by price effects, and by diversion of resources towards MV areas. Now non-MV areas correspond closely to high-MS areas. Most of sub-Saharan Africa and much of Eastern India suffer from unsure water supply in ways that impede the spread of MVs, especially to poor farmers. Potentially attractive MVs can induce Governments to support the spread of irrigation to such areas, as in Japan, Korea, Sri Lanka and the Philippines [Hayami and Ruttan, 1971, p. 22; Abeyratne, 1973, p. 6; FAO, 1971, p. 25]. Even short of this, measures to reduce moisture losses from evaporation, seepage and erosion – physical steps often complementary with, and sometimes more feasible than, MV breeding for reduced plant transpiration and better root uptake[37] – can raise water use efficiency in rainfed (as well as irrigated) areas. The international research centres' great expertise in crop-related research would benefit from much closer links to the physical sciences.

Apart from these four considerations affecting MS resistance breeding for *farmers as a whole*, three elements affect its usefulness to *poor people*: its success in adjusting the gene structure of individual plants; in increasing the safety and profitability, in 'bad years', of plant populations; and in enhancing 'food entitlements' [Sen, 1981, 1986] in the wake of

MS. The first element, on which researchers have over-whelmingly concentrated, is perhaps not the only useful approach, especially for the poorest farmers with several varieties or even crops in the field. This mainstream approach uses breeding to strengthen each plant, in its environment, against MS by selecting MVs with genes that remove its weak points. Such genetic changes may alter optimal planting dates, maturing periods, etc., either (i) to *avoid* overlaps between periods of greatest local risk of MS (drought or flood) and times of greatest vulnerability to MS in the plant's life-cycle (pre-flowering with maize, later with most other cereals and pulses); or (ii) to help a plant to handle unavoidable overlaps; this latter breeding approach seeks to *adjust* each plant to handle a given MS better. These latter genetic changes may select MVs such that a plant's roots seek out water more effectively [Swaminathan, 1974, p. 29, for millet and sorghum; IRRI, *Ann. Rep.* 1984, p. 93, for rice]. Or MVs may be selected so that leaves transpire in a more desired fashion (desired transpiration change under MS may be crop-specific: more for beans, less for cassava – see pp. 60–1). In practice, however, most breeding against MS selects varieties for yield in the relevant MS circumstances, and explains the results afterwards – as is almost inevitable, until more is learned about the genetic determination of resistance to MS.

The above, single-plant, route for MS-resistance breeding has proved difficult, especially for maize. The problem with *avoidance* is that most local rainfall data are either insufficient to spot the times of high risk, or show many such times – though quicker maturity does usually avoid some risk (at the cost of yield), and has produced many MVs of rice and wheat more resistant than TVs to overall MS. The problem with *adjustment* is that root length and mass, osmotic transmission, and leaf-water behaviour are (i) controlled by several or many genes, not by a single readily manipulated one; (ii) greatly affected, in each plant, by 'genotype-environment interaction' [Simmonds, 1981a], i.e. by the effect of environmental factors such as terrain and soil compaction upon the way in which a plant's genetic potential for, say, root mass is expressed in its environment in its specific physical form (phenotype); and (iii) therefore heritable only to a rather obscure and limited extent

– especially in plants with much heterozygosity, notably maize. Maize unfortunately also combines general vulnerability to MS with smallholders' rainfed cultivation in areas of high rainfall risk. Such cultivation has been increasing, because modern hybrids have provided the incentive of improved average maize yields.

Moreover, while rice and wheat MVs have been much more responsive to the plant improvement approach to MS, there are major gaps. One is *flood tolerance*: despite much effort at IRRI, deepwater rices – a main crop for one in ten rice farmers, among them the poorest, in Thailand and Bangladesh – show only very patchy spread of MVs that combine higher yields with either better tolerance of submergence or better capacity to elongate when the water rises. A second gap is MS tolerance in *hill farming*, where both rice and wheat have proved hard to stabilize under MS by genetic plant improvement.

These experiences, perhaps, should turn plant scientists' attention towards protecting poor people from MS by other methods, as well as by single-plant improvement: by concentrating on safety in plant populations, or in food entitlements. A poor farmer seeks a plant population showing a high, safe return over cost. Breeders tend to confine 'population improvement' to open-pollinated populations, because both inbreeders and the other two outbreeding populations (clonal and hybrid) are bred for highly homogeneous MVs – one plant is genetically very like all others (Chapter 2, d). Even the genetic diversity bred into 'improved' open-pollinated populations, e.g. via composite crosses of maize or millet, aims to produce fields of plants that – while distinct enough to protect one or other subset of plants against each of several main sources of MS or pest risk – in many respects, e.g. maturity to harvest, 'look alike" and behave alike (but produce a population with roughly similar gene mix in the next generation, after random mating). But the farmer's traditional route to security where there are high risks of MS is to plant different varieties, even different crops, preferably at each of several sites with differently distributed water risks.[38]

Current MV strategy against stresses is directed to the selection of a homogeneous MV (or, exceptionally – and hardly ever in respect of MS resistance – a 'controlledly' heterogeneous population) of a single crop, better able to withstand some forms of MS than rival varieties, and considerably higher yielding. The high yield and somewhat improved MS characteristics *for that crop* encourage farmers to abandon their former 'diverse population strategy' against MS and to replace it by a 'tough, uniform plants strategy'. Moreover, since breeders usually do much better for some crops than for others, on-farm crop diversity is discouraged, not merely varietal diversity for a particular crop. Then, if MS strikes at the 'wrong' moment – or if the MV's protection proves insufficiently heritable, especially in a new environment – the farmer suffers: the poorest most.

The case should not be overstated. IR-20 rice and Mexipak wheat proved tough enough to offset the loss of diversity that they induced, and made substantial areas safer from MS than previously. However, in many cases, perhaps breeders should now learn from poor farmers' own strategies against MS, and should aim more at diversely protected, high-yielding plant *populations*, with appropriate mixtures of varieties (and even crops) – each offering a distinct protective pattern against various MS risks – on one farm or in one field. Plant scientists are much more interested in research, including breeding, for mixed systems now than ten years ago [Monyo, 1976]; for example, some of ICRISAT's varietal selection aims at complementarity between selected MVs of millet and of chickpeas. However, this approach has not yet spread far into national research systems, nor even into some of the other IARCs. The main reason is that it is extremely complicated, and complexity has costs.[39] However, minimally, one would expect researchers to ensure that pure stands of an MV, prior to release, are tested by farmers in various crop-mixes, and drawbacks – especially under MS – reported to researchers for future action. This is not much done, partly for the bad reason that professionals resist or resent some requirements of the 'diverse population approach'. It requires close co-operation between (i) breeders and agronomists, who have very different training, procedures, and conceptual backgrounds, and (ii) agricultural scientists and poor farmers, with the latter teaching rather than learning.

Apart from enabling single plants of a stabilized MV to avoid (or adjust to) moisture stress, and apart from plant population improvement, there is a third approach to MS: an approach in which plant breeders, and agricultural researchers more generally, have not so far claimed their proper place. This approach aims to remedy a threat to the poor seen as mediated via the vulnerability to MS not just of single plants, nor even of mixed plant populations, but of at-risk people's entitlements to food. If all poor people lived entirely off more or less equal smallholdings, each growing only food for family consumption, then the third approach would be identical with the second.[40] However, it is a central theme of this book that MV research for 'the poor' needs to adjust to a new reality: that a dwindling proportion of the poor are farmers. An increasing proportion of people in *severe* poverty – i.e. so poor that they, and especially their children, eat so little as to be at significantly increased risk of death –. derive their scanty livelihoods mainly from rural employment, not from farming. In Bangladesh, Eastern India and Java, such 'landless and near-landless labourers' are already a majority of the severely poor; in many other places, including Kenya, they soon will be. These people suffer when MS denies them chances to work and earn, and are then almost always the main famine victims [Sen, 1981].

Apart from this, many poor farmers and pastoralists depend, for food entitlements, on selling their products to buy cereals and roots. MS hits these people by rendering their staple food expensive, just when they are all trying to sell their products, and when these are of poor quality (e.g. cattle damaged by drought and hunger).

Assorted proposals have been advanced for helping such 'poverty groups' – and also the urban poor – to overcome 'temporary food insecurity'. Some targeted food subsidies, some integrated nutrition interventions, and some public-works schemes (such as Maharashtra's Employment Guarantee Scheme), have shown considerable practical success [Reutlinger, 1986; Berg, 1987; Alderman and von Braun, 1984; Edirisinghe, 1987; Dandekar and Sathe, 1981]. However, MS remains a major cause, perhaps the main cause, of such insecurity – less, perhaps, because it reduces food supply than because it reduces employment opportunities in the later part of the crop season. Yet poverty-orientated agricultural research, notably into MVs in the IARCs, although centrally concerned with reducing the impact of MS, has done almost nothing about such issues. It is not at once obvious what *breeding* priorities would emerge from a determined attempt to reduce poor labourers', or non-food farmers', vulnerability to caloric shortage as a result of MS. It is clear, however, that such priorities should be examined, and would be different from those that seek to protect 'poor people' from MS on the assumption that 'poor people' means 'small farmers'.

*   *   *

Yield levels are currently low – and hard to raise – in environments at serious risk of unanticipated and major changes in water timing (including flood and drought). Therefore, a dilemma faces IARC research. Concentrate on those environments, and expect lower returns to research, lower increases in food supply, and hence dearer food, with the worst effects on the poorest consumers. Or neglect those environments, and (as extra food supplies from favoured districts glut the markets) impose losses on the many very poor producers, often immobile, living there [Brass, 1984]. In a sane world, given the huge returns [Pinstrup-Andersen, 1982], there should be ample cash for IARC research on food production in all major farm areas, and for making such work useful via national adaptive research. In the real world, the agonizing strategic choice remains: how to allocate inadequate resources between the poorest producers and the poorest consumers; between better water-use efficiency in unreliably watered places, and more food output where water is not a major problem?

Of course, the dilemma can be softened by persuading the few excellent researchers now doing work demonstrably unhelpful to the poor as producers *or* consumers to change tack (see the discussion of protein research in Chapter 5). But can something more positive be said?

Hunger – whether due to power-structures, to imbalance between population and food availability, or to 'entitlements' failures (i.e., to inadequate access by poor households to income that can buy food, to food grants, or to other legal claims) – cannot in most areas be tackled sensibly without major improvements in the farmer's capacity to maintain water supply to the plant population. We have argued that many MVs often outyield TVs somewhat, even under severe MS. However, the extra yield, from a shift from water-insecure TVs to rather insecure MVs, will seldom suffice either to feed growing populations of poor consumers, or to get adequate command over basic necessities to poor producers. Without external water security, e.g. via micro-irrigation, farmers may not risk fertilizers; without fertilizers *and* MVs, neither food supply nor rural employment income can often keep up with 3 per cent annual population growth. Usually, the strategy of MVs, water control, and fertilizers is the only game in town, and in the countryside too.[41]

This may sound surprising. Many people would argue that the 'water security' approach to poverty reduction, in areas prone to severe flood or drought, has been largely discredited. Vast irrigation (and flood protection) schemes, at forbidding and rising cost per acre, have typically featured bad water management, and often no integration between water planning and the crop-mix, let alone varietal research. Less costly approaches have also been unsuccessful, often involving paternalist and under-researched technocratic efforts to persuade or compel farmers to alter their planting dates, crop-mixes, or entire soil-water management systems, in the vain hope that what is technically feasible in the research plot (or sandbox) will prove safe or profitable in the field.

It may be that the IARC system can offer a third and better approach to external water security. Some aspects of IITA's development of *fadama* (valley-bottom irrigation), or of ICRISAT's micro-watershed development (if the technology can be made more profitable), may point the way. The centrepiece has to be substantially increased water security, and this will usually involve farmer-controlled micro-irrigation, typically a well or low-lift pump system. Into this context might come something like Sri Lanka's 'minikit-production kit' approach, in which two or three poor but 'progressive' farmers in a village first try out different combinations of MV and fertilizers in a tiny Latin square, then select a combination for larger-scale use. Ultimately and on a wide scale, of course, this is a job for national research and extension. But the IARCs should at least 'seed' the process, in conjunction with water management experts and national extension systems and after prior consultation with small farmers, in a few trial areas.

Food security requires water security. The world's rural poor, especially in Africa, require approaches to MS that combine water management with MV breeding, the latter probably aiming at total plant populations (and at stabilized entitlements to labour-income to buy food) rather than simply at a crop variety resistant to undifferentiated, or else to over-specific, concepts of MS. The separation of MV research from work in irrigation and water management will have to cease. A study team on Kenya's agricultural research system [Taylor *et al.*, 1981] was surprised at the absence – or segregation – of expertise in hydrology and water management. Yet exactly the same situation characterizes international MV research, in the IARCs. The poor suffer needlessly from MS, in part, because of arbitrary dividing lines among rich people's academic disciplines.

(h) Living enemies of crops

A crop is a plant cultivated for human use. Crops face two threats to such use. One is from inappropriate but lifeless conditions, such as those we have reviewed – wrong amounts or timings of light, water, or nutrients; these we call Obstacles. The other threat is from non-human but living enemies of crops, including weeds, rats, and many birds, insects, bacteria, viruses and fungi; these we call Pests. Pests and Obstacles have much in common (both as threats to poor people's welfare and as priorities for MV research). Both impose three sorts of cost, all especially harmful to the poor;[42] MVs represent a 'low-input' bid to cut these costs by wrapping up protection (against both Pests and Obstacles) in the seed.

First, both Pests and Obstacles reduce harvests (and therefore employment), raising prices to consumers, but seldom by enough to compensate growers – especially if they eat what they grow – for reduced output and higher unit costs.[43] Second, both Pests and Obstacles require the farmer to pay cash, or to forego otherwise promising options, in order to select, adapt or protect the crop against risks; in seasons when the Obstacle (say MS) or the Pest (say bacterial leaf blight) does not materialize – or when the threat (or any other misfortune) overwhelms the farmer's defences – those costly precautions against the threat are wasted. Third, both Pests and Obstacles add to general background risk [Lipton, 1979], increasing rural poverty in two ways: by impelling rural people to divert resources from investment, into carrying more reserves of food or cash in case things go wrong; and by discouraging rural people from profitable enterprises, because these may unacceptably raise an already high level of risk.

These similarities between Pests and Obstacles have often trapped researchers into overlooking the essential difference between these threats to crops (and poor people). The difference is that *most Pests can adapt to meet human counter-measures. Obstacles cannot adapt; but Pests, over several generations, can usually change or evolve.* Research strategy for MVs, therefore, must anticipate the responses of Pests – not simply fight them as if, like Obstacles, they could not adapt.

Of course, it is not only (i) MVs that, once selected to fight a pest, may cause it to breed a new race that negates the improvement, or worse. (ii) Chemical controls, while still often a necessary part of the armoury against a pest, can also teach it to adapt – perhaps while weakening its less flexible enemies, or starving them while the pest is latent or temporarily absent. 'When chemicals were applied [to rice] both brown planthopper and its natural enemies died. BPH populations recovered rapidly, but natural enemies took longer to increase. . . Spraying may even give pests an advantage over [them]' [*IRRI Reporter*, Mar. 1985, p. 1]. Sometimes, more of a crop is lost to pests than in the pre-pesticide era, even if the farmer now buys ever-increasing optimum doses of more and more pesticides; this 'pesticide treadmill' has financially exhausted many cotton farmers, and has many environmental risks [Bull, 1982]. (iii) Biological controls, such as parasites that kill crop-eating insects – and (iv) integrated pest management, combining

MVs with chemical and biological pest controls, and often with (v) farmers' traditional monitoring and agronomic controls [IRRI, *Ann.Rep. 1984*, pp. 194–5] – also have an important part to play in the fight against pests. They are, rightly, part of several IARCs' research programmes [e.g. *ibid.*, pp. 194–216]. However, just like MVs, such methods apply pressure to the pest to select or evolve fitter races. Indeed, they can damage some local ecologies more drastically than a shift from a TV to an MV. Both things have happened in the wake of biological control of a major pest, the rabbit, through the disease myxomatosis.

But the drawbacks to other pest control methods are not the only reasons for favouring the MV breeder's approach. Some pests are bad at some kinds of adaptive evolution, and prove unfit to survive under subtle or well-varied attack; the appropriate attack seems to have come from breeding strategy in the case of grassy stunt in rice [Ou, 1977; but see below, p. 82]. Even if the MV provides only temporary protection, the gains to poor people while the crop resists the pest – i.e. before it evolves a 'stronger' race – may outweigh the subsequent losses from that race; by the time it arrives the potential victims may have climbed out of poverty, and be able to afford a pesticide, or to take the crop losses in their stride. This can work for any form of pest control,[44] but is likelier with MVs (which raise farmers' incomes) than with pesticides (which cost farmers money). The MV strategy, moreover, builds up a research capacity that can maintain and vary a crop's responses to the evolving pests; the improving record on multiple disease resistance breeding [ICRISAT, *Ann. Rep. 1984,* p. 98; CIMMYT, 1985, p. 7] does not suggest that MV breeding involves risks of pest build-up comparable to the risks in the 'pesticide treadmill'.

So the ability of pests to adapt or evolve is not a good argument for reducing research into MVs that fight them. But it *is* an argument for subtle tactics in that fight. To explore the issues, and especially the selection of those tactics to help the poor, we ask three questions. Is MV research targeted at the 'right' pests? What tactics of pest research are appropriate? Can they preserve the MV strategy's proven capacity to provide medium-term help against specific pests (by selecting

tougher varieties), without threatening the genetic diversity
that constitutes a crop's best long-term protection against pests
as a whole? Nurseries to breed selected MVs 'clearly have a
tremendous potential to spread new genes', tough against
pests, but also 'the potential to spread the same genes'
[Hargrove et al., 1985, p. 9].

We shall conclude that most – not all – international crop
research is alert to disease risks and the need for diverse plant
populations; that yield (though not cropping intensity) and
toughness against diseases are usually allied, not rival, goals of
MV research; but that the tactics to approach these goals need
much amendment – and incorporation of 'unfashionable'
pests – to guard the interests of the poor.

(i) 'Unfashionable' living enemies of crops

Pests are found among[45] insects; fungi (airborne like wheat
rusts, or soil-borne); bacteria, such as leaf blight of rice;
viruses; and eelworms [Simmonds, 1981]. Much less
researched than these five living enemies of crops – although
probably more important sources of loss to typical farmers, in
'normal' seasons and places – are larger pests: weeds, birds,
rats, and in some circumstances other wandering animals, wild
or domestic. Post-harvest pests – while taking 3–8 per cent of
grain crops [Greeley, 1982, 1987], not the 10–30 per cent
freely alleged – also receive less research than their share of
food losses might suggest.

There are four reasons for the research neglect, especially
among MV researchers, of large and/or post-harvest pests.
First, many researchers believe that 'the plant breeder can do
nothing about' such pests as pigeons and quelea birds [Sim-
monds, 1981, p.243] – despite the success of some traditional
rice farmers in selecting bird-resistant (awned) rice TVs
[Richards, 1985]. Second, the main responsibility for control
of such pests, which are more readily detected in good time by
farmers than are most disease sources, is often placed on
agronomic practices – bird-scaring; weeding; timely planting
to minimize pest attacks; dry storage – which are seldom
costed. Third, a typical large pest (such as rats) might typically

destroy 5 per cent of all crops everywhere, in an accepted, undramatic way; but a species of virus or insect, taking only 1 or 2 per cent of crops nationally, often destroys half or all of a particular crop in a particular area, leading to articulate complaint.[46]

However, the underlying reason behind all the above explanations is that research, training, and scientific fashions and rewards are self-confirming. Once they are steered, for whatever reasons, towards the selection of MVs resistant to fungi or insect pests – and away from breeding MVs good at competing against major weeds, or unattractive to birds or rats – appropriate scientific specialists are appointed. These specialists' training, their honest preference, and their self-interest all lead them to insist that big and growing portions of research and teaching funds, publications, and prestige are directed towards their own specialisms. Largely because these specialisms have been favoured in the past – because (say) rust-resistant MVs have been researched far more than rat-resistant MVs – more familiar specialisms have scored more successes than have the neglected lines of research. That makes it easier for successful and sincere specialists in breeding against 'fashionable' pests to obtain support from politicians and administrators: to argue that scientifically unfashionable pests are not so important; or are more suitable for non-MV research (e.g. into better weedicides or rat poisons, which have their own research lobbies); or are not promising research targets at all, and should be managed in traditional ways. Such arguments, backed by the weight of scientific opinion, then determine what sort of research (and researcher) gets paid, published and promoted.

Self-interested arguments, however, may be valid. We cannot prove (though we firmly believe) that weeds, birds, and rats deserve more attention from MV scientists, relative to fungi, viruses, bacteria and insects. To prove it, we would need to know (i) the proportion of crop losses attributable to various pests, (ii) the relative harm done by such losses to the poor, (iii) the research costs, and probabilities of success in reducing losses for various periods of time, in respect of each main pest, associated with alternative 'mixes' of research strategy among MV breeding, chemicals, and biological control.

It is a serious criticism of both agricultural economists and agro-scientists – and of the politicians and organizations that finance them – that almost no evidence, transcending anecdote,[47] for any major food crop in a developing country exists on (i) crop losses due to weeds, rats or birds – let alone (ii) or (iii) above. The existing priorities and fashions in MV and other crop research reduce the thrust to obtain knowledge about the relative incidence, damage, and prospects of treatment for unfashionable pests (weeds, birds, rats, etc.). We can form reasoned judgements on incidence and prospects only among the main targets of *past* breeding research to breed MV resistance. Thus we know that airborne fungi are 'the most damaging group of diseases [and] have claimed by far the greatest part of plant-breeding attention [with] some successes but many failures', and are followed in importance by soil-borne fungi and viruses, both with considerable breeding effort and major successes [Simmonds, 1981, pp. 243–4]. Such rankings, plus estimates that 'damage or loss due to small-grain cereal diseases [averages] 10–15 per cent of [world production]' [Prescott *et al.*, 1985], are useful – though one wishes they included detailed information on losses, by pest, for particular countries and crops. However, by assessing losses only from the 'target pests' that have in the past received high priority, such estimates help to confirm distrust of research into other, less fashionable pests. That they cause massive losses is readily dismissed as an unquantified hunch. It then becomes natural to direct future funding for MV breeding research against pests towards the specialisms that have received major funding in the past.

The many excellent publications of the IARCs therefore contain very little about weeds. The occasional contributions are pessimistic, stressing shortage of resources and/or absence of interesting prospects for breeding research [ICRISAT, 1983, pp. 4, 83; IRRI, 1983]. Moreover, weed research seems to have more quality-control problems than research into more fashionable pests. It is rare to find careful distinctions – e.g. demonstrations that wheat yields vary inversely with populations of some weeds but not of others [as shown in Saunders and Hobbs, 1984]. Often, weed researchers' experimental write-ups do not specify which MV or TV – let alone which farm system or size – is fighting which (competitive?) weeds, when, or in what soil–water regime. Where weeds are generically very different from the crops with which they compete for nutrients, light and moisture – not (as with many barnyard grasses among cereals) so similar as to impede selective non-manual control – and where uses (e.g. as livestock feed) can be found for weeds, the research seems more systematic, and the outlook more hopeful.[48]

Is there a 'poverty-orientated' case for downplaying research into conventional, above-ground weeds? Poor *farmers* usually have readier access to on-the-spot family weeding labour, per acre operated, than do rich farmers, and may find it easier to 'do without' weed-tolerant MVs or weedicides – but have few alternatives to tough MVs (except for costly chemicals) in fighting insects, fungi, or virus pests, and might therefore 'vote' for research into these rather than into weeds. Poor *labourers*, especially women, may often rely for much of their income on weeding for hire. However, the case is unconvincing. Poor *consumers* lose unequivocally if major sources of crop loss are neglected. Moreover, the poorest farmers in marginal conditions, e.g. cultivators of upland rice, face the most acute competition between crop and weeds for scarce water and nutrients; and seasonal labour scarcities, especially in parts of Africa, impede weed control by family labour. In such circumstances, neglect of research into MVs, and linked agronomic practices, that improve weed control 'throws' it back to chemical methods, which often only rich farmers can afford, which can depend on precisely timed and thus educated application, and which (environmental hazards apart) are probably more labour-displacing than MVs selected to fight weeds more effectively, e.g. through denser crop canopies.

Anyway, the under-researching of interactions between MVs and 'conventional' weeds goes far beyond the point where it might represent a pro-poor allocation of research resources. Also, it is not motivated by any such aim. This latter is confirmed by the fact that the great majority of MV-related research that *is* done by IARCs relates to screening of MVs for performance under herbicides – a direct, usually free, service to help herbicide manufacturers in displacing poor weeding labourers. In rich countries, where weeding labour is scarce even despite urban unemployment, 'breeding for herbicide tolerance' is a growth sector; such breeding (especially in view of the work of large companies seeking to develop and patent hybrids that will resist specific herbicides: see below, p.372), certainly should not be an aid-supported thrust in research for poor countries. Similar doubts about possible 'displacement of poor workers via aid' apply to IRRI's and CIMMYT's [1983, pp. 89–91; 1984, pp. 77–9] work on screening interactions between weedicides, specific weeds, and yields – especially since hand weeding is apparently at least as effective in weed control [IRRI, *Ann. Rep. 1984*, p. 218]. The impact on crop yield of chemical controls of some weeds may well be offset by increased growth of other competitive weeds;[49] such findings require to be supplemented by comparing the effects (on yield, cost and poor people's incomes) from herbicides and from hand-weeding. The latter is often better at correcting such effects if, but only if, weeds can be identified in time.

Even seasonally inadequate weeding labour can produce good results; although 'labour was scarce, two properly spaced weedings helped boost yield to 77 per cent of that obtained from a cleanly weeded field of cassava' [CIAT, 1984, p. 28]. Given that MVs really tough against 'weeds' that are similar plant types are also a tough research proposition, at least MVs – where appropriate, in mixes similar to those in farmers' fields – should be screened, before selection, for economic response to hand-weeding. Good response is usually, from poor people's standpoint, better than similar economic response to costly and labour-displacing weedicides. However, limited and selective use of herbicide – as in IITA's experiments – can make sense during seasonal labour shortages.

We revert to these issues from the farm labourer's standpoint in Chapter 4. Here, we point out only that – since the emphasis of the IARCs' weed research has in practice been labour-saving – one can hardly justify its small volume by arguing that, since poorer farm families have (and poor labourers supply) weeding labour anyway, MV-linked research should concentrate on pests less responsive to labour inputs. Still less does this argument apply to the neglect of research into interactions between the choice of MVs, agronomic practices, and losses to birds and rats. 'In Africa, at least, the two biggest problems of sorghum growing are *Striga* and bird pests, yet these were dealt with in only three [of 34 topic-specific sorghum-related] papers' [Jones, in ICRISAT, 1982, vol. 2, p. 720].

Farmers select varieties, crops, and mixes and rotations so as to keep down crop losses to birds. Yet we could find only one major MV programme, that for millet in the Sudan [ICRISAT, *Ann.Rep. 1984*, p. 122], where resistance to birds was a stated objective (following resistance to drought, diseases and insects). The setting of goals of stable 'multiple resistance'[50] – to diseases and insects, but *not* to rats or birds or weeds – definitionally downgrades research to make MVs resistant to the latter pests.

(j) Breeding against pests: tactics for toughness

If the aim is maximum benefit to the poor, what tactics should breeders use to develop MVs that are tough in face of living, and therefore adaptable and evolving, enemies? The unscientific term 'toughness'[51] is used here, in order to encompass a number of characteristics of an MV that improve its performance in the presence of living enemies, and to separate controversies about these ways.

(1) *Avoidance or battle:* One can sometimes pick a variety that escapes battle against a pest by avoidance. Such a variety is more briefly exposed to pests (and to Obstacles) because it is quick-maturing, and/or – usually because planted at an unusual time – reaches susceptible stages when the pest is absent or weak. Avoidance is often infeasible; quick-maturing varieties are normally, other things equal, lower-yielding; and climate largely determines planting dates and/or subsequent crop history. Mostly, a variety must engage the battle with the pests and hope to win.

(2) *'Resistance or tolerance:* In that battle, a MV may be selected for 'resistance', i.e. capacity to reduce either infection by or growth of the pest to far below the levels in susceptible varieties; or (more rarely) for 'tolerance', i.e. capacity to accept

pest attack with little reduction in output or quality. Tolerance is often carelessly used to cover the sort of horizontal and moderate resistance conveyed by the genes for slow rusting (to be incorporated into most of the advanced CIMMYT wheat MVs by 1988: CIMMYT, 1985, pp. 17–18). The term 'tolerance' 'should be avoided unless precisely used' [Simmonds, 1981 p. 261], – but does apply to rice MVs that 'yield normally in the presence of an ordinarily damaging [BPH] population' without affecting its growth [Panda and Heinrichs, 1982, p. 13]. Because tolerance does not pressure pests to evolve virulent pathotypes, it has obvious advantages over resistance – though, as Simmonds points out, tolerance for one pest can be undesirable if it carries another (as BPH carries tungro virus, in rice fields).

(3) *Horizontal or vertical resistance:* Horizontal resistance (HR) works against all races or types of a pest; it is, in the great majority of cases, associated with several resistance genes in the crop; and it slows down the spread of an epidemic, rather than delaying or preventing its start. Vertical resistance (VR) is conferred by one major gene of a crop; it is resistance to one (or a few) races of the pest; and it prevents or delays the start of an epidemic, reducing infection more or less to zero if – but only if – 'virulent races' of the pest, able to initiate infection despite the 'resistant' gene, are absent (or nearly so) [Simmonds, 1981, pp. 257–9; FAO, 1986, pp. 3–15].

(4) *Moderate or near-immune resistance*: Suppose that the severest (and most harmfully-timed) conceivable attack of a pest, upon a susceptible variety of a crop in the field and with no counter-measures, reduces its output at maturity by x per cent, as compared with a pest-free but otherwise identical field. Quite commonly x might be 80 or even more. Immunity would reduce x to zero; near-immune resistance (NR), by perhaps 90–100 per cent of x (e.g. from 80 to 0–8 per cent loss); and moderate resistance (MR), by perhaps 75–90 per cent of x (e.g. from 80 to 8–20 per cent loss).[52]

Only vertical resistance is likely to confer near-immunity. A strong case can be made for the classic, VR-orientated approach. First, even the early MVs – though attacked for a 'notoriously low threshold', for being 'susceptible to disease and. . . insects', for being 'sensitive' in proportion to 'potential' [Whitcombe, 1973, p. 199; Griffin, 1975, p. 205; Palmer, 1972, p. 23] – have often proved, not a 'museum of insect pests' like the notorious (but in 1965–72 widespread) TN-1 rice [Fernando, 1973, p. 2], but surprisingly tough. IR-20 rice replaced TVs in many parts of Bangladesh and South India; proved more resistant to all major rice pests and diseases prior to BPH; and has lasted over fifteen years in the field. Among the original Mexipak wheats – conservatively estimated by their discoverer, Norman Borlaug, to have a life-expectancy of 4–7 years as the rusts selected virulent pathotypes – *Sonalika's* range of VRs stood up against major rust attack so well as to dominate North India, Pakistan, and NW Bangladesh for almost twenty years (though new rust races now cause increasing concern).

Second, VR breeding can correct past errors. The story of IR-22, eradicated by tungro virus in the Philippines in 1972 but replaced within a year by the tungro-resistant IR-26, shows not only the instability of VRs [Borgstrom, 1974, p. 17] but the speed with which this can be remedied by a country lucky enough to have access to a sophisticated breeding programme that anticipates and pre-researches such trouble. The IR26–IR36–BP56 rice sequence in Indonesia, in which successive VRs were embodied to handle three successive BPH races, tells the same story [Herdt and Capule, 1983]. Moreover, as MV work progresses, higher yield thresholds are achieved partly by selecting varieties to resist low-level pest attack – and this often involves VRs, which thus render higher yield and pest resistance more compatible goals.

Third, if VR fails, chemical control – or perhaps field tolerance – is available as a back-up. Finally, VR is no longer simply reliance on a single major gene. Different varieties, with distinct single VRs, can be combined in one field – mixtures or multilines. Or the varieties, and the gene providing VR against a disease, can change each season; such gene rotation may outwit the pathogen's adaptive skills, and offer hope of 'eradication of the pathogen' [Crill *et al.*, 1982, pp. 143–4]. Precise multilines (as opposed to mixtures) are very rare, however, and it seems risky to challenge a pest to develop an especially virulent new pathotype by these procedures.

However, the apparent success of major-gene resistance against grassy stunt virus[53] of rice [Ou, 1977], and the growing sophistication of VR approaches, might suggest that durable VR is feasible against many pests.

However – looking at the evidence from the standpoint of poor farmers and labourers – we believe that, when breeders have a choice, they should seek avoidance or HR first, and VR normally as a last resort only, and with safeguards. The VR breeder cannot tell in advance whether, if the research succeeds, the 'target' pest will be good at adapting. If it is, the risks from a new, virulent pathogen are not the breeder's but the farmer's. Such risks hit worst at poor farmers and labourers. These can tolerate small crop losses from imperfect HR or tolerance – but not, if a virulent race overcomes VR and is not swiftly replaced, the consequent unemployment, or costly 'access to all kinds of agribusiness' [Buddenhagen and de Ponti, 1983, p. 1] – e.g. for suddenly needed, scarce back-up pesticides. Indonesia and (backed by IRRI) the Philippines may have sophisticated VR-based breeding programmes for rice, swiftly responsive to articulate farmers hit by new biotypes of tungro or BPH; less high-prestige crops, smaller and poorer research teams, and less articulate farmers may be less fortunate, especially as MVs spread into countries with less experienced research systems and/or less close links to IARCs. The life of VR is usually shorter than is suggested by IR-20 rice and Sonalika wheat. Sophisticated attempts to extend it by (in Simmonds's term) 'aping HR' with multiple VRs, too, have major problems, especially very stringent requirements for research and its links to field management [Jennings *et al.*, 1979, pp. 122–3][54] – and for outreach to poor farmers. The breakdown of leaf-rust resistance in Nacozari-76 wheat in Mexico in 1981 exemplifies their problems – in obtaining sufficiently rapid access to newly resistant MVs, fungicide, and even up-to-date extension [CIMMYT, 1984, p. 5]. It is even possible, if only better-off farmers can obtain such back-up, that pests excluded from their fields will concentrate on the vulnerable crops of their poorer neighbours.

Certainly, some MV scientists seem to place a degree of confidence in VR, and in 'elimination of the pathogen', that may endanger the rural poor. Single-gene VR remains the major approach to millet rusts [ICRISAT, *Ann. Rep. 1984*, p. 97]. 'Perhaps bread-wheat breeders were (and possibly still are) doing the wrong thing by staying with the "boom-and-bust" VR treadmill' [Simmonds, 1985, p. 16].

Rice blast provides an instructive example. A standard review opposes breeding for HR and relies on rotation of major-gene VRs backed up by systemic fungicides [Crill *et al.*, 1982, pp. 143–4]. Yet both Japanese experience [Buddenhagen, 1981, p. 407, 410] and work in IARCs so far suggest that VR against blast is not durable; that pathogen variation causes repeated resistance breakdown [Ou, 1977, p. 282]; and – as with many other pests – that blast is most serious, and most adaptable in face of single-gene VR, in upland and sparsely-watered conditions [Buddenhagen, 1981, p. 410; IRRI, 1985]. There, presumably, pressure to adapt is severest; and VR strategies may well be most vulnerable, because the pathogen must adapt or perish – there being few or no suitable alternative hosts, timings, or places. But poor people, in such conditions, are also most exposed if VR strategies fail.

Where vertical resistance confers near-immunity, the selection pressure on the pathogen – to develop new pathotypes or perish – is very strong. Hence VR is a much riskier strategy against wheat rusts (where breeders are authoritatively criticized for 'staying with the boom-and-bust VR treadmill' [Simmonds, 1985, p. 16]) and BPH (since the early 1970s probably second only to blast as a rice disease) than for green leaf-hopper, where VR in rice MVs has tended to be moderate only.[55] However, VR seems to have been the main strategy in research against BPH, keeping new major-genes resistances in reserve as new BPH biotypes were selected, but fortunately – as with IR-46 – also seeking reserve 'field tolerance' to avoid catastrophe in case VR breaks down [IRRI, *Ann. Rep. 1981*, pp. 64, 70–1; Khush, 1977, pp. 302–3, 307–8; Jennings *et al.*, 1979, pp. 157, 160].

This, however, points to a major problem with VR, and a major reason why critics such as Simmonds regard it as such a dangerous research target (especially, we would add, for the poor): the problem that, as long as it is successful, VR obscures the presence of HR.

- This is partly because [Day, 1973] major-gene resistance must be dominant (except in annually reissued hybrid seeds), and thus occludes recessives that *might* otherwise be selected, through inheritance from both parents, and confer tolerance, as they emerged under selection pressure on the crop from new pathotypes.
- More generally [Simmonds, 1981, pp. 265–7], VR makes it very difficult to know to what extent, if any, the variety has HR. Only a new pathotype, in the relevant field conditions, can adequately test that – and, by definition, while VR lasts in those conditions, no such pathotype has emerged to do the testing.
- Most seriously – quite apart from the difficulty of *recognizing* presence or absence of HR (or MR or tolerance) in a variety with near-immune VR – so long as the crop is attacked only by the pathotypes to which VR is present in a particular variety, the crop population in which that variety looms large is under little or no genetic pressure to select for HR. Hence 'HR will tend to be, on average, low in varieties bred for VR' (unless, in gene positions *other than* those selected in such breeding, by luck, dominant resistance to the pest exists), so that 'the failure of VR, when it happens, will be correspondingly dramatic' [Simmonds, 1981, p. 260, gives examples of this so-called 'vertifolia effect'].

That (moderate) HR is safer than (near-immune) VR – because it does not ambitiously seek to 'eliminate the pathogen' and therefore does not pressure it into new virulence – seems fairly clear. Controversy remains, however, on two issues. Does durable HR exist? If so, is it fully consistent with the large rises in average yields sought by MV breeders (and necessary to feed and employ the world's poor);[56] or does the 'tailoring' of several genes for pest-resistance greatly raise the chance of 'diverting' at least one from goals conducive to higher yields in conditions with few pathogens?.

As for the existence of durable HR, projects of the Food and Agriculture Organization of the United Nations (FAO) in Brazil and Zambia report considerable progress towards HR against major wheat diseases. On a wider geographical base, wheat research has been especially successful in generating HR against *Septoria* [de Milliano, 1981; Beek, 1981, p. 384;

Simmonds, 1985, p. 15]. Moreover, the breakdown of rust resistance in *Sonalika* [Saari, 1985] spotlights the vulnerable VR base of rust resistance in South Asian wheat MVs. As for rice, HR against blast may have been identified already [Hawkes, 1985, p. 21], and is almost certainly feasible [Buddenhagen, 1981 and Simmonds, 1987, pers. comm.]. Against BPH, tolerance (i.e. low rice loss despite many surviving insects), whether achieved via one or many genes in the MV, has been the main supplement to VR. Alternatives to the VR-NR strategy exist, at least for the two main 'green revolution' crops.

But are these alternatives consistent with rapid yield improvement? The long survival of most traditional crops in their environments suggests that HR against many pests has kept many 'minor' pathogens minor [Simmonds, 1985]. However, *if HR to each pest tends as a rule to be moderate*, then HR to a wide range of pests adds up to major yield suppression by them jointly. The real reason why the naive critics of MVs as 'more susceptible' to pests and diseases are wrong is precisely that one main way to raise yields is to *reduce* short-run losses to pests (because 'sub-clinical' infestations in TVs are key yield depressants: CIAT, 1984, p. 27). Unfortunately, that tends to be done by NR–VR breeding strategies that may occlude, or even (via vertifolia effect) dangerously weaken, the older, long-run HR protection.

Nor is this the only way in which the search for very fast growth can dangerously shift researchers to VR. Yield as such, if achieved by fertilizers or irrigation – both of which are strongly induced by MVs – is not an enemy of plant safety, although increased nitrogen is often associated with more leaf disease (because it provides a microclimate more conducive to fungal growth). Fertilizers increase resistance to *alternaria* in wheat and to rice tungro virus [Saari and Wilcoxson, 1974; IRRI, 1985]; irrigation reduces losses due to rice blast.

However, increased cropping intensity (double or multiple cropping) and uniform cover (reduction of crop-mixing or monocultures over large areas) do increase disease risks, because they reduce, respectively, *seasons* and *places* when the pest lacks its preferred food. Unfortunately, higher yields – and higher profitability – are normally achieved much more dramatically, in a given micro-environment, for one crop than for others. Therefore the economic pressures increase to grow that crop (e.g. rice) for several years in succession, often for two or three seasons a year, and to displace other crops that might act as 'firebreaks' to the spread of a pest. This, in turn, increases the period over which exponential build-up of a crop pest is possible; hoppers and borers receive year-round homes [IRRI, *Ann. Rep. 1973*, p. 74]. Higher intensity, not MVs, increases the incidence of pests such as yellow stemborer [*ibid.*, 1984, pp. 386–7] – but it is the success of MV rices that encourages higher cropping intensity.

How does this affect the HR–VR choice? 'The more continual the growth or presence of a crop' – the more pure stands, or the higher cropping-intensity – 'the greater the selective pressure on the pest to produce a new pathotype' [Simmonds, pers. comm.]. To accentuate this pressure by VR is thus even more risky. However, HR too is weakened by multi-crop monocultures. HR/MR relies mainly on reducing the rate of spread of the pest (whereas VR/NR depends on inhibiting infection). More intensive cropping or near-monoculture deprives the farming system of spatial or temporal gaps, in which the host crop is absent and the pest build-up therefore falls sharply. Thus less is achieved by reduction in the pest's rate of spread via HR, since year-round build-up continues even at the lower rate. HR, too, therefore appears less attractive under multicrop/monoculture – though it becomes terribly risky to add further to the pressures on the pathogen by seeking VR to prevent the onset of infection via VR.

Hence the loss of crop and varietal diversity in the field – due largely to the very success of MVs in raising yields, not least via VR – creates risks to farmers. This is not only because one crop (and especially one MV or a genetically similar group of MVs) is more exposed to all hazards than is a farming system with several distinct crops. It is also because the new opportunities for year-round and area-wide growth, offered to pests by a successful variety of a particular crop, tempt researchers to seek radical preventives via VR, rather than suppressants via HR. All such risks, all conflicts of yield with safety against pests, harm the poorest most. And they all appear ultimately traceable to the reduction of diversity.

(k) Safer MVs, less safe varietal sets

We now approach the central problem in interpreting the impact of MVs' physical characteristics on poor people. The problem does not mainly concern physical obstacles to MVs' performance.[57] Breeders select MVs with characteristics that increase the efficiency *and robustness* of the process by which nutrients, light, and water are converted labour-intensively into inexpensive food. With few and temporary exceptions, this should help the poor at least as much as the rich, unless (in general implausibly) breeders are incompetent, or are sub-orned by a group that believes it can gain by diverting them away from pro-poor directions. If MVs have not helped the poor sufficiently, the problem lies with the socio-economic process, not with the characteristics of MVs in respect of yield or toughness against physical obstacles.

There is, however, a central problem for poor people in regard to the impact of MV breeding upon pests. This problem is not, as sometimes alleged (Chapter 2, h), that many MVs are worse than the varieties they replace at fighting local pests. Increasingly, MVs offer multiple resistance against major pests and diseases [ICRISAT, *Ann. Rep. 1984*, p. 98, for millet; Lynam, 1986, pp. 10–11, for cassava outside Latin America; CIMMYT, 1983, p. 7, for wheat]. MVs themselves almost always offer the farmer and the consumer greater effective stability than the varieties they displace (Chapter 5, f).

Yet MVs do bring a problem, best summed up in an apparent paradox: the MV is almost always *more stable* in face of pest attack *than its predecessor variety*, viz. than the TV (or older MV) it replaces; but the *set of MVs* – unless it can be fairly rapidly and regularly turned over in ways that vary the genetic base of VR, broaden resistance from VR to HR (or tolerance), or 'soften' resistance from NR to MR – will probably in the long run prove *less stable* in the face of pest attack *than its predecessor set*. The very success of a narrow genetic range of MVs – success not least in pest resistance – tempts farmers to use this range only. A single wheat variety, Sonalika, did so well in South Asia (by 1982 covering over half the wheat area in Uttar Pradesh, Bihar, and West Bengal, and over 70 per cent in Bangladesh and Nepal) partly because of its rust resistance – so that by 1983 its susceptibility to a new race of leaf rust [Saari, 1985] was a serious issue.

The race is on between (i) researchers' skill in reducing risk, in face of adaptive pests, by adapting individual MVs; (ii) the price of that success, in developing a set of MVs that lacks genetic diversity yet comes to dominate an environment. Since – except for irrigated lowland rice and wheat – appropriate tropical MV innovation is rather localized, the outcome of this race depends heavily on the competence and speed of *local* plant breeders, pathologists and entomologists in re-screening or breeding new, locally suitable MVs to meet new pathotypes – and of extension systems in reporting their attacks – as in (i) above; and on the skill of IARCs, and of local researchers, in maintaining or incorporating varietal and crop diversity, despite (ii) above. Especially with a VR strategy, few national research systems can meet the first challenge without IARC support.

The spread of MV successes to difficult regions of Asia, to Africa, and to crops other than wheat and rice, is extending the problem, especially if (temporarily?) robust varieties or their genetic near-neighbours, selected for one sort of environment, prove so tough as to displace other varieties and crops over a much wider area, as happened with IR-20 rice and Sonalika wheat [Rochin, 1973; Lowdermilk, 1972]. The decline of varietal diversity, especially if associated with VR strategies, now seems most serious for Indian and Mexican wheats, and perhaps Indonesian rices. That decline, and the associated vertifolia effect (p. 85), may well become much more of a danger in other countries, with much less developed adaptive research systems.

How serious is the reduction of field diversity, and what counter-measures are likeliest to help the poorest to fight threats from pests? On Simmonds's account [1981, pp. 126–7, 262–9], until about 1880–1900 (before modern plant breeding) – and to some extent still, in such crops as sorghum, in developing countries – farmers of inbred crops relied on 'land races'. These are not simple mixtures of TVs, but combinations of plants with (i) high 'combining ability', such that the few per cent of outbreeding plants[58] would reproduce something close to the original mixture, and (ii) considerable HR to the major local pests. Outbreeders were entirely open-pollinated populations, not 'hybrids'. So diversity was preserved in field populations (though at low yields), usually reinforced by mixed cropping and a fallow (zero-crop) season, and often by longer fallows of 1–2 years or even – in shifting cultivation – 7–12 years. Hence agriculture, before single-cropping of pure line stands, was characterized by 'diseases. . . always present but rarely epidemic' [*ibid.*, p. 262].

Or was it? Is the protection afforded by traditional diversity (and hence HR, potentially strong if challenged by a pest) anecdotal? It is not statistically testable. Recurrent epidemics of ergot in rye, destructive over wide areas and long periods, were features of agricultural history in NW Europe for many centuries. Today, *striga*-sick soils reveal the lack of any form of toughness against this pest in TVs of a very wide range, often all, of locally grown cereals (though not of some exotic crosses) in several parts of Africa [ICRISAT, *Ann. Rep. 1984*, p. 51]. Even where diversity (and hence HR) eventually produces tough populations, complete absence of VR can cause destructive epidemics of staple crops first, with terrible human hardship – e.g. during the Irish potato blights of 1846–8. The potato is mostly of South American origin; blight is Central American [Simmonds, pers. comm.]; the result of their Irish encounters confirms that new crops, or new diseases, introduced from other successful areas, bring challenges that the TVs' field diversity and HRs are unequipped to meet – or, at least, to meet fast enough to allow the poor to avoid disaster. 'The temptation to suppose that [in peasant agricultures] epidemics were always followed by successful adaptation in the hosts should, perhaps, be resisted; who knows what crops may have been destroyed. . . because they could not adapt quite quickly enough?' [Simmonds, 1981, p. 263].

With people, crops and pests all moving and adapting, traditional field diversity is unlikely to protect optimally against epidemics. But the loss of that diversity in search of just one MV, optimal over wide areas in respect of its gene mix for increased yields – even if that implies reduced pests and diseases over the next five years or so – increases the 25-year risk of epidemics. The main thrust of breeding is towards such uniformity; and of farming, towards adoption of the best uniform variety available. It is unwise to rely on the offsetting capacity of researchers to meet new pathogens by quickly breeding in new sources of resistance, especially if diverse sources of appropriate genes have been lost.

In bread-wheat breeding, the risk is perceived as so great that CIMMYT's top priority – ahead of higher yields in less productive areas, and stabler yields in highly-productive areas – is to 'avoid the narrowing of variability in both [anti-]pathological and agronomic characteristics' [Rajaram *et al.*, p. 10]. Long-run vulnerability to pests is increased by any increase in genetic uniformity of a population, not only by reliance on one major gene for VR to a specific pest.[59] Thus in rice breeding the main threats to diversity probably arise from (i) the single semi-dwarfism gene in the vast majority of IRRI-based and other semi-dwarf populations of standard inbreeders, and (ii) the fact that, for technical reasons, semi-dwarf parents in breeders' crosses are usually female, and cytoplasmic inheritance (unlike the usual form discussed in Section c above) is entirely in the female line [Hargrove *et al.*, 1985, p. 9]. Female-inherited cytoplasmic sterility (introduced into maize hybrids to prevent field crossing, seed retention, and loss of hybrid vigour) was linked to susceptibility to a new race of Southern corn blight (*H. Maydis*), lending to disastrous losses in the USA in 1970. Almost all rice MVs in farmers' fields contain uniform female cytoplasm inherited from a remote Cina ancestor. Also, a single wild rice plant, Wild Abortive, is the female-cytoplasmic ancestor of almost all Chinese and US hybrid rice (being the source of its male sterility) [Hargrove *et al.* 1985; Chang, 1984, p. 254]. So it is wrong to blame uniformity solely on the search for major-gene VR.

One possible approach towards maintenance of diversity is to limit research into MVs, either by area or by season. (i) Very sparse MV rice coverage by SE Asian standards – or, more positively, substantial survival of local TVs or even land races, especially in the monsoon season and in drier regions – is associated with relatively few epidemics in rice in 1970–83 in Thailand, Burma and Bangladesh [*ibid.*, p. 255; Hawkes, 1985, pp. 22–3]. (ii) A related approach is to reduce the exotic, highest-yielding genetic component of local MVs. In Indonesia and Thailand, nationally produced varieties have almost all contained both local and IRRI germplasm [*ibid.*, p. 23]. (iii) A third 'MV-limiting' approach used for rice was to develop and extend robust intermediate-yielding varieties based on local germplasm – the H series in Sri Lanka, Mahsuri in many parts of South Asia – before introducing top-yielding, but less robust and diverse, IRRI varieties. All three approaches reduce the threat to diversity, but in some cases at the cost of confining the gains from MVs (greater VR as well as higher yields) to better-favoured farmers or areas.

A second approach is to amend MV breeding strategy to obtain higher diversity. In this context, Simmonds [1981, pp. 267–9; 1962] usefully distinguishes diversity within varieties from diversity between varieties; one might add diversity between crops. Outbreeding crops such as maize populations, and even inbred MVs of rice or wheat with a 'surviving' few per cent of outbreeders – but not hybrids – maintain substantial diversity *within* a variety, thanks to heterozygosity (Chapter 2, c). However, the replacement of outbred populations by hybrids, the beginnings of hybridization among inbreeders, and the emphasis on pure and uniform seeds, all tend to sacrifice long-run diversity for (often big) yield gains now; that points up the need for collections, whether ex-situ or as 'genetic base populations' [Simmonds, 1987].

As for diversity *among* varieties, successful breeding for yield and VR tends to displace genetically remote varieties that are less successful, both in the field and on the research station's lands. Also, there are often economies of scale if research resources can be concentrated upon a narrow genetic range of varieties suitable for many environments, and on widely grown crops.

This threatens inter-crop diversity (and hence 'firebreaks' against crop-specific pathogens). Concentrated research is unlikely to produce advances at similar rates for different crops. Thus there is a danger not merely that one (or a few closely related) *varieties* of any given crop will come to dominate production over large areas and in all seasons, but that one *crop* will drive out others. This has advantages for poor people in areas where the MV cereals are less attractive, if these areas can grow the crops abandoned in the MV regions (Chapter 3, i). However, if MV regions become almost monocrop *and* monovarietal – and for some rice areas this is no exaggeration – deep trouble threatens. If the dominant crop is an inbreeder, there will be some (1–5 per cent?) residual outbreeding, but more and more seldom either with progeny of impure seeds from the same field or with genetically distinct varieties from other fields; thus the tendency towards homozygous and homogeneous crop populations is sharpened. This is good while the environment, to which this genetic fixity (barring mutations) has adapted, is stable. It can be disastrous if a pest changes that environment by developing a new pathotype, especially in the large majority of crops where natural mutations are generally very rare. If the dominant crop is an outbreeder, it has naturally excellent prospects for diversity; but the shift to hybridization (especially via male sterility), or to clonal reproduction, removes them once again, in the interests of yields and VR via genetic stability, fixity, and control.

*   *   *

If we assume that the very success of breeding will continue to drive out land races and cultivated crop relatives like red rice – even though in Africa that risk still looks remote, except for maize – there remains one background or reserve strategy: 'diversity within the actual breeding population' [*ibid.*, p. 268]. The literature is full of accounts of efforts to diversify sources of dwarfing [e.g. CIMMYT, 1984, pp. 1, 124–7; 1981, p. 4], of VR, and even – ironically, since it is itself such a threat to diversity (Chapter 7, i) – of cytoplasmic male sterility. Many such efforts succeed, but such success is often temporary. For the farmer, good seeds drive out bad. The seeds that best meet the requirements of cultivation are usually adopted, even if diversity is thereby lost. For the breeder (especially with a VR approach), the movement of pathogens – and the evolution of new pathotypes – often compels a search for more and more single-gene resistances in each plant – leading to ever tighter, and hence less diverse, genetic specifications.

Thus 'diversity of breeding population' must refer mainly to the deliberate conservation of *reserves*. There are two sorts of reserves. Natural or *in situ* reserves comprise areas in which are maintained the ever-scarcer land races; the residual pure-line TVs (still pervasive, and diverse, in many rainfed areas); and/ or wild relatives (such as species of *Oryza nivara*, the wild rices). Breeding or *ex situ* reserves consist of managed stocks or genebanks, containing many classified groups of viable seeds.[60] In either case, it is essential that records be readily available of the relevant characteristics of each variety (e.g. extent and type of toughness in face of many pests and obstacles; yield; height; duration), preferably cross-classified by major environments, and ideally also by socio-economic characteristics and farming systems of main growing communities, if any. It is also essential that enough viable seed be retrievable for breeding in emergency.

There is much disagreement about the relative merits of *ex situ* and *in situ* conservation. In favour of the *ex situ* approach:

- The IARCs have the largest and most diverse collections ever made of the main cereals.

- Such collections are substantially used – sometimes, as with development of grassy stunt resistance, by incorporating *O. Nivara* genes into IRRI varieties [Ou, 1977], showing the feasibility of *ex situ* methods even in circumstances where the importance of resistance sources in wild races would appear to argue for *in situ* approaches. Genebank accessions accounted, from 1975 onwards, for the distribution to breeders of 10–15 per cent of sources of NR or tolerance, covering blight, tungro and BPH as well as grassy stunt [Hawkes, 1985, p. 22].

- In many cases, there are no TVs or land races adapted to recent environments of a crop; tropical wheats effectively resistant to *Helminthosporium* are invariably breeders' inventions [CIMMYT, 1984, p. 145], not obviously suited to *in situ* 'naturalism'.

- *In situ* conservation, if perceived as 'preservation of primitive agricultural ecosystems, crops, stock, and presumably people included', is readily dismissed as 'socially inconceivable' [Simmonds, 1981, p. 325; compare Ingrams and

Williams, 1984, p. 165].

However, this does less than justice to the prospects of, and need for, partial *in situ* approaches:

- Reservoirs of land races of composite crosses with diverse parents can be maintained; and the 'gene park' approach has attractions [Bennett, 1968; Simmonds, 1962; Browning and Frey, 1979].
- Farmers themselves choose to preserve many TVs – in Thailand, Sri Lanka and Nicaragua, the 'farmer-curator' idea has been suggested [Mooney, 1985].
- While plants sit in static gene banks, pathogens evolve *in situ*.
- Massive *ex situ* collections have several weaknesses. They are often divided into several sub-sections (delaying the process of discovering the right seeds, if available), or are badly maintained [Mooney, 1983, pp. 75–8; Myers, 1983, pp. 22–3]. The genebanks are not highly valued by many breeders. Some *ex situ* stocks have no 'second copy' elsewhere, and would disappear if badly mismanaged or destroyed. Some are too inadequately referenced (too few, or wrong, plant descriptors) to be safely consultable in time of need [Duvick, 1984, p. 180; Frankel, 1984; Goodman, 1984, p. 365; Holden, 1984, p. 271; Smith, 1981, p. 32].
- Some major collections, notably for maize, have been notoriously inadequate. Indeed, the outbreeders present horrendous problems to classifiers of 'varieties'.

Much of the 'ex-in' argument may miss overarching issues. Proper descriptors, classification, maintenance, breeder information, and access are as necessary for *in situ* plants as *ex situ*, and arguably more difficult, especially for outbreeders. The problem that a gene bank, as with rice, may be able to provide access to plants with particular desired characteristics but not with desired combinations [Chang, 1982, p. 40] is at least as serious for an *in situ* source – and identifies a task not for curators but for breeders, who would be redundant if all desired combinations were available from (*in* or *ex*) existing reserves! The fear that private companies can somehow damage poor farmers by appropriating germplasm, or knowledge about it, is to some extent misconceived, especially for inbreeding crops – but much less so now than when it was first voiced (Chapter 7, i); however, public-sector (and especially international) gene banks may represent at least as good a defence against such appropriation as, say, farmer-curators, who could more easily be 'bought' by private research interests.

(l) Physical features of MVs: afterword

From the controversy about *ex situ* versus *in situ* collections to maintain diversity, as from the controversy about VR versus HR, three conclusions follow, if the top priority is to advance the welfare of the poorest. First, the apparently obvious finding – that, because the poor can least afford a major downturn in food availability, HR and *in situ* approaches are always safer and therefore better – is over-simple: HR is unavailable against many pests; *in situ* collections are inapplicable in the many environments where only specially bred MVs (e.g. tropical wheats) can survive. Second, 'commonsense compromise' is not really applicable: VR always occludes, often (via vertifolia effect) impedes evolution of, HR; *in situ* and to *ex situ* approaches compete for scarce cash and expertise.

The third conclusion from the 'collections' and 'resistance' controversies, trivial as it looks, has wide-ranging implications – mostly positive, partly negative – for MVs' impacts on poor people. It is Pope's: 'whate'er is best administer'd is best', if 'administration' covers not only competent management, but also stimulation and reward, of MV research directed to poverty reduction. Properly maintained, quickly accessible collections, indexed by disease responses and environments – plus local research and extension that can swiftly combine desired characteristics and spread varieties with the necessary toughness – matter more, to poor people's prospects of surviving epidemics, than apparently strategic *ex/in situ* choices. Researchers singing in chorus to achieve durable robustness, rather than heroic tenors seeking star status for eliminating pathogens (or doubling yields), protect the poor better than an apparently strategic, but in fact infeasible, universal HR/VR choice across crops, diseases, and environments.

Can so obvious a truth have implications, positive or negative, for MVs' impact on the poor? Positively, the application of scientific breeding to redirect plant and pest evolution – if, as with MVs, constrained by efforts to cut risk, and to produce things mostly made by the poor (i.e. labour-intensive) and mostly consumed by the poor (i.e. low-cost foods) – should normally produce better results than 'unassisted' evolution. Just as farmers' seed selection adds to and thus improves on natural selection, so scientific screening and breeding should add to and improve on farmer-plus-natural selection. To deny this – to confuse conservation and conservatism – is to show *fear of knowledge*. That is understandable in the era of Bhopal, Seveso and Chernobyl, but misplaced if the biophysical results of the scientific endeavour are increased yield *and stability* of poor people's crops. (But we shall have to be rather sure about stability.)

'Fear of knowledge' in respect of MVs is generally misplaced, because the adaptations of a plant genotype and hence of its architecture and biochemistry, sought by breeders, improve the efficiency of plant populations in converting sunlight, water, and nutrients into less pest-exposed human food. Improved conversion efficiency at all relevant levels of intake is the aim – not just dramatic improvement when conditions are favourable (regarding, say, fungus infestation or water availability). With regard to Obstacles – unfavourable amounts or timings of sunlight, water, nutrients, or heat and cold – we can say, almost unequivocally, that the physical characteristics of MVs improve stability as well as 'normal' performance. With regard to Pests, there is a caveat. MV *plants* are normally selected for, indeed raise yield partly by, fighting pests more efficiently; but that very success leads to growing uniformity in MV *populations*. This uniformity, in turn, increasingly induces pathogens to evolve races that leap that population's set of barriers (even if a complex and subtle one). In the long run, that endangers stability. This danger can outweigh the effect of the toughness bred into individual plants in MV populations.

'You may fear too far.' 'Better than trust too far.' Scientists recognize the need for MVs with appropriate resistance, and for varietal collections, as conscious safeguards against the new threats to diversity [Dalrymple, 1979, p. 37; IRRI, *Ann. Rep. 1973*, pp. 64, 82; CIMMYT, 1982, pp. 124–7]. Consciousness, scientifically mustered into MV breeding, should, if 'best administer'd', improve the physical and biochemical impact of plant populations on poor people as compared with pure evolution, just as farmers' seed selection did.

The apparent neutrality of the adage that 'whate'er is best administer'd is best', however, hides the negative implication, for poor people, of the central physical fact of MVs: the fact that, while MV research aims at crop characteristics advancing the mutually supportive goals of robustness and stability for crops mainly grown and eaten by the poor, that very aim can harm the poor by reducing diversity. Intelligently pro-poor management is the main need if such research is to help the poor, not a blanket formula about types of resistance or of seed collections. However, scientific incentives, rewards and promotions are seldom neutrally 'managed'. They are largely determined by four main factors. The first is talent in advancing the goals of a scientific enterprise. The second is the social and scientific 'agenda' that sets those goals, and finances work towards them. The third and fourth, underlying these, are the norms created by the ruling scientific paradigm [Kuhn, 1962], and the economic and political pressures on research priorities and financing.

Only the first factor clearly helps to convert the 'pro-poor' implications of MVs' biochemistry and plant physiology into 'pro-poor' crops and methods in the field. The other three factors can either help or hinder. For example, the agenda of a food strategy seeking to displace imports will direct research towards crops – and varieties – satisfying demands now met by imports, especially for fine grades of wheat, even where poor people normally grow and eat coarse wheat or maize (Chapter 5). Or the ruling scientific paradigm may unduly reward researchers for generating maximum yield potential, high but temporary VR, or gene-transfer wizardry, rather than for achieving localized, moderate, durable HR. Economic and political pressures steer researchers towards crops, varieties, and priorities favouring urban groups, bigger farmers, or particular sorts of effective cash demand; they may, for example, lead researchers to emphasize the development of

MVs that can readily be combine-harvested, in order to 'save labour' for big farmers and to reduce the price of grain sold to cities, even in economies with unemployment, overwhelmingly rural poverty, and scarce land.

In particular, pressures and agenda can interact – even where administrators and politicians want MV research that responds to poverty – to conceptualize poverty wrongly. For example, most MV research is for 'small farmers' though the poorest increasingly are near-landless labourers. Science does not respond only to such demands and pressures, professional or socio-economic. In particular, basic discovery – such as that of Mendel, Darwin, or Crick and Watson – cannot plausibly be explained in such coarsely materialist terms. Even if it could, it has produced new scope for MV breeding for many decades afterwards, i.e. long after the social and scientific 'pressures' have drastically changed. However, the direction of *applied* research – especially at national and local levels – is subject to peer-group pressures to conform.

Such pressures make it especially important to increase non-conformity in the IARC system (and also in larger national research systems). They also suggest a need for incentives and career structures that encourage researchers to set, think through, and steer their research by, pro-poor breeding priorities. Finally, they underline the importance of socio-economic and historical awareness, and linked field studies, in realizing the undoubted pro-poor potential of MV breeding. MVs increasingly tend to improve crops' conversion efficiency and toughness, and hence their 'worst-case' performance, even at the low input levels affordable by the poor. But the sequence, especially in regard to crop response to pests, is not automatic.

*   *   *

The physical features both of semi-dwarf wheat and rice, and of hybrid maize and sorghum, in the great majority of varieties, 'ought' to help the poor as labourers, consumers, and growers.

- Because such MVs are bred mainly for yield enhancement – short stalks, erect leaves and dense roots, so as to improve

per-acre use respectively of nutrients, sunlight and water –
MVs usually raise labour requirements per acre, and thus
employment.

- Because MVs produce grains that loom largest in poor
people's consumption (and because the breeders' priority
for high grain weight tends to reduce fineness, etc., and to
cause most MVs to stand at a price discount), they should be
especially important (i) in restraining poor people's cost of
living as consumers, and (ii) in the output-mix of poorer
farmers; for these, the MVs' high ratio of marketing-costs
to grain-value is less of a deterrent than for commercial
farmers, because small family-farmers eat most of what they
grow.

- Because MVs are increasingly bred to resist or tolerate pest
and disease attack, they should – if due regard is paid to the
maintenance of diversity – specially benefit poorer growers,
who are more damaged by downside risk than richer
farmers, and less able to afford chemical controls.

Yet the systems of science and of society into which MVs are
inserted often thwart these pro-poor elements; and
researchers need to gear their work more towards varieties,
practices, and inputs designed for the poor in the various total
systems, social as well as economic and environmental, where
MVs are used. We return to this issue in Chapter 6.

Notes and references

1 Human foods are *prima facie* likelier than other crops to attract
insects, birds or animals, and to need to evolve genetic protection
from them – yet cotton is probably more pest-prone than most
food crops. Most cereal straws are used for other purposes than
food, viz. feed, thatch, etc.

2 Simmonds lists uniformity of plant and product; aptitude for
rooting or grafting; economy of harvest (in tree crops); uniform
stand (for sugar beet); and elimination of vegetative prickles.

3 On sustainability, he devotes much space to the risks of genetic
uniformity in modern farming – because its effects include less
varietal difference, less heterozygosity, more double-cropping,
and less crop-mixing (or, one could add, rotation) in a field or

even a big area – and to ways to conserve genetic variety. On stability, he stresses horizontal, polygene resistance to disease, and the dangers of alternative methods. See Sees j–h of this Chapter.

4 There is some recent research on intercropping, notably at ICRISAT. But such research, while exploring how varietal choices and farm practices affect crop competition and complementarity, cannot simulate the varied, complex nature of numerous farmers' systems of mixed cropping, especially in Africa.

5 Another important example: breeders' priorities promote uniformity in a crop, raising average yield but in some circumstances also raising risk [Simmonds, 1981, p.127] and risk avoidance is a high priority for *some* poor farmers, e.g. those with insecure water supply.

6 Market value, especially where consumers have very unequal purchasing power, is often most cost-effectively increased by research to improve palatability or convenience, but many poor consumers (including many deficit farmers) will lose more from the resulting increase in cost-per-calorie, than they gain from the more luxurious product. This is especially so if the researcher is *diverted* towards palatability at the expense of yield; or if other food crops are not readily available to poor consumers; or become more expensive alongside the crop whose palatability has risen (i.e., if substitution effects against the latter are outweighed by income effects raising its price, as can easily happen).

7 Relative prices of inputs and outputs might seem to tell us what is 'wanted' and scarce. But relative prices vary over place and time, and are often different for rich and poor people. Also, many value-judgements – including ours – claim that, 100 rupees' worth of extra crop is 'worth more' if it raises poor people's income rather than rich people's.

8 The expected value of an event is the sum of (the value of each outcome × the probability of that outcome).

9 See above, fn.4. Farmers often achieve research-type results by trying the same variety in different conditions, or two different varieties in the same conditions [Richards, 1985], but this on-farm biological inventiveness is possible only with inbreeding crops (such as rice in Richards's case), or if an outbreeder has been stabilized in a research station first (annually reissued hybrids or composites).

10 A few maize or millet plants, in a typical field, inbreed, though a big proportion of offspring then die, because they inherit deleterious recessives from both male and female parts of the parent plant. A few rice or wheat plants cross-pollinate, though

the much greater ease of self-pollination, plus its adaptation to their local environments, keeps the proportion of outbred offspring to a few. Sorghum and pigeonpea can readily inbreed or outbreed [Simmonds, 1981, p. 4]; some such plants, notably primroses, evolve varieties that inbreed in environments where that is advantageous (i.e. where the primrose is ecologically very well adjusted), and different varieties that outbreed in other environments [Ford, 1965, pp. 48–9].

11 The small percentage of outbreeders, even in these inbred crops (and of heterozygosity, even in a supposedly pure line), nevertheless means that farmers must buy fresh seed every few years, to keep quality from declining. Also, it may make commercial (but in LDCs more rarely socio-economic) sense to breed rice or wheat hybrids. See Chapter 7, f.

12 Except for some impacts on backward regions; in recent years, on some farmworkers; and perhaps via adding-up effects. See Chapters 3–6.

13 Occasional mutations; the presence of say, 1–5 per cent of outbred offspring; destruction of some plants; homozygotes in particular gene-pairs that involve inheriting the same 'deleterious recessive' from each parent; 'linkage' of some homozygote gene-pairs, themselves not harmful, to similar deleterious recessive pairs on the same chromosome [Simmonds, 1981, p.80]; and 'heterozygote advantage' (see below).

14 Those receiving the recessive effect – say h in a windy environment where tall plants fall over – from both parents. For the maize stand to survive, most such offspring would probably have to die young (so as not to compete with fitter maize plants for nutrients). In other words, outbreeders are adapted to their (inevitably considerable) heterozygosity.

15 Strikingly, agronomists' reports on interaction between yields of a crop (such as maize) and weeds, water conditions, terrain, farm practices, or other 'agronomic' variables, often specify neither the type of plant population, nor the variety or varieties concerned.

16 Recent legislation, imposing high standards of uniformity on commercial breeders, strengthens such trends [*ibid.*, p. 133] and is dangerous, in the long run, for poor farmers with high risk-aversion.

17 (i) In this, hybrids are unlike the other three plant populations. PBRs are almost unenforceable in a poor country, where smallholders far outnumber agriculturally sophisticated policemen. (ii) We use 'hybrid' only to mean 'inbred varieties, crossed for heterosis by breeders, and not reproducible by seed retention'.

There is a more general use, to cover all crosses, which is not employed in this book.

18 So deeply entrenched is this belief that it is sometimes stated, with expository diagrams, alongside the honest presentation of evidence that the MV outyields the TV even without fertilizer [Wright, 1973, pp. 59–60; Hayami and Ruttan, 1971, pp. 43, 83, 193].

19 The timing of a plant's uptake of nutrients, however, sometimes varies with their source. Experimental work on rice [IRRI, *Ann. Rep. 1984*, p. 158] strongly suggests that N uptake is more gradual if there are no inorganic sources.

20 Bread wheats, though not durum, are an exception in many Tunisian soils – helping to explain why smallholder adoption of MVs there has been much slower for bread wheats than for durum [Gafsi and Roe, 1979, p. 126].

21 This process will be slow, or absent, in very rich soils that were in fact *under* exploited by TVs, as with the rice soils of IRRI's continuous-cropping experiment [IRRI, *Ann. Rep. 1973*, p. 100]. The process is faster in multiple-cropping systems. The effect of slash-and-burn is ambiguous (see below).

22 It is in the long-run interest of (i) the landlords and (ii) *all* poor tenants and potential tenants jointly, to inhibit soil mining by any one tenant. But this interest is hard to enforce, partly since soil-mining practices – not restricted to the planting of MVs with little fertilizer in marginal soils – are not easy or costless to observe or prevent. High borrowing rates on loans, especially for the poor, raise the attractiveness of quick income, and thus also reduce 'concern for the future' in using MVs.

23 Some economists [e.g. Beckerman, 1974] would argue that the concept of 'sustainable farming' is much too static: that environmental threats, e.g. via soil mining, generate research responses, and hence new ways to farm, that lower the costs of averting the threats. Of course, such responses exist. But dust bowls, desertification, and irreversible salinization reveal their limitations. Burgeoning US and EEC food surpluses, moreover, do not encourage costly research into how farmers – even Asians or Africans – can sustainably grow yet more grain.

24 See the work relating applied N to N uptake, and each separately to grain yields [IRRI, *Ann. Rep. 1984*, pp. 413, 418].

25 The relative neglect of this matter may well be due to the long period when the search for MVs was mainly a search for robust dwarf and semi-dwarf varieties, i.e. for better partition efficiency. Its interactions with conversion and extraction efficiencies, which

were thus pushed into the background as concerns for applied MV research, are complex.

26 Exceptions arise; ample supply of nitrogen fertilizers, once their use is made profitable by a new MV, can occasionally deplete a soil of zinc. Agronomists, economists, pedologists, and breeders need to combine to identify where this risk needs special action – and where it does not. This illustrates how the best returns from MVs are likely only in areas with good local research systems.

27 Placement of slow-release fertilizers in the root zone of individual plants is labour-intensive, but sometimes greatly increases nutrient conversion efficiency. Azolla preparation and application share these characteristics too.

28 This discourages farmers from growing a coarse MV if they produce for the market. If they produce for self-consumption only, the deterrent is less, but still exists, first because the price discount reflects most consumers' taste preferences (including most farmers'), and second because inexpensive MV foodstuffs could be acquired by growing and selling costlier TVs.

29 Extra N can raise the yield losses per day from moisture stress. Even 20 kg/acre of N did so, by 5 kg/acre per day of stress, at IRRI [Wickham *et al.*, 1978, p. 227]. Also, adequate water is needed to enable nitrogen-fixing organisms to operate [Swaminathan, 1984a].

30 This is partly because much feedstock comprises natural gas, a given source of which 'switches into' (or out of) consideration as a direct fuel source at only one (usually high) oil price; and mainly because fertilizer prices are determined to a large extent by waves of investment, fuelled by past peaks of demand, and coming onstream jointly, pushing prices down because they do so ahead of (growing) demand.

31 Thus MVs are criticized as 'giv[ing] higher yields only [with] extra quantities of water' [Borgstrom, 1974, pp. 14, 17], as 'less resistant to drought' [Griffin, 1975, p. 205], as 'requir[ing] controlled irrigation' [Falcon, 1970, p. 699], or as 'more prone to suffer yield losses' unirrigated than MVs [Palmer, 1972, p. 51].

32 The standard – and powerful – CIMMYT argument for open-pollinated populations, however, was that poor farmers could retain the seed and hence get the improved maize more easily – not that it outperformed hybrids, even under MS.

33 All this typifies how MV breeders have to run faster just to stay in the same place. Success in *one* area (but not in others) reinforces the thrusts towards uniformity based on genetics (Chapter 2, k). Breeders need to make their MVs especially resistant to MS

merely to compensate poor farmers for the increased risk due to crop-mix effects.

34 Even this assumes 'gross output per plant' correlates perfectly with 'net value added per unit of resources, weighted by their scarcity for poor people'.

35 See Chambers [1983]. Irrigation engineers and maintenance officials, unless very recently trained in exceptional places (such as some Indian 'Water and Land Management Institutes'), usually know little about crop-water requirements in general, let alone those of assorted MVs. And international – and even national – research centres working on MVs know little of the practical issues of water management confronting irrigators. One hopes the new International Institute for the Management of Irrigation (in Sri Lanka) can help to bridge these gaps.

36 ICARDA's work on barley is of special importance here, because – like triticale – it has good MS tolerance. Phosphorus fertilizer increases this, by advancing maturity (permitting escape from some MS) and by increasing water use efficiency.

37 It is not clear to what extent water uptake is improved by root *mass* (as against *depth*), nor to what extent root characteristics are heritable, rather than due to general plant vigour in specific environments. Nevertheless, breeding for root characteristics against MS appears promising for both rice and beans [IRRI, *Ann. Rep. 1984*, p. 93; CIAT, *Ann. Rep. 1985*, pp. 14–15].

38 For a conscious and sophisticated example – in the context of varietal selection by 'illiterate farmers'! – see the behaviour of the Mende of Sierra Leone [Richards, 1985].

39 To develop a MV of pearl millet that performs well under a specified MS is not easy. But it is much harder to develop a group of MVs of pearl millet that, mixed with a farmer's many possible types and densities of intercropped legumes, will produce a *farm population* of varieties of crops that perform well under a range of MSs. Since hard tasks take longer, researchers who seek quick benefits for farmers understandably try to solve simpler problems first. But, as we have explained, their very success, in breeding a high-yielding MV of a particular crop that resists MS, may undermine diversity and actually weaken plant *populations'* resistance to MS.

40 If each of these smallholdings grew a single crop in just one variety (or in just one fairly stable mix of an open-pollinated crop), then both approaches would be identical with the first, viz. to improve the crop's handling of MS by shifting to the correct MV.

41 This is not to denigrate the value of farmers' own 'indigenous technical knowledge' [Chambers, 1979]. This should inform and

complement formal MV-watercontrol-fertilizer research. But it would be naive to expect even the brightest farmers to generate, unaided by formal research, output growth sufficient to remove African hunger, or even to keep up with 2–4% annual population growth.

42 A given cost or risk is greater for those with fewer discretionary resources to meet it. Poor people, moreover, are likely to face both greater overall risks (to health, water supply, etc.) than others, and higher costs in borrowing or insuring to avoid each unit (say $100) of risk. Finally, poor people both earn higher proportions of income from growing food, and spend them on consuming it; thus costs and risks involving food loom larger for poor people.

43 Harvest (and subsequent) costs are cut back more or less as fast as output, if pest attack strikes. However, costs at all stages prior to pest attack, plus extra costs to deal with it, must be spread by the grower over a smaller number of units of output as a result of the harvest shortfall. Employees, of course, get less work at and after harvest, if a pest has reduced the crop.

44 If interest-rates are high it could 'pay' to launch an inexpensive broad-based pesticide onslaught that raised cotton output to 125 (from a base level of 100) for ten years, and then – by inducing the evolution of tougher races of pests – cut it back by 1 or 2 per cent below base level (i.e. to 98–99) for ever.

45 In each of these genera the great majority of species is not harmful to crops. Some (e.g. bees) are beneficial. Others are neutral, but if destroyed can cause complex and often harmful shifts in the predator-host balances of species in an area. Yet other species help some crops, but harm others. Thus a scatter-shot chemical attack on, say, insects or fungi can involve serious side effects. Since few chemicals – whether from pesticides or secreted by bio-engineered soil bacteria (Chapter 7, g) – are specific enough to kill only the organism attacked, this is an important part of the case for dealing with it via specific, tough MVs instead.

46 Since social insurance seldom exists in poor countries, is a virus that destroys all the crops of 1 per cent of farmers – because it thereby threatens the survival of some of them, or of their undernourished children – perhaps 'more serious' than rats that destroy 5 per cent of the crop supplied by each and every farmer, but appear not to endanger life or health? However, if the rats thereby raise food prices, the appearance is deceptive; some babies will be squeezed into severe malnutrition. Such babies are much less articulate or identifiable than are farmers in a region ruined by, say, a new biotype of brown planthopper.

47 That such anecdotes grossly misreport losses is shown by Greeley [1982, 1987] for post-harvest pests. A careful sample of rat-burrows produced an estimate for paddy theft by field rats in an Indian region – but it includes a large but unknown portion later retrieved from burrows by humans [Boxall *et al.*, 1978], and excludes losses in storage to house rats, usually of different species and habitat.

48 Parasitic pathogens such as *Orobanche* (of legumes) or *Striga* (of sorghum, millets, and maize) are not really comparable with 'classic' weeds, and are more popular research targets. ICARDA's work on *Orobanche* offers real promise. The root parasite *Striga* – perhaps the second most important pest (after quelea birds) in Africa, and spreading to India [Clarke and Clay, 1986] – is better resisted by appropriate maize MVs than by TVs in Nigeria [IITA, *Ann.Rep.*, 1985, pp. 71, 75–6]. MVs are a much more promising approach to *Striga* control for poor farmers than are costly *Striga* seed germinators [Ramaiah, 1983, p. 53; Roger *et al.*, 1983, p. 86].

49 'Bentazon applied nineteen days after emergence appeared to promote *cyperus* growth [in wheat] . . . possibly by reducing competition from broadleaf weeds' [Ransome, 1985].

50 Whether set for composited populations from which MVs are later to be selected [ICRISAT, *Ann Rep.* 1984, pp. 77, 107] or for specific MVs [*ibid.*, p.98; CIMMYT, 1985 p. 7].

51 The evolutionist's technical term 'fitness' is not quite the same thing.

52 These vague numbers are worrying, especially since the degree of resistance is often estimated from very small samples, yet is subject to much environmental variation (especially in field conditions). The technique of monoclonal antibodies (below, p. 373) will greatly improve the precision of MR–NR specifications.

53 In 1986, there was alarm that the only known source of resistance to GSV – from a single gene, bred into MVs from a NE Indian wild rice – was breaking down. However, a new virus now appears to be the problem, not a new pathotype of GSV [Buddenhagen, pers. comm., 1987].

54 *Gene deployment* requires continent-wide, co-ordinated planing decisions. *Gene pyramiding*, of several major-gene resistances, leaves few breeding resources if the pathogen selects a super-race. *Multilines* require long, complex research (though biotechnology may help here), and are of limited use against unanticipated pests [Day, 1973; Barrett, 1981], and against highly variable pathogens,

especially airborne fungi like rice blast and the wheat rusts. All these strategies may be 'things that plant pathologists tend to write about, not practise' [Simmonds, pers. comm.].

55 This, however, in turn creates a serious risk. GLH is the vector for grassy stunt virus, to which only one MV defence currently exists – a vertical NR. So far, GSV has proved bad at adapting to this VR. If this changes, the reservoir of GSV in GLH, which to the extent that MR has been selected is still present in significant amounts in rice, will prove a major problem. MR, then, is not so obviously a low-risk strategy if applied to a pest that is also a vector of another pest to which HR or MR is absent.

56 For the argument that higher yields in Africa and Asia are not needed because of the (excessive) productivity of cereals in NW Europe and North America, see Chapter 7, k. Briefly, the argument mis-specifies the problem of hunger as one of global food-population balance; in fact, it is mainly a problem of getting command over food to hungry people [Sen, 1981], largely by increasing their own yields or cheapening their local foods.

57 Threats to diversity of plant populations, however, can sometimes mean that with an MV an Obstacle obstructs all, or almost all, of a farmer's or employee's income sources, instead of only some as with TVs. (i) If one MV of one crop comes to dominate an area, and its peak water requirements are not met because of irrigation failure, there is much more damage than if several different varieties or crops face peak requirements at different times. (ii) Demand for labour is also concentrated on a few peaks, increasing the incentive to labour-displacing investment for such peaks. (iii) Diversity also encourages exchanges of labour among farmers facing different seasonal peaks; MVs, if they reduce diversity, challenge the communal organizations for such exchanges. All these threats from reduced diversity in MVs are worst for rice, owing to its unusually large varietal diversity among the (displaced) TVs, and to its general absence of mixed cropping [Bray, 1986, pp. 16–17, 20, 25, 29, 44, 120, 174]. However, once an MV is developed to solve such problems, they stay solved; obstacles, unlike pests, do not adapt to overcome varietal improvements.

58 In almost any inbreeding population, a few plants will outbreed, preserving some diversity.

59 This is because of genetic linkage, together with the normally polygene nature of HR.

60 Such stocks are kept at some major IARCs such as IRRI, CIMMYT and ICRISAT, and at several universities. Overview of their management and policy is in the hands of the International

3. MVs and Distribۥ
Among Farmers

(a) Small farmers in MV areas: an over-researched issue?

The effect of modern varieties of staple food crops on poverty depends, in the longer term, on how the vast changes they bring are inserted into the structures of society, of ownership, and of power (Chapter 6). More swiftly and tangibly, the effect depends on four things, in decreasing order of importance. First is the impact of MVs on income from employment, especially from hired farm labour (Chapter 4). Second is their effect on prices, availability, and regularity of staple food supplies to the poorest consumers, especially in towns (Chapter 5). Third is the effect of the big rises in output of food staples, due to MVs in a few regions, upon poor people in rural areas – and in nations – where such MVs cannot usually be grown (section i of this chapter). The final major effect of MVs on poverty is via their direct impact on small farmers and tenants in regions where MVs have spread widely. We judge that this fourth effect of MVs on poverty is relatively less important, because (i) areas dominated by MVs contain less than a third of rural people in LDCs; (ii) such areas are less likely to be poor than less progressive areas; (iii) farm operators, especially in MV areas, are less likely to be poor than landless workers.

Yet well over two-thirds of the published research into 'what MVs do to the poor' relates to this, the fourth or fifth most important issue.[1] The literature asks: do small farmers and

nts in MV 'lead areas' – areas like the Indian Punjab
wheat and rice), Central Luzon in the Philippines (rice),
Sonora in Mexico (wheat) or Trans Nzoia in Kenya (hybrid
maize) – adopt MVs, and if so soon or late (section b of this
chapter)? Over what proportion of farm area (section c)? With
what backing from other inputs, such as fertilizer (section d)?
What prices do poorer MV users pay for their inputs, and
receive for their crops (section e)? As a result – bearing in mind
the impact of MV activity in raising demand prices for farm
inputs, and in lowering supply price for the crops produced –
how are smaller farms and tenants in 'leading areas' affected
by MVs in respect of yield, efficiency and income (section f)? If
they do benefit, will they be taken over or evicted (section g)?
Apart from these effects of MVs on small farmers' average
income in 'lead areas', how are such farmers affected by the
greater or lesser risk, associated both with MVs and with the
consequently[2] greater dependence of farmers on commercialization (section h)?

These are all interesting questions. The discussion below
suggests that, on the whole, we now know the answers. But the
questions and answers tell us less than one would expect about
how MVs affect poverty, even among farmers in MV lead
areas, because the overlap of 'small farms' with 'poor farm
families' is rather bad. Five reasons for the bad overlap follow.
They are largely neglected in the massive – and in itself
interesting – 'size-adoption-yield' MV literature reviewed later
in this chapter. They imply a new agenda for research into
what MVs do to farmers in poverty in MV lead areas.[3]

(1) Land quality: Even given the crop, the region, and the
inputs per acre of labour, fertilizer, etc., 'size' gives an
incomplete indication of a farm's capacity to generate income.
A farm's slope [Colmenares, 1975, p. 21], terrain [Cutié, 1975,
p. 23], above all irrigation and drainage, can make a vast
difference to its net income[4] before MVs; to the farmer's
decision on how much land to plant to them; and to their
impact on net farm income afterwards. Hence 'poverty rankings' of farm households by *farm size* and by *net farm income from
the MV-affected crop* differ hugely. So do the interactions of the
two rankings with the adoption of MVs. An outstanding

Mexican study showed that adopters, despite having slightly less land per person than non-adopters, had significantly higher land value per person [Burke, 1979, p. 148].

(2) Crop-mix: On most farms, several crops help to produce net farm income. Yet most studies of MV adoption, yield, etc. on 'small vs. big farms' ignore what the MV innovation does to non-MV crops – cash-crops like cotton, and less-progressive food crops like millets. 'Poverty ranking', by farm size or even *net farm income from the MV-affected crop*, often tells us little about how poor farm households are in terms of *net farm-system returns*[5] – let alone about how MVs affect these returns. Research in China and SE Asia shows that, where rice is mainly a subsistence crop, its intensive production (e.g. with MVs) raises income mainly by enabling poorer farm households to reduce rice area, maintain rice output for home consumption, and divert land and labour to other crops, for profitable sale [Bray, 1986, pp. 131–7].

(3) Non-crop income: Even *net farm-system returns* do not capture all the effects of MVs on a poor household's *net income from all activity* – farm, non-farm, and off-farm – even if most of its income is from its own farm. (i) MVs of subsistence crops often enable small farm households to diversify into non-crop activity, often complementary with the intensified crop [*ibid.*, pp. 113, 117]. (ii) However, their increases in income from adopting an MV are offset by losses, if their labour and other resources are diverted to the MV: from other crops; from 'off-farm' employment by others, and from own-account 'non-farm' activity. In rural studies in eight countries, such 'non-farm' activity alone accounted for one-third of net rural household income [Chuta and Liedholm, 1979]. In Matlon's painstaking study of three villages in Northern Nigeria, off-farm income sources accounted for at least 20 per cent of total income and 40 per cent at most [Matlon, 1977]. (iii) Income and information from off-farm activity, apart from making many small-farm households non-poor, help even poor households to take risks, to earn income, and thus to succeed with MVs. In two traditional Colombian villages, innovators had much more contact with cities than did non-innovators [Rogers and Svenning, 1969, p. 298]. The proportion of days spent off the farm has been strongly linked to a household's

technical efficiency, and hence to yield, in MV rice farming [Herdt and Mandac, 1981, p. 394]. (iv) On the other hand, a poor household that relies for income mainly on farming, but also on several small-scale non-farm activities, can readily use up its access to credit (and labour time) in order to finance their profitable expansion. Then it cannot borrow to finance fertilizer purchase (or hired labour) in support of MVs (or may become reluctant to over-extend itself by doing so; Barry and Baker, 1971). This greatly affected MV adoption among forty-two small-farm households in Bicol, Philippines, in the early 1980s [IRRI, 1984, pp. 353–4]. (v) If small-farm households spend time working on other farms and these adopt MVs, then the MVs can affect the households' income as employees, not just as farmers.

(4) Household size: A household's *net income from all activity* gives a very imperfect indication of its *income-linked poverty*, because of differences in size among households. An income ample for a small household can mean dire poverty for a large one. Households with 'low net income' are a mixture of (i) small households that are not poor, (ii) poor households that are small – but exclude many (iii) large and poor households.[6] Households with high total income, farm income, or farm size tend to be big households; yet in total populations, even if we do not hold farm size constant, bigger households tend to be poorer [Lipton, 1983a]. As for MVs, a farm of a particular size has fewer problems of management and labour search if it has lots of family workers – but also fewer gains-per-person from each extra ton of produce. There is some evidence that households with a high ratio of consumers to workers are under the most effective pressure (given available resources) to adopt MVs, and work more to earn more [Harriss, 1982, pp.173–5; Low, 1986; Hunt, 1979; cf. Chayanov, 1966].

(5) Income and poverty: Even the effect of MVs on *income-linked poverty*, as measured by net income per person or per consumer-unit, is a very imperfect indicator of their effect on *real poverty*, absolute or relative to the local norms. There are several reasons for the difference between income-linked poverty and real poverty – and MVs can affect these reasons. Here it is necessary to look at the family in its social setting and in its life extending over several seasons. (i) Stocks of grain, not

just flows of income, can be affected by MV adoption. (ii) So, on occasion, can access to social services. (iii) Above all, MVs affect poor farmers' obligations. To adopt MVs (or the fertilizers that often make MVs attractive), debts are often increased, especially by poor households, at least in initial seasons, when the extra purchases (to make MVs pay) precede extra incomes. Yet, in circumstances of excess demand for official and formal credit, it is usually the poorer adopters who must go to moneylenders and incur high interest obligations. These reduce such adopters' future gains from MVs, even if they adopt and farm better than richer farmers. (iv) Informal bribes and obligations to petty officialdom are bid up by the need to acquire MV-related inputs or favours. This looms largest for poor farmers [on Chilalo, Ethiopia, see Cohen, 1975, p. 354]. (v) In adopting MVs, it is poorer farmers who are likeliest to rely mainly on family members for extra labour. This extra effort is seldom fully counted into production costs. However, it requires extra dietary calories, cutting the true net gain of poor people from MVs.

*   *   *

Very many pieces of research have sought to assess the impact of MVs on poor farming households by measuring MV adoption or performance in small farms, mostly over one season. However, there is evidence that the above five effects cast doubt on such assessments. First, only in the better-watered areas do agricultural households with owned or operated holdings of seven acres (or thereabouts) have a much better chance of avoiding extreme poverty than households with one or two acres, or even landless labour [Lipton, 1985b]. Second, even authors who emphasize the recent evidence that size of operated holding is seldom linked to eventual MV adoption point out that 'when the farmer's wealth or economic resource base is considered, those with higher incomes tend to be the main adopters' [Herdt and Capule, p. 37, citing micro-studies from India, Bangladesh and Korea]. Third, other income sources sometimes radically improve the household impact of MVs on the poorest farmers [Swenson, 1976, pp. 8–10]. Fourth, larger families are significantly more likely to adopt MVs in three out of the five areas studied – and less likely in none [Herdt and Capule, p. 32; Malla, 1983; compare Harriss, 1982, pp. 173–5]. All this underlines the weakness of inferences, even if limited to MV lead areas, from 'MVs do X to small farm households' to 'MVs do X to poor people in farm households'.

Research on the impact of MVs on poverty needs to be radically reorientated. It should move from MV lead areas to other areas. It should shift from effects via production to effects via employment and consumption. It should also extend over more than one planting season. However, to the extent that research continues to focus on poverty among farmers in MV lead areas, that focus should shift, away from further replication of studies linking farm size (or tenure) with adoption, yield, etc. – and towards innovative studies linking MV innovations causally to changes in incidence and severity of 'poverty'. This is best indicated, not by a household's farm size, but by its real income or consumption, per person or consumption-unit, from all sources (net of all production costs, debt obligations, etc.), at various stages in the diffusion of MVs and linked inputs.

(b) Adoption, farm size, and tenure

In reporting on the impact of MVs on poverty in MV lead areas, however, we must use the evidence that exists. This is overwhelmingly about the effect of MV innovation on the affected crop in small-farm households. That effect depends on (i) their adoption rates, (ii) the proportion of land that 'small-farm' adopters plant to MVs, (iii) their capacity to saturate MVs with other inputs such as fertilizers, (iv) the prices they pay for MV-linked inputs and receive for outputs, (v) their yield and efficiency, (vi) their ability to keep their land, and (vii) the effect on their income stability. These are dealt with, respectively, in sections b–h of this chapter.

The questions of whether, when and how 'small farmers' adopt MVs remains loosely relevant to equity, if not tightly correlated with impact on poverty. Apart from such places as Bangladesh and Java, where most really poor villagers depend mainly on employee incomes, widespread adoption of improved technology is especially relevant to a rural society's social cohesion and 'parity of esteem': to the sense that all classes advance together towards higher levels of income and technology. The extent to which MVs are adopted successfully by small farmers tell us less about poverty impact than was once believed, but a lot, perhaps, about rural societies' long-run prospects of coherence and stability.

*　　　*　　　*

In the early years of MVs, until about 1974, the evidence that larger farmers were adopting more, sooner, seemed overwhelming. A still-classic study for the Indian Planning Commission concluded: 'For all five crops and in each of the three years [wheat, rice, maize, millet, sorghum; 1967–8, 1968–9; 1969–70, there was] a strong positive linear relationship between the proportion of farmers adopting [MVs] and the farm size' in the great majority of villages in Indian MV areas. Also 'in 17 of [20 case-studies by Agro-economic Research Centres this] relationship was statistically significant' at 5 per cent, and no case showed small farmers likelier to adopt [Lockwood *et al.*, 1971; see also Schluter, 1971, and Dasgupta, 1977, p. 226]. In Bangladesh, where wheat was introduced late (and almost wholly in MVs), it was initially likelier to be adopted by larger farmers [Directorate of Agri. Marketing, 1977]. Early evidence for other countries was similar [summarized in Herdt and Capule, 1983, p. 33].

It is usually the case that, by the mid-1970s, small farmers had ceased to lag behind in adopting MVs. In thirty villages surveyed by IRRI, small farmers even appeared to have adopted somewhat more and/or earlier than large. However, careful inspection shows that this is a fallacy of aggregation; only one village showed small farmers readier to adopt than big ones, but showed this very strongly [IRRI, 1978a, esp. p. 94]. All the same, in India, the link between *large* size and MV adoption had disappeared by the mid-1970s for wheat, and for most states for rice; it was doubtful for maize [Dasgupta, 1977, pp. 227–8; Barker and Herdt, 1984]. Wheat MV adoption also appeared to be widespread among farms, irrespective of size, by the late 1970s in the Pakistan Punjab, NW Bangladesh, and NW Mexico [Byerlee and Harrington, 1982, p. 3].

What has changed and why? Are there exceptions? What are the lessons for research and policy?

The first change is that, in many places, big farmers adopted first and small ones caught up [Prahladachar, 1983, pp. 929–30, and Harriss, 1977, pp. 139–40 for India; Burke, 1979, for Mexico; Ruttan, 1977, p. 17]. For Kenyan hybrid maize, early adoption was strongly related to size (and to no other variables tried) – but 'mature' levels of adoption were not [Gerhart, 1975, p. 42]. In an apparently typical South Indian village, the same pattern for MV rice was traced to bigger farmers' better access to reliable information, credit and water; this made early adoption safer for them [Harriss, 1982, pp. 162–72]. This means that big farmers obtain the 'innovators' rent' [Anderson and Pandey, 1985, p. 8; Dalrymple, 1979, pp. 720–1; Binswanger, 1980, p. 180]. They do so because, during early adoption, the staple food in question is often scarce and expensive, at least locally; but food prices are held back – or even pushed down – by the success of the MV (and hence the early innovators' extra sales) by the time the poorer, later adopters are ready to sell their MV-boosted crops. At the same time, extra demand for farm inputs – also due to MVs – has been pulling their cost above what the early innovators had to pay.

Big farmers' access to innovators' rent – their capacity to act first, and avoid this *cost-price squeeze* – is partly due to their cheaper, safer, and better access to inputs. However, innovators' rent is also partly a reward for bearing risk. Early innovators also incur the costs of failure when the risk goes wrong, as when downy mildew hit the early hybrid millets [Binswanger and Ryan, 1977, p. 224]. Moreover, in one important case in Western India (where poor farmers consumed much of their extra output), the effect of the cost-price squeeze upon the returns to late adoption did not give smaller farmers any 'enduring and self-reinforcing disadvantage' [Shingi *et al.*, 1981]. And adoption of MVs is a repeated process, with small farms at less disadvantage in subsequent adoptions; where 77 per cent of new MV seed is procured from neighbours, and where 1 in 3 farmers plants more than one MV with the stated aim of avoiding risk (as with rice in Nueva Ecija, Philippines, in 1984) [IRRI, *Ann. Rep. 1984*, p.

194], small farmers are unlikely to be late adopters on account of either high risk or low access to seeds. In general, though, late starters finish last.

Moreover, small farmers' adoption rates are not catching up everywhere. Of twelve quantitative studies for rice MVs in Bangladesh, seven show a positive size-adoption link, and only one a negative link [Herdt and Garcia, 1982, p. 3]. In India, while both wheat and rice MVs featured the usual pattern that bigger farmers adopted earlier but smaller ones followed, there were signs that small-farm MV rice adopters were prone to revert to TVs by the early 1970s [Lockwood *et al.*, 1971] – though the later, safer rice MVs may well have overcome this. With maize hybrids, the small farmers, reliant on timely distribution of small amounts of seed each year, may suffer long-term adoption delays, abandonment, and re-adoption lags, especially since small farm size is usually linked to absence of extension visits [Colmenares, 1975].

Whether small farmers adopt early, or catch up, depends on policy; there is no universal law. In India the spread of MVs to the poor falters or fails in areas of greater initial inequality and institutional inadequacy or bias; catch-up is thus 'by no means automatic (which seems to be suggested in the evaluations of over-zealous enthusiasts of the green revolution)' [Prahladachar, 1983, pp. 930–1]. Co-operative services are an institution which can enable small farmers to share savings, thus structuring their farm capital away from buildings and towards larger, jointly managed irrigation assets. This made a major difference to MV adoption among small Mexican farmers, favouring *ejidatarios* over small private farms [Burke, 1979].

Apart from smaller farmers 'catching up' using old MVs, new MVs of some crops may be getting more 'smallholder-friendly'. New wheat and rice MVs frequently outyield local varieties, even if both MVs and TVs are exposed to low inputs, disease risk, and some moisture stress – as early MVs, such as TN-1 rice, certainly did not [Byerlee and Harrington, 1982, pp. 1–2]. Where maize and millet composites replace hybrids, small farmers are less damaged by the problem of annual seed replenishment; this is a reason for caution about smallholders' capacity to adopt and sustain the upcoming rice and wheat hybrids.

More generally, whether smallholders find successive MVs 'friendly' depends on policy (just as policy largely determines the presence or absence of institutions helping smallholders to adopt a given MV early on). In Tunisia, higher smallholder acceptability of MVs of durum wheat, compared to bread wheat, is related to lower risk, especially at low input and management levels [Gafsi and Roe, 1979]; but research policy helps decide whether breeders seek safety or yield potential first, and rural policy helps decide whether rural institutions help small farmers to manage risk.

As a rule, bigger farmers take the early risks, adopt MVs first, and get innovators' rents; smallholders catch up, and gain less from MVs, but still something (Chapter 3, f). The breeding of MVs for lower initial risk, like the H-varieties in Sri Lanka, can accelerate smallholders' adoption. So can the encouragement of institutions that ease access, credit, or extension advice for small farmers.

Small size of farm is linked with slow, or no, adoption in many early studies; but some have argued that the delay is not due to small farm size, but to something for which it is a 'proxy', and which must be held constant in correlating size and adoption. Hold constant (i) a farm's topography and willingness to farm pure line stands [Cutié, 1975], (ii) the access to credit [Colmenares, 1975], irrigation, fertilizers, and (iii) the farmer's education and off-farm income [Perrin and Winkelmann, 1976]; and behold! the effect of farm size in impeding adoption vanishes. But this may be misleading. Small farm size, as we have stressed, is a bad indicator of poverty; but they *are* correlated. Poverty both brings farm size down, and impedes education, off-farm earnings, and access to credit and farm inputs. Through these impediments, poverty delays adoption, and ties that delay to small farm size. It is a loose connection; smallholder-friendly institutions, and lower-risk MVs, can break it. But to deny the (usual) connection, or that it (usually) harms the poor, would be unhelpful.

Tenant farming is even less clearly linked to poverty than 'small farming'. However, in many places – Bangladesh, Java, Eastern India, the Philippines – *sharecropping* tenants lack security of tenure, surrender perhaps half their crop as rent, and thus are likelier to be poorer than owner-farmers. Do sharecroppers also lag behind in adopting MVs? There are good theoretical grounds to expect that (i) a pure sharecropper and a pure owner-farmer will manage the same land in much the same way, but that (ii) a farmer with both share-cropped and owned land will manage the latter more intensively, with higher inputs, work, and output. These expectations are strongly confirmed in North Bihar, though not specifically for intensification linked to MV adoption [Bell, 1977].

Farmers who own all their land do not show systematically different MV adoption rates from pure tenants [Herdt and Capule, p. 37]. One reason is that it pays landowners to vary the terms of tenancy in order to raise the crop on a share-cropped farm, e.g. by offering to share in costs if the farmer adopts MVs [Bray, 1986, p. 188]. A major exception is Bangladesh, where the institutions, especially those for credit, gravely disadvantage tenants [Herdt and Garcia, 1982; Shahid and Herdt, 1982]. There, too, owner-tenants are likelier to sow MVs on their owned land than elsewhere [Hartmann and Boyce, 1983, p. 211].

(c) Risk versus access: land in MVs, smallness, and policy

Do small farmers adopt MVs later, and thus lose much of the benefit, mainly because their access to inputs is worse or costlier than for larger farmers? Or is it mainly because of factors associated with risk? Obviously, policies to reduce risk *and* to improve access are desirable, but are costly to governments: they 'cost' scarce administrative skills (and perhaps also offence given to the powerful) as well as cash. Should governments, seeking to spread MVs faster among small-holders, stress the provision of access or the reduction of risk?

Poor people often give uncertainty as a reason for delaying or refusing adoption. First, although many MVs reduce risk objectively [Roumasset, 1976], smaller farmers are especially likely to know less about them than about TVs (or older MVs), and in particular to know less about how the MVs will perform if rainfall or pest attack is unfortunate; smaller farmers everywhere enjoy less extension advice than their bigger neighbours in similar circumstances; later, when a neighbour has been seen to succeed with MVs, smaller farmers acquire more confidence to adopt them. Second, small farmers – to the extent that they are 'poor people' – are especially likely to face high 'background risk', e.g. from ill-health, and thus to reject extra perceived risk from a little-known MV [Lipton, 1979a]. Third, given 'background risk' and perceived levels of risk and uncertainty from an MV, poor people have higher risk-aversion than others, although the differences observed in empirical work [Binswanger, 1981] are surprisingly small. Work in Gujarat, India, suggests that risk is a more serious constraint on smallholders' adoption than access to credit for inputs in unirrigated areas – where risks are especially high – but less serious in irrigated areas [Schluter, 1974].

The choice of policy priorities between 'risk-reducing' and 'access-improving' strategies to accelerate smallholder adoption of MVs is illuminated by the following 'paradox of proportions'. Although smallholders are slower to *adopt* MVs, the *proportion of land* that adopting smallholders sow to MVs is frequently higher than for adopting larger farmers [Herdt and Garcia, 1982; Asaduzzaman, 1980; Dasgupta, 1977, pp. 229–32]. Several explanations are possible. One is that smallholders eat a much bigger proportion of what they grow, and market much less of it; therefore, given a decision to adopt an MV at all, its normally lower sale price (relative to marketing costs per bag, which will be the same as for a TV) is less of a deterrent to smaller farmers. A second explanation of the 'paradox of proportions' may be that medium and larger farmers often prefer to retain, say, 10–15 per cent of their land in a specially tasty TV, such as a *basmati* rice, which may otherwise not be readily available locally – a luxury that poorer farmers could seldom afford.

However, a third explanation of the 'paradox of proportions' – almost certainly a part of the truth – casts light on the policy question, of whether reduced risk or improved access best helps small farmers to adopt MVs. Suppose you have very little land. You have to incur some fixed costs in learning about, and obtaining, MVs and associated inputs. These costs can be justified – alongside the possibly greater subjective risk due to a change in the variety planted – only if they are spread over a certain minimum area and output, i.e. by planting most of your land to the higher-yielding MV. A farmer with more land could have recouped those costs, i.e. made a profit out of switching to the MV, while planting it on a smaller proportion (but not a smaller acreage) of his land – and leaving some of his land for a traditional tasty variety for household consumption.

This is consistent with the argument that risk-aversion impedes adoption only if – and to the extent that – there are 'fixed costs of adoption' [Feder and O'Mara, 1981, pp. 60–1]. Otherwise, a highly divisible input such as a new MV might be tried out on a tiny handkerchief of land – a negligible risk even for a small farmer – and spread further as its performance was monitored. So the 'paradox of proportions' is important, in regard to policies for spreading MVs to small farmers, in two ways. (i) Directly, the paradox suggests that, once small farmers have adopted an MV, they rely on it more than bigger farmers. That is good for yield, but bad for diversity and hence (unless the variety is very tough) bad for long-run risk among small farmers. These are relatively ill-equipped to bear greater risk, and likely to respond to it by conservative (i.e. less profit-seeking) actions in other aspects of farm enterprise. (ii) Indirectly, one plausible resolution of the paradox confirms that small farmers' risk aversion, and hence their delays in adopting MVs, may be effectively treated by reducing 'fixed costs of adoption'. This involves smaller packages of MVs and fertilizers; competitive small-scale rural intermediation, credit, and retailing; extension systems with incentives to reach small farmers; or robust MVs that do not require farmers to obtain too many new sorts of input or information.

(d) Other inputs to support MVs: big farmers and small

While appropriate recent MVs frequently outyield TVs even at low input levels, substantial gains in profitability (as against *modest* gains in *yields*) usually require higher input levels. Early adopters usually expect such substantial gains, in order to compensate them for the subjective risks, and for the information costs, of trying something new. This is one reason why bigger farmers usually adopt earlier than small farmers; big farmers enjoy better access to cheap credit – and economies of scale in selecting, ordering and transporting inputs – and can thus more readily afford the associated inputs required, not to render a new variety profitable, but to make the extra profit sufficient to justify the risk of innovation.

Three sorts of inputs, as the demand for them rises in the context of MV adoption, can restrain it by posing special problems for small (or rather for poor) farmers. Labour-linked inputs such as irrigation ought not to, but often do, as we shall see. Capital-linked inputs such as tractors clearly do, but their link to MVs is usually artificial (Chapter 4, f). Supportive inputs, such as credit, may pose the most serious problems.

(i) Irrigation and fertilizers increase labour-use per acre. In principle, this should mean that smaller farmers adopt such inputs (and associated MVs) more readily than large farmers. Small farmers generally have more family labour per acre, and can thus avoid or reduce the search and supervision costs of labour hire. Indeed, traditional manual irrigation techniques such as the *doon* in Bangladesh, and traditional organic manures like bonemeal in Sri Lanka [ILO, 1971, vol. 1, p. 92, citing 1962 Census of Agriculture], are more used on smaller farms than on bigger ones, and help explain their traditionally higher yields (see section f below). As for risk, better irrigation cuts *objective* risk – and, given that, so may extra fertilizer [Smith *et al.*, 1983].

However, irrigation and fertilizers can create two problems for small MV adopters. First, MVs often require more precise control, if the yield increase is to promise enough extra profit to outweigh the apparent drawbacks;[7] this extra control often means a shift to forms of input – tubewells, bagged inorganics – that require farmers to find more ready cash, and less family labour, per acre than did older input systems. Second, MVs often require a quite big rise in water and nutrient inputs, if they are to achieve sufficiently attractive (20–45 per cent or more) yield increases; this, too, requires extra cash, as family labour (even in densely populated areas) is unlikely to suffice, in peak seasons, to obtain the extra plant nutrients and water by traditional manual methods alone.

Irrigation in most parts of India is about as unequally distributed among farmers as is land. Access to dug wells is more equally distributed, to canals about as equally, but to tubewells much more unequally. MVs have greatly increased the public and private preference for costlier but more reliable tubewell-linked control, as against canal systems (in the Punjab) or *doon* irrigation (in Bangladesh). Indian work [Narain and Roy, 1980; Dasgupta, 1977, pp. 91–2] confirms the experience repeatedly reported in Bangladesh [Hartman and Boyce, 1983, ch. 19]: the inability of poor farmers to obtain credit for, or access to, tubewell water can retard their adoption of MVs. Evidence from Maharashtra, India, suggests that this is partly because the construction costs of a well increase less than in proportion to its capacity to deliver water, so that small farmers' unit costs in obtaining tubewell water are relatively high [Ketkar, 1980].

As for fertilizer, we saw that traditional inorganics were if anything more intensely used on small family holdings in India and Sri Lanka (just as in late nineteenth-century Russia: Lipton, 1977, p. 115). For MV rice in most of South-East Asia, small farmers also use at least as much inorganic fertilizer per acre [Herdt and Capule, 1983, p. 33]. However, especially in the early stages of MVs (when *subjectively* risky input costs have to be incurred before the benefits materialize), there are reports of higher per-acre fertilizer use on larger farms. This contributes to their greater access to innovators' rents. In Bangladesh [Hartmann and Boyce, 1983, p. 181], the main reason appears to be that larger farmers can better afford timely access to fertilizers. In a South Indian village study, larger farmers were better able to afford time – or peons – to seek out scarce urea, and to queue for days to get it [Harriss, 1982, pp. 167–71]. In a village in North India (in Western UP) in 1974–5, fertilizer use, per unit of area in winter wheat, was strongly and positively linked to farm size, i.e. to total area cultivated by a household; slightly less so to total owned area; and not at all to owned area per consumer-unit – suggesting that 'not. . . wealth. . . but concern and ability to be involved in agriculture' goes with intensive fertilizer use [Bliss and Stern, 1982, p. 201]; it also suggests that economies of scale in fertilizer use are helping larger farmers (though not necessarily richer ones) to use more of it.

There is reason to believe that small-scale farmers' disadvantages with fertilizers and irrigation, as complements to MVs, are transitory. In 1974–5, after ten years of MVs, a thorough survey of 1663 farm households in the Indian Punjab showed that, in the Central Region which had the highest per-acre input costs, outlay on manure and inorganic fertilizer per cropped acre rose – very slightly (from 134 to 167 rupees)[8] for irrigated wheat, much more (from 125 to 228 rupees) for irrigated rice – as between the smallest group of farms (operating below 2.5 acres) and the largest (over 25 acres). In the other two regions, there was no relationship between farm size and fertilizer use. Even in Central Region, it appears to have been mainly the rise in fertilizer prices in 1974, following the oil shock, that caused smaller farmers to cut back [Bhalla and Chadha, 1983, pp. 61–4]. As for irrigation, success with MVs stimulates both its spread (from 59 per cent of net sown area in 1965–6 to 85 per cent in 1979–80 for the Indian Punjab) and its shift towards the more reliable sources, namely tubewells and to a much lesser extent dug wells (from 41 per cent to 58 per cent) [*ibid.*, p. 12]; this reduction in scarcity is bound to spread irrigation, especially from tubewells, increasingly beyond the bigger farmers, who had initially used it to enjoy higher innovators' rents with MVs.

Of course, even if the Punjab story is typical, it does not mean that, if only they will wait, all will be well for poor farmers seeking fertilizers and irrigation to complement MVs. First, the main gains – from innovators' rents, before the cost-price squeeze – go to bigger, earlier adopters. Second, these may engross or evict small farmers, so there are few left to enjoy the remaining MV gains, even once they overcome the barriers (to access, credit, or risk-taking) that had initially kept them from using enough fertilizers or irrigation to gain much from MVs – or in some cases to feel that it was worth adopting them early. That was not so in most of India [Vyas, 1979], but eviction and engrossment in the wake of MVs cannot be assumed away; we return to them in section g below.

Despite real problems, we believe that small farmers' *long-run* access to fertilizers and irrigation, as complements to MVs, is strongly supported by market forces. It pays a supplier to get such inputs to smaller farmers, because they derive more economic gain from it than big farmers, and hence can pay more (or buy more per acre) once finance problems are overcome. The reason why small farmers, in principle, should gain more from a given per-acre application of irrigation or fertilizers than large ones, is that such inputs require more labour, especially for efficient weed control; and there are advantages for small farmers in respect of access to family labour (pp. 135–7).

For policies and markets to overcome small farmers' financial and risk problems, so that small farmers come to support MVs with at least as much fertilizer or irrigation as do big farmers, takes time. However, since small farmers eat much of what they grow, they gain even if their switch to MVs brings them higher yields only after extra outputs of the crop from early innovators (big farmers) have cut its sale prices. Breeders can help small farmers to adopt MVs early and with good input practice, by developing MVs that, even at some cost to optimal or potential yields, perform well unirrigated, or with less precisely timed irrigation; that use high *conversion* efficiency (Chapter 2, g) to substantially outperform TVs even with low nutrient supplementation; that respond well to organic manure, including azolla (Chapter 7, h); and that are best suited to labour-intensive use of these inputs, e.g. to placement of fertilizers near each plant's root zone (Chapter 2, e).

(ii) Tractors, threshers, and herbicides lack the 'natural protection', for small farmers, of labour-intensity. They normally reduce labour input per acre substantially. The machines, too, are 'indivisible', which favours big users; of course small farmers can hire [Schultz, 1964], but they get lower priority in peak periods than owner-users, and must pay more per acre, partly because it takes longer to plough or thresh ten one-acre farms (even if contiguous) than one ten-acre farm; partly because of transport costs among remote small farms; and partly because of elements of monopoly in hire markets.

Herbicides (usually requiring sprayers), tractors, threshers and reaper-binders all displace labour, especially family labour, and all increase cash requirements. Extra acquisition of such inputs thus offers deeper problems to small farmers than do extra purchases of irrigation water or fertilizer. Fortunately, these labour-displacing inputs, unlike irrigation or fertilizer, are not often genuinely complementary with MVs; they are made so by errors or biases in policy (Chapter 4, f). In thirty Asian villages in an IRRI study of constraints on rice yields, MVs were strongly associated with extra purchases of herbicides, tractors and threshers only in the twelve Philippine villages, and with tractors in the two villages in Malaysia [Barker and Herdt, 1978, p. 85].

These machines appear to be genuinely profitable in the wake of MVs only in some cases of double-cropping. For this reason, technical change based on MVs was size-neutral in single-cropped areas in Malaysia, but favoured larger farms (which could obtain tractors at lower cost per acre) in double-cropped areas; the same was found in West Java [Gibbons *et al.*, n.d., p. 221; Lingard and Baygo, 1983, p. 54]. However, most studies of tractorization show little or no association with cropping intensity, if other factors are held constant [Binswanger, 1978; Agarwal, 1984b]. Farm machinery (and hence big-farm advantage) is linked to MVs (and hence to their delayed adoption by small farmers), if at all, through policy errors or biases – especially subsidization of fuel, or of credit for tractor use (Chapter 4, f) – not through the sort of natural complementarity that links MVs to extra plant nutrients and better water control. Also, although MVs' short stature assists weed growth, this is usually at least as well managed by extra labour, which favours small family farms, as by extra herbicides, which favour bigger farms – unless, once again, policy errors or biases supervene (Chapter 4, d).

Breeding choices, again, are relevant. Denser stands and erect leaves, if MVs' timing is right, can shade out weeds, offsetting the competitive disadvantages of short-strawed crops. Different combinations of MV and recommended plant density can favour hand-weeding as against herbicides. As for tractors and threshers, experience in Sri Lanka showed that a quick-maturing second-season rice variety, even with a considerably lower yield than a full-duration MV, is better for poor farmers – and probably for GNP – because the full-duration variety requires quick turnover between first season and second-season crops, thus encouraging costly mechanization (and the quest for subsidies for it).

(iii) Input support via credit and extension may be more 'difficult' for small farmers seeking to adopt MVs profitably and safely, in the long term, than either irrigation and fertilizer (which are ultimately labour-using, thus helping farmers with family labour) or tractors, threshers and herbicides (which, although labour-saving, are not normally stimulated by MVs as such, though they may be associated with MVs by the facts of socio-economic power). Not only are increases in credit and extension genuinely associated with MVs (as is the case with fertilizers and irrigation, but not herbicides or tractors). It is also true of credit and extension (as of farm machinery, but much less so of irrigation or fertilizers) that their supply and use may genuinely be easier or cheaper for large farm units. A given volume of credit, or of extension information – in support of inputs used with MVs – is genuinely cheaper and easier to supply to ten 25-acre farms than to 250 one-acre farms, especially since bigger farmers 'know the ropes' and can afford to buy literacy, numeracy, and bus fares to rural credit agencies. In addition, where credit is subsidized and scarce, it is especially likely to reach mainly powerful people, i.e. to miss out small farmers [Harris, 1982, pp. 168–71 and fn. 4].

However, extra subsidized credit for the rich gets loaned onwards to the poor, at a profit of course, but eventually cutting the price of credit all round. And extension workers' information, if not much else, does trickle down. In Yaqui Valley, Mexico, initial biases towards big farmers in the spread of both credit and extension were very important early in the process of diffusion of MVs, but much less so later [Hewitt de Alcantàra, 1978; Byerlee and Harrington, 1982].

*   *   *

Overall, the findings on MVs suggest that in the long run small farmers can support them with as much input-per-acre as big ones, and with more labour-per-acre. Unfortunately – unless successive MVs are tailored to small farmers' needs for lower risk, for higher conversion efficiency of purchased inputs, and for good response to simple and labour-intensive inputs – innovators' rents from MVs will accrue mainly to bigger farmers, who can use both MVs and inputs before the cost-price squeeze reduces the gains from them.

(e) Prices for inputs and outputs

Three effects, somewhat different in timing and policy implications but often confused, tend to cause small farmers to receive worse prices for extra outputs, induced by MVs and then sold, than big farmers. First, to the extent that bigger farmers adopt MVs sooner and saturate them with inputs earlier after adoption – and in view of the fact that almost all such farmers' extra output will be sold, not eaten at home – prices for small sellers (and big) will be reduced *after subsequent seasons*, as the adoption process spreads to small farmers.[9] The policy implication here is to emphasize lower-risk, lower-input (but not extractive) MVs, especially of 'poor people's crops' such as millet and cassava, so small farmers can adopt early; institutions of credit and extension to help them do so; and – a less expensive approach – concentration of extra purchasing power with poor people, including some small farmers, who use a large part of their extra output to raise family consumption, thus reducing the fall in output prices.

Second, in a given season, smaller farmers often need to sell quickly after the harvest in order to meet costs and repay loans, while bigger farmers can hang on for price rises. However, such rises seldom reflect more than the cost of storage (including the 3–6 per cent of grain lost in store: Greeley, 1987) and normal profit to allow for price risks. For millet and sorghum, the ICRISAT village studies show only small price differentials of this type among farm-size groups [T. Walker, ICRISAT, pers. comm.]. However, in Thanjavur, South India, the need to sell paddy quickly meant that farmers operating below 20 acres obtained about one-fifth lower prices than bigger farmers [Swenson, 1973, pp. 77–8, 113]. The policy implication here is for the state to ensure ample and competitive – not subsidized or monopolized – rural credit to poor farmers in MV areas early in the adoption process.

Third, even when big and small farmers sell at the same time, big ones often get better prices. Buyers at the farm gate face much lower costs if they acquire 100 tons of grain from one farmer, rather than one ton from each of 100 farmers – and are naturally prepared to pay the big farmer for this saving. Small farmers are much likelier than big ones to face one or two buyers, rather than strong competition; and to have to contract their grain sales forward in order to obtain timely credit or inputs, in which case the merchant can insist on a lower grain price, in part as a reward for risk-bearing. Here, the remedy is for competitive marketing and input supply, with control over cartels, especially in remote rural areas. In the Indian Punjab, by the late 1970s, this remedy meant that small farmers were getting as good prices for their marketed output as big ones [Bhalla and Chadha, 1983, p. 47].

These factors interact in the wake of MVs. Between 1965–6 and 1970–1, as MVs spread all over Thanjavur, seventy farmers operating below 2.5 ha. saw their price per kilogram for paddy sales rising by only 17 per cent – less than inflation – while nine farmers with over 20 ha. achieved a 48 per cent rise. Much evidence from elsewhere confirms that larger farmers obtain better sales prices in rice areas [Swenson, 1976, p. 3; B. Harriss, pers. comm.].

However, small farmers consume much of their extra MV output. This helps them to escape some of the effects of the moderation of farm-gate cereal prices – effects that hit bigger sellers – as MV expansion takes hold [Cordova *et al.*, 1981; Deuster, 1982]. Another output price effect favouring small farmers is that, as big ones switch to MVs, they may leave premium varieties (e.g. *basmati* rice) to smaller sellers [Chaudhry, 1982, pp. 176–7]. Such varieties have, of course, an even greater price advantage over MVs than over TVs. This encourages small farmers to stick with them, especially when (like kichili rice in North Arcot district, Tamilnadu, South India) they seem less risky [Harriss, 1982, p. 66].

Price advantages for larger operators are clearer for inputs, especially *fertilizers*. Such advantages may decline as fertilizer use increases and spreads; by the later 1970s there was no evidence that poor farmers paid more in the Indian Punjab [Bhalla and Chadha, 1983, p. 64]. However, it is quite costly to 'debulk' fertilizers at retail into the 5–20 kilogram packages suitable for really small plots. Unavailable or costly small packages have proved a major problem in Sri Lanka, Kenya and Zambia, perhaps retarding MV adoption by the poor. *Credit*, too, is affected; in Thanjavur, interest on loans 'for paddy production' fell steadily from 13 per cent for holdings below 2.5 ha. to 9 per cent above 20 ha. [Swenson, 1973, p. 184] and differentials on informal-sector loans are often much larger than that.

Finally, if tenants are usually poorer than owners, rising *land* prices in the wake of MVs normally harm the poor *relatively*. Sharecroppers must hand over 30–60 per cent of the output to landlords. This share is in effect a price for land services, and tends to rise when MVs raise the profitability of farming on own-account – and thus the 'opportunity-cost' to the landlord of renting out. In general MVs tend to raise land prices and rents sharply [Cohen, 1975, pp. 350–1].

Price *movements* and smaller farmers' price *disadvantages* in the wake of MVs add up to a 'distributional' version of the famous 'cost-price squeeze' (Indian terminology) or 'agricultural treadmill' (Anglo-American terminology). Extra MV-induced output of a crop – on a world scale, or in a country or region 'protected' by trading policy or by transport costs – pushes its price down; extra demand for water and plant nutrients pushes unit production costs up.[10] These effects of MVs have, over much of the past twenty years, been reinforced by world market trends: as regards production costs, rising oil prices have driven up the costs of nitrogenous fertilizers and many other agrochemicals, and of agricultural transport; as regards crop prices, massive agricultural subsidization in Europe and Japan (and to a lesser extent in North America) has provided incentives, not only to farmers to glut world grain markets *each* year, but to researchers to meet farmers' demands by helping them to grow even more grain (per unit of cost) *every* year,[11] pushing world grain prices

down. Income changes since the advent of the 'green revolution' (1963–86) have reinforced the effects of MVs and
developed-country farm policies upon world crop prices.
Growth of income-per-person in Europe, America and Asia,
especially because it has been increasingly distributed towards
the better-off within most countries, has reduced the proportion of income used to buy grain,[12] and raised demand for
industrial products. Decline or stagnation of the standard of
living among Africans has reduced their absolute food purchases per person, and shifted them towards near-free
imported food (food aid), reducing the impact of this demand
(and of Africa's rapid population growth) on market prices for
food.

How has this cumulative cost-price squeeze, itself partly due
to MVs, affected the poor? Some of the main effects are
regional (Chapter 3, i), upon workers, or via consumption
(Chapter 5, b), but a little can be said about the effects on
'small/poor' farms in MV areas. Plainly, if they tend to join the
MV game late, they will be worse affected by the cost-price
squeeze than bigger and earlier adopters. However, their
greater reliance on non-market inputs (e.g. family labour,
instead of weedicides; organic manures if gathered rather
than bought,[13] instead of fertilizers), and on self-consumption
of outputs, immunizes smaller farmers to some extent. Moreover, the later adopter, typically a smaller farmer, can often
skip the first couple of stages of MV research and move
straight into adopting the latest, highest-yield stage; and the
impact of the cost-price squeeze is obviously less serious if
yields go up really sharply (that is why very small rice farmers
in India's Punjab, but probably not in North Arcot in
Tamil-nadu – where the 'old' MV IR-20 still predominates,
and where unreliable irrigation also renders yield increases
smaller – have gained more *as rice growers* from MVs than they
have lost *as input buyers and rice sellers*).

Three more 'price facts' need emphasis, before further
policy conclusions are drawn: the first two unfavourable to
small and late adopters of MVs, the last favourable. (i) The
cost-price squeeze multiplies up the static cost to small farmers,
from the fact that they enjoy less favourable input and output
prices (and sales timings) than do large farmers for MV-linked

activity in a given season. (ii) Also harmfully to small farmers (unless growing a crop purely for subsistence), they cannot usually[14] opt out of the cost-price squeeze by staying with TVs; MVs pull up demand prices for inputs whatever the crops on which they are grown, and push up supply (lowering market prices) for all varieties of the crop concerned, MVs and TVs alike. Ironically, however, (iii) the impact of MVs upon the poorest small farmers is improved by the fact that they remain in food staple deficit, so that they gain not only from the extra yield from their own MV (which is eaten), but also from the restraint due to MVs upon the price at which they buy their remaining necessary purchases of the staple.

Where the great majority of poor people in a country or region produce grains substantially for family consumption, but are still hungry, they will eat much of their extra output. Then, the squeeze on output prices does much less to damage small farmers as MVs spread. Also (as compared with commercialized farming) the input cost squeeze affects a much smaller proportion of such adopters' total costs. However, if a country or region specialises in Crop X, sells most of it, and imports the inputs needed to grow it, small-farm growers are especially vulnerable, if they adopt MVs late, to the erosion or reversal of yield gains by price effects. This is not an argument for local self-sufficiency in a specific staple, even with MVs. But regional specialization and interregional trade, especially if combined with polarization of a region into big early adopters and small late ones, makes it more difficult for small farmers to gain from MVs when they adopt late, in face of the problems with input costs and output prices reviewed in this section. If specialization by regions is sought, and if small farmers are to gain substantially from an MV despite the cost-price squeeze and their own relatively unfavourable prices, then – except in places of very fast yield growth, such as the Punjab – small farmers' gains depend on the tailoring of MVs and supporting inputs, technically and institutionally, for early adoption, especially by deficit farmers.

That requirement is underlined by the 'price conflicts' around non-subsistence uses of MV crop expansion. Even in the case of a poor people's crop, the use made of it is important in estimating how its price is likely to move in the wake of MVs.

Thus elasticities of demand, in response to both price and income, are quite high if cassava is exported for animal feed; low if eaten locally by townspeople; but, in the latter case, capable of major increase if appropriate MVs are selected for urban cassava-cereal flour mixes [Lynam, 1986, pp. 17–19]. In other words, in planning MVs or institutions, the choice is between price effects that bring (i) good results for *poor cassava consumers' nutrition* (via urban consumers of tapioca – not of bread, or cassava luxury products [Falcon *et al.*, 1984]) but bad price prospects *for poor cassava growers*; or else (ii) few nutritional gains for poor cassava-eaters (only for European cows), but better price prospects for poor cassava-growers. This unwelcome choice is avoided if cassava MVs are used for extra self-consumption by poor farmer-growers. Unfortunately it is in most of Africa – where poor people are still largely grower-eaters of cassava, millet and sorghum – that progress with MVs faces the greatest problems and is most steered towards largely non-subsistence crops. We return in Chapter 7 to this African challenge, of how to breed for poor people rather than for price conflicts.

(f) Farm size, yield, efficiency: impact of MVs

Chapter 2 showed that most MVs increasingly improve plants' efficiency and safety in transforming sunlight, water and nutrients into food – and, despite serious dangers to crop diversity, in withstanding pests. All this (i) cuts risks; (ii) needs more labour (Chapter 4); but (iii) in raising yields and supplies of staple food crops, restrains their prices (Chapter 5). Poor and small farmers in 'lead areas' should gain: (i) they are concerned to reduce risk, more so than big farmers; (ii) they have more family labour per acre; but (iii) they eat much of the extra food they grow. This chapter has confirmed that they indeed adopt MVs widely, over large areas, with supporting inputs – but has demonstrated that they suffer severe adoption delays, access problems, and price disadvantages. Do small and/or poor farmers in lead areas, on balance, gain much from MVs?

This requires two conditions. In this section, we establish the first: that small and poor farmers use land, and resources in general, at least as productively as large and rich ones,[15] both before and after the MV technology arrives. Section g examines the second condition: that small, poor farmers in lead areas can hang on to (or get compensation for) land-rights as MVs spread.

*   *   *

One of the most debated topics in the whole of agricultural development is whether 'small farmers' (or poor ones) 'perform better or worse' than others, overall or in particular physical and technical environments. We lack the space (and the effrontery) to try and settle this debate. However, we must clarify it enough to draw out the implications for our main theme: how MVs affect poor people.

The very strong balance of evidence, especially in potential MV lead areas, is as follows. First, *all* major groups of farmers act sensibly to further their goals with the resources at their disposal. Second, larger and/or better-off farmers do so with less labour, and more of other resources, per unit of land than smaller and/or poorer farmers. Third, the latter therefore produce *more* gross output (and add *more* value to purchased inputs) per year from each acre – the celebrated *inverse relationship*, (IR) – but *less* per unit of labour.[16]

The evidence for the 'inverse relationship' is weaker in four cases. The first is when even the 'bigger' farmers have only a few smallish (usually garden) plots near home, and can thus, just like small farmers, enjoy the advantages of family overview of cultivation [Bliss and Stern, 1982, p. 198; Hill, 1982, p. 171]. The second is where shortage of seasonal labour (rather than of usable farmland) constrains some families to farm tiny holdings of bad land at low intensity, because it pays them to concentrate on off-farm earnings [*ibid.*, pp. 168–9].[17] A third possible exception arises in some semi-arid areas, where there are few available techniques for small farmers to use, in order to apply their high ratios of family labour to land so as to intensify grain farming [B. Harriss, 1982, p. 40;[18] Berry and Cline, 1979, p. 46, for the exceptional 'Zone C' in Brazil]. Even in these three exceptional circumstances, the balance of evidence still suggests an IR, though a weak one; and the second and third circumstances are removed, almost by definition, with the arrival and spread of MVs. The fourth possible exception is more serious: it may arise because, in some circumstances, big farmers' more favourable access to credit and inputs destroys the IR because it outweighs small farmers' advantages.[19]

* * *

Before assessing how new inputs such as MVs affect the IR, we need to explore the *nature* and *size* of small farmers' alleged 'advantages', to see if MVs are likely to strengthen or weaken them (and hence the IR). As for their nature, three points are central. (i) First, small farms are supposed to be good at achieving, not necessarily high yield, but high output (or high value-added) per acre-year. (ii) Second, this is due to smallness and compactness of acreage in several senses: per unit of supervision or enterprise; per unit of *family* labour or effort; and per unit of consumption requirements. (iii) Third, *any* persisting advantages (or disadvantages) for small (or big) farms have to be due to imperfect markets.

(i) A small farm's greater endowment of supervision, or of lower-cost family labour, per acre per year often cannot be reflected in bigger yields (or higher value-added) per acre, in a season, with a specific crop. Improved techniques to achieve this may be absent, risky, labour-displacing, or more readily available to richer farmers (though MVs should eventually lessen all these dangers). Small farmers' advantages, however, can be expressed in other ways than via crop-specific yields. They often leave a smaller proportion of land fallow [Berry and Cline, 1979, pp. 31–6; ILO, 1971, pp. 91–6]. Or they double-crop a larger proportion of their land [*ibid.*, pp. 45, 92; Berry and Cline, 1979, p. 14] – an activity that MVs render more rewarding, and sometimes (via shorter duration) more feasible. Or they select a higher-value crop-mix, often involving vegetables, or replacement of pasture by crops[20] [*ibid.*, p. 60; J. Harriss, 1982, p. 153]. Most of the time, in most environments, however, smaller farmers obtain somewhat larger yields even with specific crops and seasons [S. Bhalla, 1979, p. 154].

(ii) A persistent minority of careful scholars [most notably Hill, 1982] has always questioned the IR – but usually on the basis of reports from very under-endowed regions, where the poor can get little extra output from higher labour-intensity in farming, and are likely to use their spare labour to escape from it; or from transitional areas, while only the lead adopters have moved towards improved methods. Even those in this minority, however, usually give evidence showing that the IR survives in a substantial majority of cases [Chattopadhyay and Rudra, 1976; Roy, 1981]. Most observers, and the great weight of evidence (though based on data of mixed and sometimes doubtful quality), support the IR.

But its *implications* depend on its *causes*. Does the IR mean that redistribution of land, MVs, or inputs to poorer or smaller farms would help output?

Yes, if the IR rests on poorer or smaller farmers' 'better' use of *given* land and inputs. No, if it rests on their initial good fortune in operating land of higher quality [Khusro, 1973] – better soils, irrigation, etc. – perhaps because parents with good land subdivide it among more children [Sen, 1984]. More often than not, however, the IR cannot be traced to differences in the quality of land;[21] even where it can, such differences may well be the result of bunding, levelling, grazing, or micro-irrigation, carried out by poorer farmers (because their slack-season family labour has lower income expectations) – not of any good fortune in the sense of readier natural access to better-endowed soil-water systems.

If small or poor farmers obtain more annual output per acre (though less per hour worked), what exactly are the alleged advantages of smallness? They are of three types, relating to smallness per worker, per consumer, and per decision-taker. First, smaller land area *operated per unit of family labour* – if such labour is committed mainly to the farm – means that, in order to work the land, a smaller part of the total labour force needs to be hired. As compared with bigger commercial units, more work is put in per acre on a small family farm, because the workers are not deterred by the costs of seeking work, or by the risks of a fruitless search; because it pays the worker to farm in ways that preserve the capital value of the land; and because family farmers can avoid some costs of supervision (leaving much of the workforce, family-style, to supervise themselves), and of screening workers for particular tasks since they know who does what task best. We should not romanticize families (with their exploitation by age and sex), nor oversimplify small family farming (with its huge variety of arrangements, and, invariably, partial commercialization). However, the avoidance of costs of labour screening, search, and supervision means that, for small family farms, higher levels of labour-per-acre 'pay' than for large commercial farms. Such 'labour market dualism' is certainly the main reason for the IR [e.g. Bhalla, 1979, pp. 161–8]. MVs, by further raising labour use per acre, should increase the importance of this family-related advantage of smallness. However – especially if MVs spread alongside tractorization – they may also increase the relative importance of hired labour, so that it pays both big farmers and workers to learn more about labour-market opportunities outside the traditional ambit; such learning would erode the advantages of the family farm.

Second, farms with little *land per mouth to feed* (and not much off-farm income) – small, or more accurately poor, farms – have another 'advantage', pushing them to saturate land with family labour. Each extra unit of farm income is worth more to a poorer family. It will therefore increase labour-input, as compared with a richer family on a bigger farm (or a smaller family on the same farm), because it values extra income more and will therefore tolerate more 'drudgery' [Chayanov, 1966].[22] In the many rural cases where operated land is largely owner-farmed, and where land ownership is the main source of family income, this process will also mean more labour-input per operated acre on farms with little land per consumer-unit. The process has been demonstrated in Mbere, Kenya [Hunt, 1984], Northern Nigeria [Longhurst, 1984] and probably in North Arcot, India [J. Harriss, 1982, pp. 173–5]. This greater application of labour may well account for the greater 'efficiency' with which purchased inputs such as fertilizers are used by families owning less 'land per standard family member' – a quite good measure of wealth – in the heavily MV-affected North Indian village of Palanpur [Bliss and Stern, 1982, p. 290].[23] The lash of poverty is the least desirable advantage of smallness; MVs weaken it in absolute terms, but probably not relative to larger farms.

Third, farms with little land per *decision-taker* face him or her with relatively low costs of overviewing, co-ordinating, and deciding. In industrial economies, an analogous advantage of smallness probably sets a limit to economies of scale [Kaldor, 1934]. In tropical farming, it helps the head of a small family farm to respond quickly to the changes and complexities of managing MVs. Of course, this advantage is reduced if the household's activities, requiring management and co-ordination, also include a good deal of off-farm and non-farm work – just as such work reduces the access of the 'small farm' to family labour, while the income from such work reduces the pressure on the 'small farm' household to work more in order to survive. However, reduction, as the persistence of the IR shows, is not the same as removal.

All three advantages relate not only to smallness, but also to compactness. Smaller farms have fewer plots (even if, frequently, *more* plots relative to their total size). They are thus less time-consuming for workers or decision-takers to get around. And a smaller proportion of 'drudgery' is wasted on doing so, encouraging more of it to be put into the business of feeding hungry mouths, than on an extensive or scattered farm.

(iii)   As indicated, the advantages of smallness and compactness can persist only if, in some sense, markets are imperfect; but they always are. If *land markets* were perfect, owners of 'over-large' holdings would rent or sell some or all of their land, or have it managed, in appropriately small units. However, transactors are imperfectly informed about others' land, farming skills, etc.; and big farmer-owners value security, control, hedges against inflation, and land-related power or prestige. Even if such things deterred dealings in land – for sale or for rent – perfect *labour markets* could also destroy the IR [compare Binswanger and Rosenzweig, 1981a]. With perfect markets for labour, it would move from 'over-farmed' smallholdings to 'underfarmed' big holdings until its marginal contribution to output, and its reward, were equal in the two cases, and underfarming vanished alongside the IR itself. But we have seen how costs of search, screening and supervision (and imperfect knowledge, which is costly to improve) deter both farmers and farmworkers from adjusting fully in this way. Finally, perfect *product markets* could partly compensate for imperfect markets in both labour and land.[24]

Probably, however, the processes of modernization, commercialization, and market integration that accompany MVs tend to reduce market imperfections – to improve workers' knowledge of off-farm or distant jobs, farmers' information about possible tenants, etc. This process is probably part of the reason why MVs somewhat weaken the IR. But the process is slow and incomplete – and subject to countervailing forces, as new opportunities create new special knowledge, and new market power.[25]

*   *   *

Even if the IR is general, how important are small farmers' advantages, given the quality of land and the technology, in using land (and other non-labour inputs) more intensively? *First*, 'the graph linking [gross or net output per acre per year] and size of farm holding is a curve, not a straight line, which may slope upwards as well as downwards for part of its length', e.g. (in an area of reliably irrigated land) between 1/4-acre part-time farmers and committed 1-acre farmers. *Second*, the curve 'is bound to vary. . . from time to time and from place to place' [Hill, 1982, p. 169]. The enormous advantage of farmers with below one acre, in annual output value per unit of operated land in Pakistan in 1959–60 (or per quality-adjusted unit in Muda, Malaysia, in 1972–3) [Berry and Cline, 1979, pp. 80, 117], is exceptional. A gradual IR, typical of most other studies, would show annual output value per operated acre above the all-farm average by perhaps 5 to 25 per cent among farmers operating the equivalent of .05 to 1.5 acres of good irrigated land (say 0.3–10 acres of inferior, but still arable, dry land). Third, and above all, 'as there are so many variables other than holding size which affect yield, no curve linking those two variables would be devoid of detailed kinks and bumps' [Hill, 1982, p. 173]. Thus, the 5–25 per cent advantage, claimed above as a normal IR, for 'small' over 'typical' farmers, would apply to 'average' managers within each size-group; management quality (including personal relations, farming ability, attitudes to risk, and willingness to work at farming) is likely to vary greatly *within* any group of, say, 5000 farmers with about the same land quantity and quality.

So we should not expect the inverse relationship – although it survives MVs – to have massive direct implications on its own for policy towards plant-breeding (or other things). That is because the predicted output gains from concentrating resources on poorer farmers are small, relative to the gains, *within* each size-group, from concentrating on more efficient farmers – or from raising technical efficiency to best-practice levels. Nevertheless, economists have not been simply 'academic' or irresponsible in devoting so much attention to the inverse relationship, especially in the context of MVs, for two reasons.

(i) Even if small farms produce only a *little* more net output (per acre per year) than large farms, it normally implies that typically 'poor farmers' produce a good deal more output per acre-year than 'richer farmers'. Households, cultivating farms of a given small size, tend to be *poorer* if they have to support more working and family members.[26] But the large number of family workers not only deepens poverty; it also allows the farm to enjoy large advantages of access to family labour per acre (p. 137). And the larger number of non-working family members, apart from increasing poverty, thereby increases (up to a health limit: Lipton, 1983) the pressure to work the family farm harder (pp. 137–8). At the other end, big *and very well-off* farms have rather smaller numbers of family members, working and non-working, than less well-off big farms. Thus yield can be expected to show a much stronger IR to 'household freedom from poverty' – or, given land quality and off-farm income, to 'land size per family member' – than to 'land size per household'.[27] Hence, output should increase even more if inputs (or lands) are steered to poor farm households, than if they are steered to small farms as such.

(ii) Suppose there is only a weak, or even a zero, IR of farm size to annual output value per year (especially if net of purchased input costs). Suppose, too, that such a relationship applies to each technique (i.e., more or less, to each 'given' set of non-land and non-labour farm inputs, conditions, and methods). This still justifies steering MVs as a *new* technique, with other inputs and institutions in MV lead areas, towards smaller and poorer farmers on ethical grounds; for no output is sacrificed by so doing. At least, the implication holds for a

'Rawlsian' morality [Rawls, 1972] if we can assume that people who tend to be even poorer than these 'smaller farmers' (among local labourers, among consumers, or among farmers or labourers outside MV areas) will not suffer as a result.[28]

* * *

Thus the static (but largely traditional and pre-MV) inverse relationship, at least across all except the smallest 5 per cent or so of farmers, appears to be clear, not usually very powerful, but important. What about the post-MV situation, and (since the relationship tells us nothing about dynamics), what about the transition to it? Can we still safely conclude that, if social conditions permit, the average small farmer – and, even more so, the average poor farmer – can obtain rather better yields? And will efficiency advantages follow yield advantages, if any?

It has been claimed that MVs, by giving advantages to those who can better afford credit or inputs or tractors, remove or reverse the IR. Initial doubt is cast on this claim by recent data, subsequent to the spread of MVs; for example, for all but two of fifteen developing countries, standardized farm manage-ment data collected by FAO showed that 'output per unit of farm land systematically declines as farm size rises' [Cornia, 1985, p. 518]. However, Roy [1981] analyses data to suggest a reversal of the IR in some parts of the Indian Punjab exposed most intensely to the new technologies (though only in a minority of them, not including the absolute 'lead area', Ludhiana District). Another study there concluded that in 1974–5, among 1663 sampled cultivators, significant regres-sion coefficients show that the IR 'stands reversed', though only for the main MV crops, rice and wheat, in the most progressive area, Region I [Bhalla and Chadha, 1983, pp. 62–3, 70–3]. Over the border in the Pakistan Punjab, as MVs of wheat were introduced between 1960–1 and 1972–3, the IR was substantially weakened. In progressive districts, bigger farmers actually moved ahead of smaller ones in output per cropped acre – though this was outweighed by smaller farmers' greater ratio of cropped area to operated area [Berry and Cline, pp. 91–7]. In an area of Nueva Ecija in the Philippines, the yield advantages of small farmers in wet-season rice – very large in 1960 – had shrunk by 1970 as

MVs spread [*ibid.*, pp. 73–7]. In Mexico, small private farms –
unless they co-operate to finance land improvements through
the system of *ejidos* – appear unable to attain the post-MV
yields of larger farms [Burke, 1979]. Do we conclude, from
these 'lead areas', that as MVs spread big farmers will overtake
small farmers in annual gross output values, or in the
efficiency of inputs used, per acre? We believe not, for six
reasons.

First, the few recorded reversals of the IR in MV lead areas
almost all refer only to the gross output of a particular crop in a
particular season – e.g. winter wheat per acre. But more crops
(and less fallow) per acre-year, and higher-valued crops, were
always the main advantages of small farmers in respect of land
use. Even in the leading 'Region I' of the Indian Punjab, small
farmers retain these advantages. They may indeed now be
enhanced by the new shorter-duration varieties. By 1975–6,
these advantages had in many areas cancelled out the temporary lead in wheat and rice yields previously gained by large
Punjabi farmers [Bhalla and Chadha, 1983, p. 75].

Second, higher yields attained by larger farmers via higher
outlays, even for a particular crop and seaon, need not mean
better private income or social efficiency per acre/year. In
Bhalla and Chadha's 'Region I', the reversal of the IR was
accompanied by a steady tendency of purchased inputs to
increase with farm size [*ibid.*, pp. 62/63, 70/73]. The evidence
of Palanpur [Bliss and Stern, 1982, p. 272] and of parts of the
Philippines [Herdt and Mandac, 1981] suggests that poorer
farmers – even where no longer achieving higher yields – were
using scarce MV-linked inputs more efficiently. In
Bangladesh income gains for small farmers, from given MV
adoption, have been in excess of gains for larger farmers
[Alauddin and Tisdell, 1986].

Third, in many of the areas where MVs were spreading
while surveys cast doubt on the IR – even in the Punjab – some
small farmers were still in transition, from lower production
functions without MVs, to higher production functions with
MVs [Lipton, 1978, p. 324; cf. evidence in Chattopadhyay and
Rudra, 1976, pp. A-109, A-117]. Lower small-farm yields with
MVs in such conditions are temporary only. As smaller
farmers follow the leaders up the 'S curve relating the

percentage of [adopters] to time. . . the earlier [IR] is likely to be established [alongside higher yield] at all sizes' – as evidence from India and Pakistan suggests [Berry and Cline, 1979, pp. 28, 105, 114].

Fourth, there is evidence against scale-economies in several MV areas. This is notable in Sri Lanka [Herath, 1983]. A large, slightly earlier farm-level survey in the MV period in India confirms the IR, both for irrigated and for unirrigated areas [Bhalla, 1979]. By late 1970s a strong reversal of the IR – which had emerged for wheat in Yaqui Valley of Northern Mexico, as MVs spread initially to bigger farmers – had been greatly weakened [Byerlee and Harrington, 1982, p. 3]. One of the clearest IRs – with 'total factor productivity' about 50 per cent higher on farms with $1-2^1/2$ acres than on farms with above 7 acres – was for a careful survey in Muda, Malaysia, which by 1972-3 was dominated by irrigated MV rice [Berry and Cline, pp. 116–20]

Fifth, the results casting doubt on the IR after MVs arrive – even in respect of gross output of a specific crop per acre in a single season, which is all that the reversal usually covers – are very special. In both a Malaysian and an Indonesian case, rice MVs were size-neutral on land operated in the main seaon, but reversed the IR (yielded more on big farms) on double-cropped land [Gibbons *et al.*, n.d., p. 211], probably because of special advantages to large farmers in obtaining access (perhaps with hidden subsidy) to tractors and threshers for quick turnaround. In Northern Ghana, large units of mechanized MV riceland were set up, registered, and given special access to inputs amid a sea of unregistered small producers [Goody, 1980]. It would be evasive to blame the results on a reversal of the IR relationship by MVs.

There is a final reason for confidence that small, poor farmers' three advantages – per acre farmed, more labour, more supervision, and more consumer pressure to produce – will survive MVs in good shape. It is that such advantages are not indicated only by annual yield-per-acre. True, if 'small farms' after MVs have lower yields than large farms (reversing the IR), market pressures will favour amalgamations. However, the financial and economic fate of small farms depends on their efficiency in transforming all inputs, not just land, into outputs. Small and poor farmers, because they often cannot earn much (or cannot easily find work) off the farm, are ready to apply more of their effort and enterprise than are big farmers to the intensive, efficient conversion, into outputs, of seeds and fertilizers (not just of land). These current inputs gain importance as MVs spread. Surveys in the Philippines and Northern India [Herdt and Mandac, 1981; Bliss and Stern, 1982], while finding few signs of yield differentials between big and small farms, shows that the latter achieve a given yield more efficiently, i.e. with fewer scarce inputs. Probably, for example, small family farmers can more readily replace some fertilizers by manure collection, and some weedicides by hand-weeding – using otherwise under-employed family labour, rather than buying scarce purchased inputs.

*   *   *

If 'small farms' are those which can supply each acre with more hands to work, heads to manage, and mouths to feed, than most MV-fertilizer-irrigation technology should increase the possibilities and incentives to generate somewhat higher yields, and much higher annual income net of cash costs [Cornia, 1985, p. 525, Table 3, columns 1 and 3], per acre on small than on large farms. However, these advantages of 'small farm' households are reduced, or even removed, if they have little access to education, credit, extension, or timely inputs. The higher person/land ratios of small farms are a fact of life; their lack of access is a policy variable. This suggests that, instead of doing more research on MVs and size-yield relationships, analysts need to investigate how MV packages can accompany appropriate institutional change to improve small-farmer access. One example is the need to remedy the failures of institutional credit in Uttar Pradesh, so that it becomes better able to respond to the uncertainly-timed cash needs facing small farmers [Subbarao, 1980]. Then, the high cost of informal credit would be less of a deterrent to the use of MVs and fertilizers. Alternatively, researchers might ask how MVs, etc., could be rendered more robust, if that access is delayed or denied. Poor farmers must delay fertilizer application because the credit arrives late. These are not, of course, arguments for generalized subsidies, but for properly costed remedies for market failures that hit the poor hardest.

(g) Can small farmers stay the course?

'Small farmers', tenants, and in general low-income farm households have the capacity to adopt MVs as much – and to manage them at least as effectively, especially for year-round net income per acre – as other farmers in the long term. However, big farmers usually adopt sooner, enjoy temporarily lower unit costs of production, and thus drive down prices – at least by the time small farmers have adopted substantially and are trying to sell their output. So can small farmers keep their land into the long term?

The two processes alleged to stop them in the wake of MVs are *eviction* of tenants, as ex-landlords find it more profitable to resume personal MV cultivation themselves, and *engrossment* as large operators buy up, or rent from, small ones. We have seen that MVs do not, as a rule, involve economies of scale. So there is no obvious reason why they should stimulate eviction or engrossment. Landlords indeed find that biochemical innovation raises land productivity and prices, but why should they not capture part of that rise by raising rents, or changing the terms of tenure [Bray, 1986, p. 188], rather than by seeking to turn themselves into farmers? The rise in the price of land, too, should discourage large operators from buying or renting in more; if they do, it should enrich the smallholders who rent out to them.

There are, however, notorious cases where MVs have been linked with eviction. In Chilalo, Ethiopia, where about half the 60,000 farm households were tenants before the MV-based development programme started in 1968, 'as of 1971 some 20-25 per cent. . . had been evicted'. Although the survivors had increased their real incomes by over 50 per cent, this was no help to the 'evictims'. One cannot blame the MVs, or the Swedish donors, for the persistence of three dubious policies of the (Imperial) government of Ethiopia: heavily subsidized mechanization (favouring very large scale); grants of big individual land ownership rights; and broken promises of land reform [Cohen, 1975, pp. 348–9]. But if MVs are introduced into such a context the effects on tenants can be terrible. It is, after all, the *combination* of MVs and the policy context that renders it profitable for landlords in such circumstances to adopt their new strategy of eviction and tractor–combine farming.

In the Indian Punjab and Haryana, the spread of MVs has been accompanied less by eviction than by engrossment, despite a steady post-1967 downtrend in the proportion of all farmland rented. Middle and big farmers stopped renting out, and middle farmers increasingly rented in from very small farmers. Operational holdings below 1 acre fell from 24.3 per cent to 4.3 per cent of all farms (i.e. operational holdings) from 1953-4 to 1971-2, while the pattern of owned holdings changed little [Bhalla and Chadha, 1983, Tables 1.6 to 1.8]. In other parts of India, however, farms operated as miniholdings have generally risen greatly as a proportion of total numbers and area [Vyas, 1979]. The Punjab's trend, amounting to a conversion of many of the poorest from micro-farmers into micro-landlords-cum-employees at slightly rising real wages [Lal, 1976; Bhalla, 1979] and rapidly rising rents, cannot be definitely linked to MVs. However, since MV adoption in the Punjab initially favoured those with better access to inputs – especially to reliable tubewell water (and more recently to threshing-machines) – a causal link is probable.

No general link of MVs to changing tenure or farm size has been established. Too many other things, from person/land ratios to land laws, are changing at the same time. It is, however, essential that programmes to introduce or change MVs be pre-evaluated in the tenurial context where they are to be introduced. Chilalo is not an isolated case, and MV planners (and researchers) do bear some responsibility for such results. In such social conditions, biochemical technologies need to be designed to favour small scale and labour-intensity. It does not suffice that innovations be 'neutral' as between such methods and capital-intensive large-scale farming.

(h) Poor farmers in MV areas: more income, but what stability?

Even in Chilalo (Chapter 3, g), MVs greatly enriched even very poor farmers, provided they could keep their land. It follows from what has been said so far that small farmers in regions where MVs do well, if they can keep on farming, will usually gain real income, and will seldom lose. Despite some gloomy assessments from the early years of MVs when few areas and very few small farmers had adopted them [e.g. Frankel, 1971], that is the general message of later work [Barker and Herdt, 1978, 1984; Chaudhry, 1982; Deuster, 1982; Blyn, 1983]. One study even concludes that such analyses 'have provided a body of evidence which proves beyond doubt that they [the critics of MVs from the viewpoint of the local poor] are wrong' [Pinstrup-Andersen and Hazell, 1984, p.11].

It must, however, be stressed that, on Indian evidence, the prospects that MVs will benefit even those who remain as small farmers are good only in those lead areas where the institutional structure is not too firmly in the hands of rural élites [Prahladachar, 1983]; for example, a comparison between rice villages in Bangladesh and South India [Cain, 1981] shows that only in the latter could poor farmers be reasonably confident that their legal rights to land would be protected by the law. Further, extra farm income from MVs – even if distributed no worse than 'in proportion to the initial landholding' where (as in the Indian Punjab) 'landholdings are. . . very skewed' compared to income – enriches the rich proportionately much more than the poor and is thus 'quite inequitable' [Bhalla and Chadha, 1983, p. 160]. Also, in any area, 'small farmers' are not the same as 'local rural poor' (Chapter 3, a) – especially as population growth increases the numbers of labourers with little or no land. A final qualification is that the real poverty problem with MVs arises in the rural areas that they leave out.

However, for small farmers in MV lead areas, the new technology does normally bring higher real incomes. Yet such farmers could still suffer if it also made that income more unstable, because they have special difficulty in handling risks and fluctuations (Chapter 3, c). MVs' effects on the physical capacity of crops to withstand droughts, pests, etc. were reviewed in Chapter 2. The effects on stability of food consumption are treated in Chapter 5, section d. Here we ask how MVs affect real income stability for poor farmers in MV lead areas.

Almost all the measurement has been of crop *output* instability (or its alleged increase after MVs). Since food-crop prices are normally higher in bad crop years, such measurement overstates the instability of *incomes* (or any effect of MVs in increasing such instability) for poor farmers who have a crop surplus for sale, but understates it for very poor 'deficit farmers', who in bad years not only harvest smaller food crops and therefore must buy in more food, but must do so at higher prices. Moreover, the way in which 'instability' is usually measured systematically exaggerates the phenomenon, and/or any increase in it that may be due to MVs.

*   *   *

(i) A commonly used measure of instability [e.g. Walker, 1984] is the *variance* over a period of years of output per year, i.e. the average of squares of differences between annual output and its mean (or, alternatively, trend) value; but if this mean (or trend) value is raised by 50 per cent – say due to MVs – while fluctuations remain the same as a proportion of it, then the variance measure gives the surely erroneous impression that 'instability' has risen by 50 per cent. Standard deviation, or square root of variance, is subject to the same objection, to a lesser degree.

(ii) The most usual indicator, coefficient of variation (CV), avoids this objection by taking the standard deviation of, say, wheat output as a proportion of the mean or trend.[29] However, if MVs raise a poor farmer's average wheat output substantially, even a quite big rise in its CV could leave 'worst-case' output – let us define this as typical output in the worst 10 per cent of years – considerably higher than before. The evidence of Chapter 2 suggests exactly this, namely that MVs typically raise output on a farm in favourable and in unfavourable years, but in favourable conditions by proportionately more. Thus both its 'worst-case output' and its CV rises! People, however, eat and sell wheat, not coefficients of variation. It is worst-case wheat output on a farm, not the widely used CV indicator, that best reflects the impact of MVs (and of everything else) on the farmer's risk in growing wheat.

(iii) Yet most analyses use variance, or at best *CV, of output* – of one crop such as wheat, or of all crops – to reflect *worst-case risk to income*. A further problem is that analysts' data are usually derived from output series for a nation, a province, or at best a district. If, as is probable, MVs increase the extent to which all farms are 'covariant' – i.e. *each* tends to produce a lot in the same years, and to produce a little in the other years – then such aggregated output series will show a higher CV, although at the same time the CV might be falling for most individual farmers (even in principle for every one of them).

(iv) Finally, comparison of instability 'before and after' MVs often overstates their destabilizing effect by attributing all the change to them. However, at least two factors, not associated with MVs, have rendered farm output less stable in MV areas in the 1980s than the 1970s, and in the 1970s than in the 1960s. Probably, weather instability has gradually increased, and has affected larger parts of the developing world. Certainly, the extent and forms of agricultural protection and subsidization of cereals in developed countries – especially EEC [Koester, 1982] but also Japan and the USA – have increased and deepened price fluctuations affecting farmers elsewhere,[30] causing greater responsive fluctuations in the areas they sow to the affected cereals.

*   *   *

Given these four factors, we must set aggregate data about CVs against the evidence that MVs normally *should reduce* single-farm-level risk (Chapter 2); *do so* in specific cases [Roumasset, 1976]; and *encourage complementary action* by farmers to cut risk further (e.g. to re-time production in the water-secure non-monsoon season, as in Bangladesh; or to invest in or purchase groundwater, as in the Indian and Pakistan Punjabs). The evidence, however, does show that aggregate CV of foodgrain output, around its trend, has risen in India, and in most of its States, during the spread of MVs [Mehra, 1981; Hazell, 1982, 1984, 1987a; Ray, 1983; Pinstrup-Andersen and Hazell, 1984]. However, there are several reasons to reject the inference, from such data, that MVs have therefore made small (or other) farmers' lives riskier in lead areas.

First, data from low-income developing countries now strongly suggest that in really 'risk-averse' populations the MV spearhead crops – wheat and rice – usually showed *declining* CVs of yield around trend [Hazell, 1986], especially in countries with major MV impact. For rice, wheat, and 'all cereals' respectively, India's CVs of annual yield in 1960/1–1970/1 (1971/2–1982/3) were 10.6 per cent (5.9 per cent), 8.0 per cent (7.9 per cent) and 6.2 per cent (4.6 per cent); Pakistan's, 9.8 per cent (4.1 per cent), 8.2 per cent (3.1 per cent) and 8.6 per cent (3.0 per cent); and Egypt's, 10.2 per cent (4.3 per cent), 5.8 per cent (3.8 per cent) and 4.5 per cent (2.3 per cent). Bangladesh and the Philippines also showed falling CVs of yields (though they rose in middle-income countries: Brazil, Argentina, Mexico, Turkey, South Korea and Thailand) [Hazell, 1987a, Table 2.4]. It was rising output variability among less-improved crops, and growing covariance among crops and regions, that sometimes caused CV of total foodgrain output to rise. This suggests that tropical MVs, far from raising output or income instability among adopters, fought a brave – but sometimes losing – battle to reduce it, in face of 'CV-increasing' effects: at home, population growth that pushed a growing proportion of farming into riskier lands, and perhaps worse weather; abroad, the 'export of instability' from the West through agricultural protectionism (Chapter 7, k).

Second, our suspicion that rising CVs of output at *national* level are masks for falling CVs at *farm* level – especially in MV lead areas, where the prospect of yield breakthroughs encourage farmers to seek safer water supplies – is strengthened by more disaggregated work. Analysis for single crops in Indian Districts – as opposed to the above analyses for all foodgrains in Indian States – showed that 95 per cent of recent rises in all-Indian aggregate production variance[31] about trend for sorghum, and 92 per cent for pearl millet, is due to extra covariance among producing Districts [Walker, 1984, pp. 6–8]; this is a much larger proportion than for States, but less than would be found if we could go below District level, to smaller units of disaggregation.[32] Nevertheless, for these relatively risk-exposed crops, MVs and regions[33] (though probably not for wheat and rice), this still leaves 5–8 per cent of extra variance that hits the individual district as extra instability in a crop's output; perhaps part of this might remain as a rise in variance at farm level, rather than being attributable to extra covariance among areas within the region. Of these remaining few per cent of extra output variance attributable to rising on-farm risk, some might, in turn, be due to MVs as such, e.g. via their tendency to narrow the genetic base [Hazell, 1984], although more evidence is required.

Third, it has been argued that risk is rising as MVs are pushed into riskier areas in India [Ray, 1983]. If correct – and the riskiness of most MV millets is worrying – this bodes ill for MVs in most of Africa. However, it is not proven that greater restraint with MVs, accepting low yields in less reliably watered areas, would have been a better policy for the poor [Walker, 1984, pp. 11–12], or even a less risky one. If yields had risen more slowly, population increase in the absence of MVs could have forced traditional farming onto even more, and more marginal, semi-arid and ill-watered uplands and thin topsoils than has been the case in most of Asia. The growing depletion of many lands – and hence the increasing risks of crop production – in East Africa and the Sahel in the 1980s, usually in the absence of MVs, owed much to this process.

Fourth, as discussed above, changes in the CV of gross product are an odd way to measure risk before and after MVs. This CV has risen in semi-arid parts of South India with the *early* spread of successful millet and sorghum MVs – yet 'worst-case' output rose too, though normal or 'best-case' output rose by a larger proportion.

Fifth, even if worst-case output in the worst 10 per cent of harvests falls due to MVs – a rare event with most of the recent MVs – there are compensations for small (and large) surplus MV farmers.[34] Usually, output will be sold for a higher price at such times (making things even worse, of course, for deficit farmers). Some inputs will be demanded in smaller amounts, especially at and after harvest, and their prices will be lower than in good seasons. We know nothing about whether such compensations have been increased or reduced by MVs.

All these facts give grounds for a rather optimistic inter-pretation of the effect of MVs on risks to poor farmers in developing countries; and this applies also to risks to poor consumers (Chapter 5, f). However, threats to plant diversity could change this (Chapter 2, k). Also, a *given* reduction or shortfall in a farmer's gross output is often worse with MVs than with TVs, because MVs usually make it profitable in a typical year to incur much higher costs for working inputs; most of these costs have to be borne even if the crop does badly. So, even if MVs increase worst-case gross output, they could reduce worst-case net farm income, especially as sale price is restrained by extra output due to MVs.

It is important for MVs to aim at robustness, partly through greater genetic diversity, and thus at lower *individual disaster-risk* for a poor producer. The authors cited in this section avoid the mistake of inferring that MVs have, so far, tended to increase such risk, even in circumstances where there has been higher *aggregate output variability*. So should we.

(i) MVs and the rural poor in non-MV areas

Even those who are most positive about the effects of MVs on poor farmers and workers in progressive areas, and on consumers, agree that 'backward' areas, especially areas with less reliable water supply, have not done well [Barker and Herdt, 1984, p. 48; Ruttan, 1977, p. 18; Pinstrup-Andersen and Hazell, 1984, p. 13]. 'In some countries optimum environments are frequently controlled by the larger and better-off farmers' [*ibid.*, p. 13], so that land is less unequally held, and landlessness is less, in villages in backward areas [Dasgupta, 1977a]. Thus the initial emphasis of MVs on better-watered areas, and on wheat and rice regions, tended to leave out initially poorer areas, which were also the very places where prospects for fair distribution of gains from MVs are best.[35] Indeed, by increasing the surpluses sold from better-favoured areas, MVs made prices for cereals sold lower than they would otherwise have been; this was outweighed in the lead areas by gains from higher yields, but in non-MV areas the restraint on cereals prices harmed farm sales with little or no such yield compensation. This reduced incomes circulating there (although also cutting the cost of food purchases to landless workers and deficit farmers).

However, the problem of 'regions left out' – while the main poverty problem of MVs – should not be over-generalized. (i) In some important cases (India, West Malaysia), inequality among rural areas is associated with only a small proportion of either poverty or national inequality [Malone, 1974, p. 16; Anand, 1984]. (ii) In other cases, some of the regional bias in benefits from MV research corrects earlier research biases towards regions suitable for major export crops, especially within West Africa. (iii) Also, to some extent, migration from non-MV areas to lead areas can correct some regional biases. (iv) At least since the early 1970s, IARCs and some national MV researchers have been redirecting priority towards the cropping systems of 'backward' and unreliably watered areas – upland rice, millet, hybrid sorghum – with some major successes, notably hybrid sorghums and finger millet in some of India's poorest regions. (v) Only if there is a price break, i.e. if prices are reduced in 'backward' regions by extra output in MV lead areas – which happens only if a national economy is big, or remote, or partly closed to trade by transport costs or protection – will more MV grain from, say, the Punjab or Central Luzon reduce absolute income in non-MV regions; even then, net food *buyers* in all regions should gain (Chapter 5). (vi) Perhaps above all, the model in which there are only *two* types of region – for instance, in India, lead areas such as the Punjab that steam ahead, and backward areas such as most of Madhya Pradesh that lose absolutely as well as relatively – while perhaps appropriate until 1975 or so, has more recently proved to be misleadingly over-simple. We shall nevertheless adhere to this model in the first part of this section, but later in this section we consider the implications of the reality that there are at least *four* types of region, three that gain and one that loses income from MVs – and that net food buyers and net food sellers, both including poor people, often experience opposite income changes in any given region.

What, then, has happened? Regional income distribution has improved in some countries in the wake of MVs. Notable is Taiwan, where most cropland is in irrigable MV rice almost everywhere. In Pakistan, with about 40 per cent of cropland in irrigated wheat, and with much spread of MVs to rainfed *barani* areas mainly on grounds of risk reduction [Rochin, in

USAID, 1973], inequality among rural regions has fallen since MVs [Chaudhry, 1982]. Even in India, if we exclude the specially difficult problem of the Eastern rice States (and also, for data reasons, Kerala), MVs have not increased inter-District inequality. A survey of 228 Districts reveals that, from 1958/9–1960/1 to 1981/2–1983/4, those with a trend of rapid annual compound growth in yields of the five main cereals – wheat, rice, sorghum, millet, maize – tended to 'start off worse' than slow-growing Districts. Of the 112 Districts with growth above 3 per cent yearly, 61 had yields below 700 kg/ha. in the late 1950s, and only 17 started at over 1000 kg/ha. Of the 54 Districts with growth below 1.5 per cent yearly, only 18 began below 700 kg/ha., and 13 at over 1000 kg/ha. [Jansen, 1988, p.41]. In such cases, it is a problem of at-least-average regions that 'become relatively poorer' because they are left out by MVs, not a problem of growing regional inequality.

However, in most MV-affected countries – and probably in India too, if the Eastern rice belt is included – poorer rural regions have lost, through lower farm-gate prices, as MVs from elsewhere raised cereal supplies, owners of land (which is in inelastic supply) bear more of the initial losses than workers (who may be able to migrate, or to shift jobs); but many poor farmers *and* workers are unable to move readily from land in poorer regions, and have lost absolutely from MVs [Binswanger and Ruttan, 1977, chapter 13; Binswanger, 1980, p. 187; Binswanger and Ryan, 1977, p. 229]. And, despite shifts in research priorities, huge areas remain exposed to risks of such loss; over 48 per cent of Latin America's 22 million hectares under rice (but under 22 per cent of its rice output) is in 'upland less favoured' or 'upland manual' systems, where on present research strategies 'adoption of [MVs] will not occur' [CIAT, 1984, pp. 3, 9].

An example of the reinforcement of regional inequality was the spread of wheat MVs in Mexico in the mid-1960s. It was heavily concentrated in the Pacific North (Sonora and Sinaloa, above all, enjoyed huge yield increases). Already in 1960, before the MVs arrived, agriculture there had enjoyed two huge advantages: over half the land was irrigated (the national figure was 15 per cent), and farm income per person was over 50 per cent above the national average [Tuckman, 1976, p. 20].

MV surpluses must have restrained cereals prices, but the resulting setback to farm incomes in Sonora and Sinaloa was far outweighed by the yield increases. These did not happen in most other areas growing wheat and corn, which were poorer to begin with. So the MV bonanza for Sonora and Sinaloa meant less real income in many poorer areas – not only for both rich and poor private farmers with a surplus, but also for poor *ejidatarios*. Such regional damage to Mexico's rural periphery was reduced by the prospects of migrating to share in the rapid growth of urban employment, especially in services, in the wake of the oil boom; this compensation for poorer regions was also available in Indonesia, but not in most countries affected by MVs.

In India, for example, small farmers (or workers) in non-MV regions, if damaged by lower-cost competition from rice and wheat lead areas, could not readily shift to non-farm work. The proportion of workers engaged in agriculture appears to have been constant between 1961 and 1981 [Lipton, 1984]. Yet the spread of MVs and associated yield improvements did leave out large areas. In the Eastern rice zone, already in the early 1960s India's poorest rural region, average yields rose by less than 25 per cent in 1960–77 – well below population growth (and there was hardly any spare land to bring into cultivation); meanwhile, proportions of riceland in MVs reached about 30 per cent, and yields reached around one ton per acre. In the SW and NW wheat–rice zones, initially the least poor, proportions of rice in MVs rose to about 60 per cent, and rice yields about doubled, reaching about three tons per acre [Brass, 1984; Herdt and Capule, 1983]. MVs also tend to increase interregional inequality by shifting the incentives towards more intensive use of other inputs further towards already-favoured regions. Thus in 1960–1, before MVs, the Indian Punjab already irrigated 54 per cent of net sown area (19 per cent was the all-India figure); by 1979–80 the proportions had moved even further apart, to 85 per cent (27 per cent) [Bhalla and Chadha, 1983, p. 12]. Meanwhile foodgrain output, which had grown at similar rates in Punjab and all-India (60 per cent) from 1950–1 to 1960–1, grew from 1960–1 to 1978–9 by a factor of 3.6 in the Punjab, but only by 1.6 in all-India [Chadha, 1983, p. 131].

The CV of *output over time*, as we have seen, is a dubious indicator of *instability* (Chapter 3, h). However, the CV of *output or income per person over areas* is a plausible indicator of *regional inequality*. The interstate CV of foodgrain output per person among Indian States, stable in the 1950s, increased dramatically in the 1960s due to the confinement of progress in MV wheat to the North-West [Krishnaji, 1975]. Lagging States did not, as a rule, compensate by relatively better non-foodgrain performance. Since about 1973, however, the take-off in rice and sorghum has somewhat reduced inter-State disparities in foodgrain output per person [Sawant, 1983, pp. 493–6]. Nevertheless the maize and millet[36] areas, the winter sorghum zone, and above all the populous and poor Eastern region – comprising Bihar (outside Kosi), Eastern Uttar Pradesh (except tubewell areas), most of Orissa, and West Bengal (outside the irrigated Northern wheat zone) – remain largely left out of the national, and MV-based, agricultural progress, as do analogous agro-climatic zones in other developing countries.

The parentheses in the last sentence give a major warning about regional impacts of MVs upon poverty and inequality. Most so-called 'disaggregated' data for Bangladesh, India, Indonesia, Pakistan and the Philippines show changing averages for vast States or Provinces, each with tens or hundreds of thousands of farmers, and with administrative, not agro-climatic, boundaries. Further, disaggregation of MV effects often produces surprises, not least for India. *Below State level*, it is well known that the MVs did much more for Western than for Eastern UP – but not, perhaps, that their spread, plus that of tractors, etc., was associated with a 1.7 per cent decline each year in per-hectare labour use per acre in Western UP, while the much more restricted spread of MVs (alongside rapid growth of the labour force) was associated with a 0.3 per cent yearly rise in labour use per acre in Eastern UP [Joshi *et al.*, 1981, p. 4]. Deficit farmers, providing most of their own food but supplementing it by work as farm labourers, may have done no worse in the Eastern than in the Western wing. As for Indian *Districts*, 81 of them, accounting for 51 per cent of Indian wheat output on 42 per cent of area in 1962–5, had each by 1970–3 raised output at over 10 per cent per year, to provide 67 per cent of output on 52 per cent of area in 1970–3 [Bhalla and Alagh, 1979, pp. 95–7].

However, despite the heavy concentration in the Indian Punjab of both high initial productivity and rapid growth of foodgrain output, there is no indication overall that Districts with slower growth of total farm yield and output were initially worse endowed with good *land* than those helped by MVs to improved performance. The laggard Indian Districts were areas of low *labour*-productivity, i.e. of low income per farm-operating or labouring worker. Bhalla *et al.* [1983, Table 17] compare 74 leading Districts (of 281 surveyed in India), which achieved 5.1 per cent annual compound rates of all-crop yield growth and 5.6 per cent growth for total farm output in 1962–5/1970–3, with 67 laggard Districts, or rather back-sliders (output −2.1 per cent, yield -2.1 per cent, yearly). Land yield in the leading districts in 1962–5 had been a mere 6 per cent above the laggards' yield (but was 88 per cent above by 1970–3), but output per male worker started 14 per cent higher and ended up just over double. The widening of gaps in *labour*-productivity means that the regional patterns of Indian growth, in a period where MVs dominated these patterns, almost certainly increased substantially the relative disadvantages of very small farm operators – depending mainly on labour (in owned, rented, or 'employing' land), not on the value of their land, for income – in poor districts, as compared with similar micro-farmers in better-off places. The absence of linkage between growth performance and initial *land* productivity, however, prevents any ready-made theorizing about just how and why this happened – and strongly suggests that the reasons did not lie mainly in any emphasis by researchers upon land that was 'good' initially. Causes and cures of regional backwardness do lie partly in research emphases. But the processes are specific to crops and environments – and seasons.

For example, the success of sorghum hybrids in the 1970s enabled parts of semi-arid India to catch up with the growth-rates (if not to reach the levels) of output and income in the best of the irrigated wheat districts. Yet this is specific to kharif (main monsoon cropping) areas in two States; the importance of kharif-season, as against other, sorghum explains 88 per cent of inter-District variations in the adoption of hybrid sorghum [Walker, 1984; Walker and Singh, 1983]. Where these kharif MVs work, they are robust: 'in [several districts] in the. . . black-soil belt. . . 100–110–day hybrids [able] to stand grain deterioration when rains occur late have taken firm root' [Rao, 1982, p. 46]. However, it is not clear to what extent such districts are very drought-prone; normally, if they were, not *kharif* but *rabi* (winter) sorghum would be grown, and for the latter there are not yet any really successful hybrids. Even the *kharif* hybrids – despite their great success in some very poor areas – are susceptible to mould and *Striga Asiatica*, and have gone to places where such reducers of yield and grain quality are less important [Walker, pers. comm., 1986]. For MV rices, the gain over traditional rices is much greater in non-monsoon, controlled-irrigation areas and at top ends of canals [Herdt and Wickham, 1978, p. 5].

MVs thus tend to confirm the regional bias that existed previously, because they usually build on prospects known to exist in certain conditions, but can do less for areas where the genetic range of crops is adapted to survival at low yields. Long before MVs, the lower level and slower progress of output-per-person (or per worker) in less-favoured areas – relying, in the above two cases, on second-season sorghum and on rainfed or tail-end rice – were familiar issues throughout South and East Asia. For hybrid maize in Kenya, too, the spread to smallholders in 'progressive' areas – defined as those with higher labour income – was in marked contrast to the failure to reach long-neglected areas [Gerhart, 1975].

Even well below district level, inter-village differences in MV offtake and yield appear much greater, notably in ICRISAT studies, than intra-village differences. Faltering yields with hybrid millet, together with the rapid advance of MV wheats, can lead to 'patchwork quilts' in areas with water regimes and topography that induce major local variation of the main staple [Sharma, 1981, on Gujarat].

Even where crops and varieties are similar, regional and local gaps in MV results persist. Should we look to the quantity and quality of farmers' education to explain these gaps? They are clearly linked to labour productivity. 'Human capital' may help determine the response to MVs among villages, districts, and Indian States such as Bihar and the Punjab [Nair, 1979]. Education, co-operation, and perhaps motivation surely vary locally, and are linked cumulatively [and statistically: Rogers, 1962] to innovation, growth or decline, whether among rice villages or computer manufacturers – and among farmers within a village [Jamison and Lau, 1982]. This last piece of evidence gives hope for MV impact on 'backward' places even in terms of human capital, since MVs do spread out from early, educated innovators, although they alone enjoy economic rents from education and early adoption combined (pp. 116–17). Hope for less-favoured regions also arises from the frequent changes in pattern that transform desperately poor and long-stagnant places – e.g. the shift to irrigated MV wheat in NW Bangladesh, or the bamboo tubewells and MV rice of North Bihar, or the MV sorghum takeoffs in much of Maharashtra.

All these facts about diversity within, and progress among, 'backward' areas also preclude explanations of MVs' regional patterns in terms of ethnic determinism. Caste [Bliss and Stern, 1982] or ethnic group, where related to MV performance, are normally proxies for 'income and size of landholdings' [Herdt and Capule, 1983, p. 32]. For paddy villages in Malaysia, 'the average total Technical Progress score is exactly identical for the Chinese and Malay villages' [Gibbons *et al.*, n.d., pp. 194, 205].

There is also great *localization* of regional disparity. Except perhaps for the legendary successes like Sonora and Ludhiana, every MV lead area has its backward villages, and almost every laggard (at least in Asia) contains villages with several successful MV smallholders. Thus it may not be useful to identify vast and disparate 'zones', like the Eastern Indian rice zone, as places where no sensible MV-based strategy is yet available, as appears to have been done at IRRI [Brass, 1984]. By 1978–9 Assam, Bihar and Orissa had respectively 23 per cent, 25 per cent and 30 per cent of rice under MVs and West Bengal 41 per cent [Herdt and Capule, 1983, p. 49], by no means all in the winter season. The spread of rice MVs in some ·rainfed areas in Bangladesh ever since IR-20 [Lowdermilk, 1972] has been impressive. *Local* areas of MV takeoff, inside regions of poor water-control, should be identified; East Asian evidence suggests that detailed local water systems and timings are crucial, and may imply that careful selection among MVs could help [Bray, 1986, p. 62]. For rice, better drainage appears to be crucial in improving poor response to nitrogen fertilizer [David and Barker, 1978, p. 178], and hence presumably in producing 'islands' of adoption within many such regions.

There is an increasing consensus that inadequate retained nitrogen, not too much or too little water, is the major constraint on yields in areas without good water control. Hence fertilizer distribution, type and placing may be key issues in broadening the MV impact on backward regions. Accessible and promising regions are often well served by competitive private fertilizer distribution, but remoter farmers, often already deterred by water risk from applying levels that maximize expected profits, may also require some public involvement [Ahmad and Hossain, 1984, p. 40; Govt. of Bangladesh and USAID, 1982]. African practice in extension and fertilizer supply frequently involves supplying a standard NPK mix (e.g. 'Compound D' in Zambia) unadapted to local soils or even crops, a procedure which not only wastes many nutrients (which poorer farmers can less readily afford), but also almost certainly discriminates against areas of higher water risk, especially for smallholders. The CIMMYT proposals to shift wheat research towards stability in marginal environments could also help here [CIMMYT, 1983, p. VII].

*   *   *

Can any conclusions be drawn about the design of a research strategy to help poor farmers in non-MV regions? We first approach this in the context of the deliberately simplified 'lead-area, backward-area' model that we have so far largely adopted. Then, we try to allow for more complex realities. Following the model, Brass [1984, p. 7] has pinpointed what might be called the Poverty-Alleviating Researcher's Regional Dilemma, or PARRD: work for the backward areas, and risk doing little or nothing for poor consumers; work for the dynamic areas, and do little or nothing for poor growers in backward areas. His example of the dilemma is the choice of regional targets in rice research. Should IRRI (and the Indian

Council for Agricultural Research) divert 'resources to the poor subsistence paddy growers of [Eastern] UP and Bihar when production and consumption needs might be better met, and the political dangers of discontented urban populations. . .warded off', by concentration on MV lead areas? Must one, perhaps, 'introduce [more] inequalities into the agricultural economy to prevent the greater evil of an inadequate food supply'?

However, it is not clear that Brass is right in blaming 'production orientation' for the delay in attending to lowland rainfed rice at IRRI (a balanced review is Barker and Herdt [1979]). He cites [IRRI, 1979, pp. 20, 45] showing that 46 per cent of non-Chinese Asian rice is 'shallow and intermediate rainfed' and received 31 per cent of IRRI's 1977 budget – hardly dramatic neglect, especially as these two categories provide 33 per cent of anticipated benefits. Shares in IRRI's research budget remain far below shares in rice output, farmland, or population for dryland, upland, and deepwater rice [*ibid.*]. The same applies to the share of African national agricultural research outlays devoted to the adaptation and selection of MVs of rainfed foods, especially millet and root crops [Judd *et al.*, 1973; Lipton, 1985].

IRRI's emphases represent à decision about research prospects. Plainly IARCs do not help the poorest regions by redirecting research towards their problems, if more such research is likely to produce few or no economic results. The African imbalance in *national* adaptive research, however, damages the impact, on poor areas, of the *international* system, notably of IITA and ICRISAT. ICRISAT, combined with Indian national adaptive research, has had major recent impact on sorghum and pigeonpea among poor farmers in neglected regions of India. In places without such national adaptive work, 'foreign' agricultural research demonstrably achieves less [Evenson and Kislev, 1976]. The work of ISNAR, so far mainly directed to improving the organization of national programmes, might do more for the impact of varietal change on the poor if it now turned to their content, especially to the appropriate crop-mix in poorer regions – both for farmers and for research emphases.

*   *   *

There are several respects in which the PARRD, as defined above, is too crudely stated, in ways that conceal possible resolutions. Above all, while in 1965–75 it made some sense to see just two economic types[37] of staple food areas in developing countries – the water-controlled 'green revolution' flatlands, growing wheat if and when moderately cool, rice if hot and wet; and the neglected remainder of Third World farmland – this simplification will not do today. There are *four* main 'economic types' of tropical and sub-tropical food-crop areas; each type, and particularly its poor, are being affected differently by MVs, and are responding differently to the combination of yield upsurge and cost-price squeeze[38] that MVs bring with them.

Type I areas, or *lead areas*, mostly comprise well irrigated, or reliably rainfed, flat wheatlands or rice paddies. It is in Type I areas that the initial 'green revolution' output breakthroughs took place. These areas were sufficiently similar that closely related genotypes – sometimes the same one – could be used over a wide range of them, and even transferred to grow rice in what had been mainly wheat areas (the Punjab) or vice versa (NW Bangladesh). MV research on such lead areas continues to show good, though probably diminishing, returns. Moreover – despite such areas' usually large internal income inequality – MVs have alleviated poverty for most poor farmers and farm labourers.[39] Furthermore, MV-induced growth of food surpluses, sold by Type I areas to urban (and some other rural) areas, has in many poor countries provided the buying regions with most of their extra food since the mid-1960s and lowered its price, thus making it less difficult to alleviate poverty there as well (Chapter 5).

Type II areas, or *backward areas*, are those for which, in 1987 as in 1960, there is little sign of substantial MV-based improvements in the production of a staple food crop that are profitable for adoption by smallholders – and where there are no major spin-offs to producers from developments in lead areas. Landless or near-landless workers, and in bad years some small farmers, are likely to be net buyers of staples; as such, they nevertheless gain in Type II areas (as elsewhere) as consumers, because extra output induced by MVs renders such staples cheaper than they would otherwise have been. However, many Type II areas feature soils, water, or terrain that confines the farmer there mainly to producing food staples, and exchanging them for all other needs. The greater supply (at lower unit cost) of MV-linked staples from Type I areas – by keeping down prices – damages the income from sales of such growers, and their employees, in Type II areas, and pulls up their farm input costs. In most Type II areas,[40] there are few rich farmers, and small growers and labourers tend to be poorer but less unequal than in Type I areas. Unless such people can leave Type II areas, or shift out of agriculture, they are likely to be absolutely impoverished, because MV research can do little for them; fortunately, however, many areas formerly believed to be Type II in fact fall into Types III and IV, below.

Type III areas are those, once thought backward, that seem likely to become *second-generation varietal breakthrough areas*. In India several kharif sorghum areas in Maharashtra and ragi areas in Karnataka, and in Africa some maize areas in Malawi and Zimbabwe, are already in this group. Suitable MVs of rice have also spread in West Africa to Type III areas from Type I areas through farmers' own experiments [Richards, 1985], and/or have been designed explicitly by research stations.

Type IV areas, which may be called *crop-shift MV-shadow areas*, represent an interesting and important, though only recently recognized, development. A steady flow of new MVs to Type I areas has usually kept wheat and rice yields there increasing fast enough to increase profitability substantially, outweighing the 'cost-price squeeze' – amply in wheat, rather less so in rice. Therefore, farmers in lead areas have moved land into wheat and rice, out of pulses and fodder crops. Since fodder crops, even processed, have high weight/value ratios, while the taste for pulses is generally rather localized to specific crops, prices are in both cases largely determined in the national, rather than world, market. Therefore, when the lead areas shift into rice and wheat, and thereby reduce the national supply of (say) lucerne or groundnut, the price – and producer profitability – of such crops goes up sharply. Type IV areas are those which have enough flexibility and security of soil, water, temperature and terrain to move profitably into the crops abandoned by Type I areas.[41] In India, the shift from rice to groundnuts and sugar in North Arcot, Tamilnadu, and from wheat to mustard, rapeseed and groundnuts in many parts of Gujarat, illustrate this move. Typically it happens where (i) some MV crops are grown, and yields do rise due to MVs, but for many farmers (and on marginal lands) not by enough to offset the cost-price squeeze; (ii) some of the potential replacement crops are also already grown. The poverty impact is complex, depending partly on the employment income created by the 'expanding' fodder, pulse, etc. crops in Type IV areas; one reason to expect a good impact is that, in face of the sharp increase in initial labour requirements following MVs, Type I farmers will be particularly likely to give up crops that need a lot of labour, at once raising the price of such crops, and leaving them (and their expanded employment income) to family-farm or employee labour in Type IV areas.

The main point in our context, though, is that – given the impact of an innovation on income distribution *within* regions – the existence of Type IV (and Type III) regions greatly softens the harshness of the PARRD. In breeding rice or wheat MVs for Type I areas, the researcher can give preference – if other considerations are roughly in balance – to varieties or practices that induce Type I farmers to use those MVs to replace crops that can then be profitably grown in Type IV areas, if possible labour-intensively and/or by smallholders. Breeders seeking to diversify away from Type I areas – especially as poverty there becomes less serious, and as new MVs are absorbed into steadily less employment-generating farm systems (Chapter 4, b) – can look to the relatively hopeful Type III and Type IV areas, or can seek to create them out of Type II areas. In effect, these strategies recognise that breeding effort for the initially most promising Type I areas eventually suffers from diminishing returns,[42] so that the cost-effectiveness of agricultural research (possibly with economies of scale) for Type III and Type IV areas can come to exceed that of yet more Type I research.

However, there is little point in doggedly seeking 'pro-poor' MVs in deeply unresponsive, irredeemably Type II places. Where the latter exist – and are not near to more promising soils, terrains, or agroclimates – policymakers will need to recognize that the PARRD bites hard. MVs in such conditions will raise yields little. The very poor often cannot readily migrate. Agricultural research (and even biotechnology) cannot develop silk-purse MVs from genetic materials, and for depleted soil-water environments, perfectly adapted to sow's-ear TVs. Crop-shift on any major scale is ruled out too, unless agro-forestry presents a major new hope [Chambers, 1983]. If these depleted environments cannot be economically and sustainably improved, an off-farm future must be sought by (or for) their poor rural victims.

*   *   *

Another regional consideration for MV researchers is the impact of their work on the distribution of benefits between urban people and rural people, including farmers. Of course there are many poor townspeople and many wealthy rural people, but rural people are on average much poorer – and less unequal [Lipton, 1984b] – than townspeople. Thus initial income rises are normally likely to benefit the poor more if rural. Part of the central motive of the international MV effort was to reduce hunger, but another important part was to remedy the neglect of the rural poor.

Yet it is not self-evident that the MVs' benefits are overwhelmingly rural. First, with any MV-based breakthrough, even in semi-arid farming systems or sorghum, a large part of its benefits tends to be transferred to input suppliers – especially fertilizer producers, in towns or abroad [Ghodake, 1983, pp. 8, 11, 15; Ghodake and Kshirsagar, 1983, p. 12]. Second, many of the gains from widespread MV-based rises in food supply – except where world prices alone determine food prices, i.e. in small, compact, free-trading, cheap-transport economies, which few LDCs are – tend to be transferred, via cheaper food, to urban and other consumers (or to their employers: Chapter 5, b). Some rural surplus farmers, especially in areas unsuited for MVs, can be left worse off [Scobie and Posada, 1978; Evenson and Flores, 1978; and see Chapter 5, a].

In many countries, therefore, MV research raises an issue of strict equity or fairness – rather than of equality, or even poverty alleviation – for poor but food-surplus farmers in 'backward' regions. So far, the Government has used tax-payers' resources – and diverted scarce skills – for research that increases the yield greatly, and real income somewhat, in rural Type I areas; greatly benefits urban consumers, and the poor as consumers in all types of rural area; but (because it restrains food prices) *lowers* returns to farming in 'backward regions'. The damaged farmers in such regions should, in equity, have some compensation, for being harmed by Government research policies that (although financed by all taxpayers, including these farmers) have helped everyone else. At least for potential Type III and IV regions, this justifies MV research outlay, even at the expense of *somewhat* lower returns – including slower progress in bringing down urban food prices than might be achieved by greater concentration on 'advanced' regions.

There are also three efficiency arguments for such a switch of emphasis. *First*, research on irrigated areas, so long the priority, must be running into diminishing returns, compared to work on neglected, generally poorer regions. This effect is likely to be underestimated, because of the heavy weight of advice from specialists engaged to work on one set of problems (those of Type I areas), and seeking honestly to reinforce the contribution of their special knowledge. Such experts – like those who study 'fashionable' pests (Chapter 2, i) – naturally tend to overestimate prospects for their initial work, as against work in other areas they know less about. *Second*, under urban pressures, national research agencies tend to favour areas which deliver food to cities. International centres, by their own area priorities, can help to correct that emphasis. *Third*, neglected rural areas do expel some poor migrants, but this process imposes heavy costs on others. Thus – while migration (partly caused by neglect of Type II, III and IV areas) has done a little to redress the damage from regional neglect (notably in the Indian case of Bihar and UP, which sent seasonal migrants to the booming Punjab: Chapter 4, g) – it creates serious problems. The cities face increasing slums and congestion. Rural-to-rural migration often brings overfarming of environmentally marginal recipient areas; movement to favoured rural areas is better, but is often temporary, in face of their increasingly capital-intensive farming, and of opposition from local labour. The whole process, too, often deprives 'backward' rural areas of potential leaders.

*   *   *

Finally on regional matters: what have MVs in developing countries done to income distribution between rich and poor *nations*? We know of no global research on this, but the impact on North-South distribution via lower grain prices, while favourable (since the Third World is a major net food importer) and measurable [Flinn and Unnevehr, 1984], must be small compared to that of the big growth in EEC and North American exportable grain surpluses under the impact of domestic agricultural research and of such countries' policies to protect or subsidize agriculture. Only in rice – where trade is a small part of total output, and where consumers' preferences over other grains are especially strong – is improved performance based on tropical MVs, above all in China, Indonesia and India, the major factor restraining world prices; here, the main losses accrue to tropical exporters, especially Thailand (though some accrue to the USA), and the main gains to rice importers, especially African countries.

More generally, among developing countries, the few grain exporters lose out from the major restraints on world prices caused both by MVs themselves and (much more) by Western policies to subsidize cereal production. However, farmers in the Third World who are geared towards food exports, and who are often poor (or employ poor workers), have safeguarded their interests, (i) by switching from MV cereals, e.g. from rice to rubber (Thailand), and (ii) by concentrating more upon premium grades of such cereals for export, e.g. on *basmati* rice exports (Pakistan). Moreover, cassava and soya are major areas of MV-linked export expansion from Third World countries, mainly for feed in the West. However, all three crops increasingly face severe restrictions in Western markets. Also, by means that are not well understood (but that apply at least as much to domestic as to foreign or multinational activity), a switch to export farming in food crops often appears to involve larger scale, greater capital-intensity, and hence less favourable poverty impact.

Africa presents a special problem in the context of the international effects of MVs on the poor. Although MVs' successes – and Western failures to reduce subsidized cereal mountains – have temporarily helped keep down grain prices for African urban consumers [Koester, 1982], Africa's poor are overwhelmingly still small farmers. Most of them live in an international Type II region; MVs have not only depressed prices for, but also failed to make much production impact on, small farmers in most poor countries in sub-Saharan Africa (Chapter 7, d). There are possible exceptions for cassava in a few places [Hartmans, 1985]; for rice (at high cost, so far) in a few parts of West Africa; and for hybrid maize in parts of Kenya [Gerhart, 1975], Malawi, and two middle-income countries, Zambia and Zimbabwe. But the concentration of agricultural research on irrigated conditions has been damaging to African farmers. Barely 3 per cent of cropland in sub-Saharan Africa is irrigated, mostly in the Sudan, and not much more is fertilized. In the main African crops (maize, sorghum, millet and cassava), which are grown almost entirely under rainfed conditions, MVs have seldom proved economic in the field, except for maize hybrids, mostly released up to about 1970–2.[43] The IARCs have spent more – per head, acre, or ton of food – in sub-Saharan Africa than elsewhere, but with limited impact, due to rather ineffective and costly local research backing which is seldom in practice directed towards poorer regions or smallholders. While the experiences with sorghum in West Africa have bred caution about adapting Asian *seeds* to African farming systems, Asian national *research* and farm policymaking methods and priorities are more relevant to developing MVs for poor African regions than has yet been recognized [Swaminathan, 1984; Lipton, 1985a]. We return to this major regional issue for MV research in Chapter 7.

Notes and references

1 Three reasons for this odd research emphasis may be (i) a natural tendency to seek MV effects above all in the productive units where MVs can be used, (ii) a wish to understand changing

poverty among farmers by including social interactions within an area (Section g of this Chapter), (iii) the relative ease of research into farm incomes *only*, in the same place, before and after MVs. See also Chapter 7, c.

2 'Consequently' because more poor farmers, when they adopt MVs, have a surplus (or a bigger surplus) to sell after meeting family food needs; and because the proportion of farmers hiring labour, and the hired/family labour ratios, are increased in the wake of MVs (Chapter 4).

3 This new research agenda has heavy data needs, but that is because the 'eaiser' data so far gathered (on the old agenda that wrongly equated 'small farms' with 'poor farm families') are not really suited to understanding the poverty impact from MVs on farmers in MV lead areas. Such researchers as continue to work on this impact should, we believe, reallocate almost all their work to this new research agenda, thereby helping to meet its data needs. However, fewer research resources should go to this issue (and almost none to its old research agenda) – and many more to the other, relatively neglected and more important, issues listed in the opening paragraph of this Chapter.

4 Net of production costs, valuing family labour in each season at its market wage (*minus* costs of job search), multiplied by the probability of finding work.

5 That is, household income, net of costs, from the whole system of household farm activities.

6 This is so serious that, in rural Malaysia, if we rank households twice – in decreasing order (i) of 'income per person' and (ii) of 'income per household' – typically over 60 per cent of households are assigned to different *quintiles* by the two rankings! [Datta and Meerman, 1980].

7 Increased *subjective* risk for early innovators; normally a 10–15 per cent price discount for output below TVs; a smaller per-acre amount of digestible straw.

8 A difference probably outweighed by the fact that a larger proportion by value of small farmers' purchases comprised manure – which is less expensive, because it requires more labour, per unit of absorbable nutrients; the small family farmer has less costly access to such labour (Chapter 4, c).

9 Local food prices are affected by local supplies, and not simply determined (even in small and otherwise open economies) by world prices. Few developing (or other) countries' governments allow world grain prices to determine local prices without intervention. Even if they do, bigger, timelier and more reliable local grain supplies induced by MVs bring prices down, partly by

cutting transport costs. Anyway, the downtrend in world grain prices is partly due to the early adoption of MVs – by bigger Third World farmers, but even more by big Western farmers who have not only captured innovators' rents, but also persuaded Western governments to guarantee domestic prices, inducing massive overproduction, and hence transferring subsequent price-cuts to the Third World's farmers: late adopters on a world scale (Chapter 5, b).

10 Work in progress by P. Hazell and C. Ramasamy shows that in North Arcot district, Tamil Nadu, between 1973 and 1983 this squeeze wiped out most of the extra profit from MV rice, despite yield growth at about 3 per cent per year.

11 This latter process would continue to increase the gluts of world grain for several years, even if agricultural subsidies in the developed world were phased out.

12 This has been only partly offset by the shift in demand for food from grain to animal products, which *increases* grain requirements because animals are inefficient convertors/filters of grain into human foods.

13 If traded, manure may well rise more in price than fertilizers in the wake of MVs, as there is less non-local competition (and manure does some things that fertilizers cannot do).

14 There are two important exceptions, favouring small farmers who are late in deciding whether to adopt: the 'basmati effect' (Chapter 5, h) and the 'Type IV region crop-shift effect' (Chapter 3, i).

15 Otherwise, they can lose from MVs, unless they eat almost all they grow. Even if their unit costs are eventually reduced, they will often be unable to compete against bigger farmers and earlier adopters in output markets.

16 A comprehensive and massive source of evidence – favouring the IR for many different countries, crop types, regions, and ways of gathering data – can be found in Berry and Cline, 1979. [For supporting data, see Bharadwaj, 1974, chapter 2; Lipton, 1974; Dasgupta, 1977, pp. 173–7; Shah, 1975, pp. 35–7; Hunt, 1984; Bliss and Stern, 1982, pp. 82–4]. A very careful recent measurement concludes that – in a village well advanced in the MV transition – gross value of annual output per unit area is independent of area cultivated, as is the proportion of fallow land to cultivated land. Thus, in this case, gross annual output per acre would seem unrelated to operated farm area (cultivated plus fallow). However, the operation of more farm area (whether fallow or cultivated) was strongly associated with more use of fertilizers – the main productivity-enhancing input cost. Thus output value, net of the cost of purchased inputs, was almost

certainly more on smaller operated holdings. Also, 'efficiency in the use of inputs was negatively associated' with owned 'holding per standard family member' – itself correlated with cultivated, and hence with operated, holding size [*ibid.*, pp. 198, 201, 290–1, 153].

17 Under such conditions 'the unattractiveness of very small-scale farming, compared to. . . other types of work' [*ibid.*, pp. 168–9] corresponds to the finding that – in very unreliably watered rural areas *only* – farm households with up to 7 acres or so are as likely to be in ultra-poverty as the landless [Lipton, 1985b].

18 This valuable work, however, in fact relates gross *crop* output per cultivated acre, not to farm size, but to the farm household's 'labour status', i.e. the extent to which it is a net hirer-in of land. The relationship between farm size and labour status is very weak [Harriss, 1982, p. 29].

19 We return later in this section to the view that MVs thereby bring a technology that in itself destroys or reverses the IR.

20 Note that 'crop area as percentage of total farm area' commonly falls as farm size rises from 1 hectare, but sometimes rises as between the 0–1 and 1–2 ha. size groups – confirming that one *can* be too poor to farm intensively [Berry and Cline, 1979, pp. 210–13; cf. Hill, 1982].

21 If anything, irrigation – especially in modern systems – is distributed less equally than land [Narain and Roy, 1980; Hamid *et al.*, 1978, p. 42] – not more equally, as would probably be required for poor farmers' land-use advantages to be due to better land quality.

22 Originally Chayanov showed this for parts of Russia where extra land was available to family farmers; he showed that, given the number of 'hands' available for work in a household, it would cultivate *more* land if there were more non-working family members – more 'mouths' to be fed. The same sort of argument applies to land-scarce economies, except that poorer (smaller, higher mouths/workforce-ratio) farms must work *existing* cultivated area more intensively, raising its output per acre.

23 This process, too, has limits. Age- and sex-specific labour force participation rates rise as income per consumer unit falls (and as mouths/hands ratios rise) – but only up to a point. Among the poorest 10–20 per cent in low-income countries, existing high levels of effort (and of ill-health) mean that participation rates do not rise further as poverty increases [Lipton, 1984a]; this restricts the 'Chayanovian' impact on the IR for *very* small farms, and leaves a group closely analogous to the group which Hill [1982, p. 169] calls 'too poor to farm'.

24 Small farmers would specialize in labour-intensive, large ones in land-intensive, product lines; and exchange between small farmers, large farmers, and non-farmers would push product prices (if not determined by world prices) *towards* the point that equalized the profit rate on cash put into large and into small farms.

25 Conversely, imperfect *capital* markets – enabling bigger farmers to modify or reverse the IR by borrowing more cheaply than small farmers to obtain improved inputs, using 'financial power and family solidarity' [Hill, 1982, p. 173] – are likely also to improve in the wake of MVs; probably faster than other markets, if MVs increase confidence that poor farmers can repay debts.

26 This is in spite of the link between smaller households and smaller farms; that link occurs because fewer people can work less land, and need less income from it. There is nevertheless a strong link in *total* rural populations between big households and poverty, because, within *each* size-group of farm households, bigger ones tend to be poorer [Lipton, 1983a].

27 Ideally 'per family member' should be adjusted to reflect the relative working power and/or consumption needs of children, women and men, as in [Bliss and Stern, 1982, p. 150].

28 This assumption is reasonable. *Local labourers*, are usually poorer than local farmers, but gain if it is smaller local farmers who get MVs; though such farmers have a somewhat larger family/hired labour ratio than big farmers, their total labour use per acre is considerably larger; and extra income from MVs also tempts smaller farmers to supply less household labour-for-hire (and thus to reduce competition against pure labourers), especially as own-farm work is made more lucrative by MVs. Poor *consumers* – who gain most if farmers use MVs to market more food (thus restraining prices), rather than eating it themselves – lose somewhat if MVs go to smaller farmers, on account of the latter's lower marketed supply/output ratio; but this is offset by the fact that smaller farmers tend to produce more per acre. As for *non-MV farmers* seeking to sell grain, the converse applies.

29 The CV of a group of successively dated numbers – e.g. wheat output in 1971, 1972. . . 1982 – around its *trend* (as opposed to the CV about the mean) is 'square root of average of squares of difference between (observation for each date) and (observation that would have been made if that observation had been on the trend line)', as a proportion of mean value of observations.

30 For example, the EEC's agricultural policies guarantee its farmers fixed prices for many crops, whatever happens to world prices. So, when there is a glut of a crop, EEC farmers are not deterred from

producing it – and the world price falls further still. The deterrent to producers elsewhere is increased; because EEC's farmers are largely insulated from making short-run reductions in output in response to falling world prices, the burden of these reductions (i) is increased as prices fall further, and (ii) falls entirely on other agricultures, notably those of developing countries. Similarly, rising world prices for a crop give no signals to EEC farmers, so that the signals (and hence responses) of output by farmers elsewhere are strengthened. Non-EEC, including Third World, farmers will suffer bigger swings of farm prices and output as such policies spread (e.g. as Greece, Spain and Portugal join the EEC and thus add more farmers to it).

31 CV would be a better risk measure, 'worst-case' output better still, and 'worst-case' income best of all; but the *proportionate decomposition* of causes of variance is useful, even if an *absolute increase* in variance means little – e.g. because it is consistent with *any* absolute rise in the divergence between each observation and the mean (or trend), even if that rise is proportionately smaller than the rise in mean output (or in its trend rate of growth).

32 Where rising CVs are found at national level for wheat, they too are largely due to rising covariance among producing places where output is increasingly concentrated. These 'places' are the MV lead areas, such as those in the Indian Punjab. Why should such places show much output variation, let alone covariance, since they are mostly irrigated? First, because such irrigation merely supplements rainfall – and, when this is good (or bad) in a handful of nearby Indian Districts, a much bigger part of total wheat output is affected now than in the early 1960s. Second, because surface irrigation – and even to some extent groundwater – is in reduced supply (as well as increased demand) when the rains are inadequate. On a *typical* lead-area farm in the Punjab, Haryana, or coastal Andhra, however, the CV of output is less than on a typical farm in a 'backward' area – but MVs have concentrated Indian farm output much more upon such lead-area farms; and among these farms *as a whole* water supply, and hence food output, tends to do well (or badly) together.

33 Regions, because they are largely semi-arid, unirrigated, and at high rainfall risk; MVs, because *early* hybrid millets were susceptible to downy mildew, and hybrid sorghums to *Striga Asiatica* witchweed.

34 These may be very poor people, selling MV rice or wheat to buy coarse grain or roots.

35 It has been suggested [Lynam, 1986, p. 18] that scale-neutrality may well apply to irrigated rice and wheat, but has 'no basis [for]

rainfed crops [if] mechanization or input use must precede varietal adoption'. One could add that the transport infrastructure was also usually much worse in these more sparsely settled areas, benefiting those who could afford their own vehicles and/or stocks. However, such factors are surely outweighed by the less unequal distribution of land-rights in semi-arid rainfed areas.

36 This is despite a substantial spread of MVs of millet. In most places and seasons, their on-farm performance has been rather disappointing.

37 The economic types are classified by forms of reaction, to MVs, of the size and distribution of income and output. Therefore each economic type covers several different cropping systems (and ranges of person/land ratio), though Types I and III below are each more homogeneous, in agro-climates and population densities, than Type II or IV.

38 See section e of this chapter. This is the effect of extra MV output in rendering cereal prices lower – and input-costs higher – than they would otherwise have been.

39 See sections (a)–(i) of this Chapter, and Chapter 4. In some countries, such alleviation is only relative; the poor got absolutely poorer in Bangladesh in 1960–75, but without MVs would have been even worse off.

40 For evidence on greater equality in 'backward' areas, see Dasgupta [1977a] and B. Singh [1985]. On grower-labourer gaps in Type I and Type II areas, see Lipton [1985]. Of course some Type II areas, notably the unreliably irrigated rice areas of Eastern India, feature very great inequality, but these are seldom the most backward areas; such inequality is often induced by the desirability of assets in land or water, and this provides hope that research may turn the area into Type III or IV, or even Type I as in much of Eastern Uttar Pradesh, India.

41 In principle, Type III areas could create a second generation of Type IV areas, as the former increasingly come to specialize in MV coarse grains. But, in practice, the agro-economic gains from intercropping – and farmers' wish, especially in rainfed conditions, to avoid risk by maintaining crop diversity – reduce this process.

42 Also, in a Chinese context, Bray [1986, p. 151] argues that rice MVs in well-watered regions are so attractive to the policymaker as to make him or her liable to enforce risky monocultures upon such regions – a sort of PAARD in reverse!

43 A new generation of maize hybrids is being released in parts of Kenya and Zimbabwe. Hybrid sorghum varieties successful in parts of India have proved unsuitable in West Africa, but are

promising (crossed with local varieties) in parts of East Africa, and
have already made a major contribution to output in Zimbabwe.

4. Labour and the MVs

(a) Economics, farmers and labour

In Africa, a growing minority of the poor relies for income mainly on hiring out rural labour, not on operating farmland. In Latin America, a large and growing *majority* of the poor relies for income mainly on hired labour, mostly urban. In Asia (excluding China), a growing majority of the poor relies for income mainly on rural hired employment. In all cases the word 'mainly' is crucial: the typical poor household in the Third World still cobbles together its income from a variety of sources (operating cropland, hiring out labour, informal self-employment in non-farm activities, remittances, mutual gifts) [Lévi-Strauss, 1962, chapter 1; Mauss, 1970]. However, it remains noteworthy that, whereas the IARCs were set up with an anti-poverty mandate directed largely towards helping *small farmers both to grow and to consume* more food,[1] the main 'anti-poverty problem' is increasingly one of helping *people who depend largely on employment income to afford* more food.[2] How can economists, sociologists, and other social scientists help agricultural researchers to bring about the overdue reorientation of their work towards the new, largely assetless, majority of the world's poor?

There is an interesting puzzle, which casts light on the scope and limits of different *types* of social science to help IARCs in directing MV benefits to the poor. The puzzle is that, although theoretical arguments from economics alone (i) probably cannot show that small farmers in MV lead areas gain from MVs but (ii) strongly suggest that labourers do, there is (iii) growing (though often overstated) evidence and consensus that these small farmers gain significantly from MVs, while (iv) net gains to labourers from the processes involving MVs in their total socio-technical setting are getting less clear, and may in a few cases be turning negative [Smith *et al.*, 1983; Smith and Gascon, 1979; Jayasuriya and Shand, 1986].

Economics alone cannot give firm guidance about the impact of MVs on the absolute and relative income of poor farmers in MV regions. Certainly the physical properties discussed in Chapter 2 – MVs' use of more labour and management per acre, their production of coarser and cheaper varieties favouring self-consumption rather than marketing, and recently their greater robustness – should be more helpful to small farmers than to big ones. But the societies into which an MV is inserted also influence the outcome. Their structure of power can divert, to the richer farmers, the benefits even of innovations whose economics appear to favour poorer farmers. Moreover, the transfer of some MV benefits to consumers and fertilizer-makers clouds the issue. Hence the limitations of economics in predicting impacts of MVs on farm poverty and inter-farm distribution within MV areas. Hence, too, the very variable outcomes for farmers in MV lead areas; variable, but in general favourable to the absolute real incomes of poor farmers.

Economic theory can apparently make stronger and clearer statements about the impact of MVs upon hired labour in such areas. MVs – via greater needs for fertilizer, water control, harvesting and threshing, and often via double-cropping – increase the *demand* for labour per acre, apparently pushing up labour's share of income [implicit in Binswanger and Ryan, 1977, p. 225]. However, in a large and growing majority of developing rural areas (and especially in irrigated areas, where MVs are especially important), the *supply* of labour is ample. It can usually respond by moving to nearby MV areas if the real wage starts to rise. However, the supply of land is restricted, and cannot rise much in response to the greater demand for it caused by the new, profitable MV farming opportunities.

So employment of labour goes up somewhat, the real wage rate does not go up a lot, and the rewards (price, rent) of land go up a good deal, probably reducing labour's share in income [these elasticity effects are considered in Binswanger, 1980, p. 283, and Anderson and Pandey, 1985, p. 9]. Although employed labourers find initially that more work is on offer, and is better spread over the seasons (though perhaps not over the years), with MVs than without, this rise in work may be outpaced by growth of the workforce, since population is growing. The real wage *rate* stays the same, or rises a little; the *share* of wages in total farm incomes falls, because rising land-rents enrich landowners proportionately more than rising employment enriches workers; but rising employment due to MVs[3] does mean that the real wage *bill* rises. Real annual earnings rise, but – because labour force is typically increasing at 2–3 per cent per year – not necessarily per worker.

As we shall see, this simple economic 'story line' corresponds reasonably well with observed facts in MV areas. Two complications, very important in other respects, do not greatly affect this story line, but a third can produce an unhappy ending.

The first complication is that a proportion of gains from MVs is transferred from producers to consumers, because extra output cuts food prices. However, labour-per-acre in the MV crops is still increased (footnote 3); and the pre-MV wage-rate in many areas represents a 'historical and moral' subsistence minimum, so it cannot fall. Therefore, the above processes still work: MVs still make the real annual earning-per-worker higher than it would otherwise have been. If prices of the MV crop do subsequently fall, producers may switch out of it, reducing the employment gain; however, farmworkers (as net food buyers) find that their wage will purchase more than before, though the evidence on real wages suggests that this effect is small.

Second, it has been familiar at least since Robbins [1930] that, as people become better off, they tend to substitute leisure for income. Recent work on family farm households' response to better income[4] shows that they too, as their income is raised by MVs, tend to take it easier. They probably switch family labour from other crops to the now more profitable MV crop in the first season (especially as the new methods may need more supervision); but they reduce *total* input of family labour, i.e. they rely more upon hired labour. This reinforces the above processes of employment expansion.

Third, and less happily for the poor, MVs are in some circumstances linked to labour-displacing inputs – tractors, weedicides, more mechanized irrigation, or (especially) threshing. The view that such inputs do *not* displace labour is either 'special case' or special pleading; a good review on tractors remains [Binswanger, 1978]. Some argue that MVs do not 'cause' this labour-displacement. Others disagree, emphasizing the incentives to mechanize created by MVs: seasonal labour peaks and the bidding-up of seasonal threshing wage rates in some areas. However, if MVs do 'cause' mechanization, etc., and are thus labour-*saving*, this is a direct result of employers' reaction to the labour-*using* effects mentioned above. Almost always, such labour-saving effects only partially offset the labour-using effects of MVs. Certainly, in the absence of the former effects, employment and real wage-bills would rise rather more, and wage-shares would fall slightly less. However, the current processes of mechanization of draft and threshing operations would in combination imply lower total labour inputs in most rice and wheat cultivation activities. There are several factors leading to this labour displacement; the relevant question to ask is whether MVs make labour displacement more likely.[5]

It has been argued in a Javanese context [Pollak, 1986] that physical features of rice MVs increase the gains to farmers of switching from (i) transplanting to mechanical direct-seeding, (ii) hand to rotary weeding, and (iii) above all, individual plant-by-plant reaping to sickle reaping. However, this is probably unusual, being partly due to the pressures from Indonesia's 'oil boom' to find substitutes for formerly rural labour that is moving to the towns – especially for the uniquely costly process of *ani-ani* reaping. Moreover, in Malaysia the transition to sickle reaping was associated with shorter-duration MVs, together permitting double-cropping of rice – creating more (and smoother) year-round employment, far outweighing the decline in main-harvest work [Bray, 1986, p. 124]. In general, the effects of MVs in raising yield, cropping-intensity, and the profitability of applying water-control, fertilizers, and therefore weed control – all labour-using processes – are likely to increase the demand for labour (and, because family farms get better off, to reduce the supply of family labour). The key question is whether pressures in hired-labour markets – pressures due partly to MVs, but usually felt more in organizational difficulties and uncertainties than in wage increases – will induce (i) reorganization of labour controls and processes, possibly harmful to poor workers; and (ii) labour-displacing mechanization, not only in farming but also in post-harvest operations (e.g. if MVs – bred for low ratios of dry matter, including husk, to grain – shatter too easily to permit conventional hand pounding).

In summary then, apart from possible mechanization effects, economic theory makes fairly strong predictions about what MVs do to labour. (1) Employment should rise significantly, especially in the short run. MVs raise not only labour use per acre-year, but also hired labour as a proportion of total labour use, for three reasons. First, farm families, unless heavily underemployed to begin with, must meet most of the extra labour requirements of MVs by hiring in. Second, as MVs enrich farmers, the 'income effect' induces them to take more leisure, and to hire in workers instead (footnote 4). Third, because MV systems are often simpler than TV systems, they tend to require less direct family involvement [Kikuchi *et al.*, 1982], at least after the first season. Moreover, (2) stability of employment should rise as MVs increase double-cropping, as has been known since at least 100 A.D. [Bray, 1986, p. 22]; double-cropping not only raises employment stability for casual workers, but makes it pay for farmers to shift their demand for hired labour from a casual to a permanent basis. (3) However, MVs will raise real wage-rates little, and (4) labour's share in income falls, due to the much greater elasticity of supply of labour than of land. Do the facts support these strong predictions of economics?

(b) Labour use, wage rate, factor share

As rural incomes rise and as labour shifts from the household to the job market – processes happening anyway, but much enhanced by MVs – two main things happen to labour. First, there is a fall in adult 'participation rates', the proportions of person-days supplied to the workforce, especially among women. Second, there is also a fall in the proportion of workforce-days spent in employment (or self-employment). Both Indian village cross-sections and analyses of the aftermath of MVs support this conclusion [Dasgupta, 1977, p. 172]. The many studies of *labour use* under MVs hardly ever separate these two effects. Nor do they often separate the role of MVs from that of other factors. Thus studies of *unemployment* in LDCs almost all show increasing rates [Lipton, 1984]. but that is because it is pulled up by workforces (growing at 2–3% yearly) faster than it is pulled down by falling *participation* rates. In fact MVs help both to moderate unemployment by requiring extra labour, and to reduce labour supply as better-off families reduce participation rates.

This chapter is confined mainly to the effects of MVs on labour use, per acre per year, on the MV-affected crops. That is partly for want of evidence on indirect effects, and partly because they probably tend to raise demand for poor people's labour anyway. (i) The greater profitability of MVs shifts land into cereals from competing crops that are usually less labour-intensive (there are a few exceptions; MV rice in Bangladesh sometimes displaces more labour-intensive jute). (ii) Extra spending by enriched MV farmers – and by consumers enriched by the effect of MVs in restraining food prices – further creates employment, though the size of the effect cannot be guessed at by looking at local linkages, or extra demand for non-farm goods, only [Hazell and Roell, 1983]. (iii) There are other employment gains as MVs drive up demand for irrigation and other inputs.

What of the important and controversial direct effects of MVs on the employment of unskilled workers?

Labour use

'Early observers' of MVs often found they raised labour requirements per acre-year by about one-fifth [Barker and Herdt, 1984, p. 38]. Village-level increases in such requirements in MV areas for 1968–73 varied from 10 per cent in

Orissa and 13 per cent in East Java to 40 per cent in a Thai region and 42 per cent for boro rice in Bangladesh. These were the proportionate rises as between TV and MV areas; the fact that many TV areas remained (due to incomplete MV adoption in these early days) meant that employment rose 'much less' [ADB, 1977, p. 60]. Thus the total impact of the new technology on employment would be much less than these figures suggest as not all adopted the technology; even adopters left some of their land in TVs. Jayasuriya and Shand [1986] reviewed evidence which indicates that in the early phases of the 'green revolution' a doubling of yields was associated with labour demand increases of 15–20 per cent, though most of the data we have seen suggest higher figures, typically in the 30–40 per cent range [see also Bray, 1986, p. 157]; where MVs made a second crop profitable, the employment impact per extra ton of grain would normally be far more than where they simply raised yields in the existing crop season.

However, these data need careful interpretation. Big farmers usually adopt MVs sooner than small ones (Chapter 3, b); they also usually have lower labour input per acre.[6] A cross-section of farms early in the adoption process, therefore, reveals big adopters who are not very labour-intensive, perhaps using not much more labour per acre than the small farmers, who have not yet adopted. Thus, when farm size is held constant (or later on, when almost all farms have adopted), the extra labour input associated with MVs is seen to be *more* than appears to be the case from the simple comparison of MV and non-MV land, early in the adoption process, in respect of labour use. On the other hand, MVs are likelier to be grown on irrigated land, where labour input per acre is higher. Thus, when the degree of irrigation is held constant, the extra labour input associated with MVs is *less*, per acre, than appears from the simple comparison [for Indian evidence from Andhra Pradesh, Tamilnadu and Orissa, see Agarwal, 1984a, pp. 23–8].

As MVs spread to less favourable environments, the yield impact fell and with it the direct crop employment effect [Herdt and Wickham, 1978, pp. 4–6]. Where rice yield rises were large, rises in total labor use per acre remained clearly linked to MVs [Barker and Herdt, 1984, p. 43]. However, the areas with lower impact – and the history of some high-impact areas – showed worrying employment trends. Among 100 rice MV adopters in a block in Eastern Uttar Pradesh, India, the big rise in labour input per acre from MVs between 1967–8 and 1972–3 had been wholly reversed by 1979–80 [D., V. and R. Singh, 1981]. Widespread evidence of recently weakening effects of MVs on employment comes from some of the very sources that earlier documented strongly favourable effects [Jayasuriya and Shand, 1986, esp. Table 1]. It must be added (i) that these worrying tendencies are strongly in evidence only for rice – though common sense suggests that they also apply to wheat, with the big spread of threshers, tractors and weedicides in MV areas; (ii) very promisingly, that the spread of MVs to new crops (sorghum, finger millet, cassava) in new areas, especially as they are associated with less unequal farming, should start up employment effects again at the favourable end. However, declining effects for rice (and probably wheat) are worrying.

Several reasons for the deterioration may be proposed. The key roles of (1) migration and (2) mechanization are treated later (sections f and g of this chapter). (3) Another general feature is that the upward pull-on costs, as the production of MVs increases demand for material inputs, may most seriously affect inputs complementary with labour (especially irrigation and fertilizer), leading farm employers to use fewer such extra inputs – and to hire less associated extra labour – alongside new expansions of MV area than they had previously done. Also, as in Central Luzon, (4) rising 'full costs' of labour-time, including search and transaction costs as hired workers replace some family labour, may be partly responsible [Smith, Cordova and Herdt, 1981] – perhaps because the spread of MVs calls into use labour of lower quality or experience, and/or higher leisure preferences, than was typically engaged previously.[7] (5) As in Java, institutional changes, under pressure of seasonal labour bottlenecks immediately after MVs, may be destroying traditional work-increasing or work-sharing arrangements or technologies [Hayami and Hafid, 1979, but cf. Hart, n.d., Chapter 7]. (6)

Larger farmers, to avoid MV-related labour costs, may eventually get together with import licensees – and, as in the case of British tractor exports to Sri Lanka, with foreign producers and aid donors [Burch, 1980] – to extract and share, at the expense of domestic and foreign taxpayers as well as of poor local workers, subsidies for labour-displacing inputs. More hopefully, (7) in Laguna [Smith and Gascon, 1979], and surely in Taiwan and Korea, growing off-farm opportunities have reduced availability of farm labour. Most worryingly in the long term, (8) the 'theory of induced innovation' means that researchers face incentives to push down unit costs of factors scarce, not simply to producers in general [Hayami and Ruttan, 1971], but to rich and powerful producers [Grabowski, 1981]. As research is internationalized, these increasingly come to mean bigger, including Western, farmers; the factor that these seek mainly to save is hired labour. But such research impinges upon LDC environments too. IARCs need to be very careful to avoid pressures to provide cost-reducing, aid-subsidized research to help big farmers in displacing labour following the spread of MVs. Such research is, unfortunately, a feature of 'weedicide screening' in several IARCs, and of the 'IRRI-PAK mechanization programme'.

Wage rates

Although the direct impact of MVs on farm employment usually remains positive, it has become less favourable. Moreover, 'except in Malaysia, the [Indian] Punjab and Thailand, [no] significant rises in real agricultural wage rates have taken place' in Asian MV areas over the past two decades [Jayasuriya and Shand, 1986, and citations there; cf. K. Singh, 1978]. Choice of particular years or seasons for comparison can be misleading, because fluctuations in real wages far outweigh trends [e.g. in Haryana in 1967–78: Kumar and Sharma, 1983]. Even in the areas of very rapid MV-induced growth, such as the Indian Punjabs, the rise in real wage-rates is very slow [Lal, 1976; S. Bhalla, 1981; at village level cf. Leaf, 1983, p. 251; Blyn, 1983, p. 719].

This stagnation of rural real wage-rates means partly that MV gains are being passed on to urban consumers – except to the extent that MVs displaced imports, or added to stocks (Chapter 5) – and landowners; and partly that farm employers displace labour by tractors, weedicides, etc., if it looks like getting more expensive (or 'troublesome' and well organized); such big farmers (on whom the urban sector relies to deliver surplus food and exportables) often have enough power to obtain subsidies for such inputs through the political system. But the main reason why wage-rates stagnate, and why wage shares decline, is that, with workforces growing fast, extra *demand* for labour due to MVs meets increasing *supply* of labourers prepared (or compelled by competitors) to work at rates barely above subsistence. Without MVs, such rates would probably have fallen further.[8] Moreover, unless all the extra food grown by MVs would otherwise have been imported, food price rises would have implied real wage falls. In view of transport costs, it is naturally in the areas where MVs spread fastest that they did most to restrain local food prices, and thus indirectly to prevent real wage falls [Jose, 1974; Parthasarathy, 1974].

Since population growth, as the mainspring of workforce growth, is largely responsible for the disappointing impact of MVs on real wage-rates and underemployment rates, it is natural to ask: do MVs affect the rate of population growth? Is it higher or lower in MV areas, or after MVs have spread? Unfortunately, these are almost wholly unresearched issues (see Chapter 7, c).

Factor shares

Farm labour's falling 'factor share' (wages × employment, divided by net total output) in MV areas has been caused mainly by (1) the rise in the ratio of rental to the wage rate and (2) the drift of extra post-MV gross farm incomes to suppliers of inputs – so that *labour*'s share in *gross* income has declined much more, in the wake of MVs, than *farm* labour's share in *net farm* income.[9] However, the latter too has declined. The rental-wage ratio in Thanjavur District, South India, doubled between 1971–2 and 1980–1 [Rajagopalan *et al.*, 1983, p. 427], though with many fluctuations; a good (bad) year usually led to much higher (lower) rents next year. A falling wage *share* in factor incomes from farming, due to the rising price of land relative to labour – alongside a rising absolute real wage *bill* as MVs pushed up labour use – is confirmed elsewhere in India [Prahladachar, 1983, p. 938] and in Mexico [Burke, 1979, p. 150].

So modest is the impact of MVs upon labour income that new farming systems for semi-arid areas are commended because labour gets even 9 per cent of extra farm income, as against as little as 1 per cent reported for previous MV impacts in the Philippines (while gross farm revenue rose by 70 per cent), and 2–5 per cent elsewhere [Crisotomo *et al.*, 1971; Ghodake and Kshirsagar, 1983b, p. 9]. In these new farming systems, the greater inputs of fertilizers induced by MVs mean that most of the projected income gains from extra production go to urban or foreign producers of inputs, especially of fertilizers. That often leaves both land and labour with a smaller *share* in gross revenue, despite a rise in the real rental and the rent/wage ratio [*ibid.*, p.12]. A similar process transfers extra gross farm incomes to producers and repairers of machines, if these are linked to MVs [Ahmed and Herdt, 1981].

Overall impact on labour incomes

With rapid population growth and scarce land, the position of landless labour is much worse without MVs than with, as village comparisons show [Hayami and Kikuchi, 1981]. MVs seldom raise real wage-rates, or prevent falls in the wage share, through their effect on incomes from the MV crop. But this effect does bring higher employment and a rising real wage bill. There are four cautions, however.

First, these benefits usually have to be shared among a growing number of labour households, partly because population increases naturally, and partly because of labour-migration to MV areas. Thus employment and hence wage receipts *per household* rise less, and may even fall. This is not usually 'because of MVs' but because of population growth. However, a part is also played by developments that may sometimes be linked to MV-induced changes in landholding structure (Chapter 3). The possibility was established in some

Bangladesh villages that due to early MV innovation 'with their extra resources and relatively increased power the village rich were [better placed] to push the poor off their land' [van Schendel, 1981, p. 245], thus increasing the supply of persons seeking hired employment. This is precisely analogous to the possibility that the greater resources and power of the 'village rich', in the wake of MVs, enable them to 'push the poor' into a worse market, or a worse bargaining position, as regards their labour, and to acquire (or to get subsidies for) labour-displacing equipment or current inputs.

Second, little is known about the effect of MVs on *wage-rates or employment on other crops, or off-farm*. In Bangladesh the shift from jute to rice – even, though less so, to MV rice – is labour-displacing [R. Ahmed, 1981; Harriss, 1978, 1979; see next section], though usually MV cereals are more labour-intensive than the crops they displace. MVs certainly increase work in post-harvest processing, if techniques remain the same, because there is more food to process. However, the growing proportion of outputs processed as urban surpluses raises the capital/labour ratio in milling, baking, etc. MVs can also push local, rural techniques in this direction: sharper 'labour peaks' encourage adoption of mechanical threshers; easy shattering of MV husks, as researchers get higher ratios of grain to dry matter, provides a further advantage to mechanical milling; and big surplus farmers' new power, as MVs enable them to displace imports as the main urban food source, helps them to secure subsidies for post-harvest labour-saving equipment or ancillary fuel and credit. Researchers should screen MVs, not usually for post-harvest loss rates (which are generally small: [Greeley, 1981, 1987]), but for post-harvest amenability to labour-intensive, capital-saving methods of threshing and milling or dehusking. Both the physical features of the plant, and its avoidance of peak labour requirements, are important here.

Third, no research has been done on the effects of MVs on *labourers in non-MV areas*. Such workers benefit as consumers (Chapter 5) and as migrants to MV areas (Chapter 4, g). But they lose as MV output cuts relative grain prices, and hence the incentive to employ them locally. Probably, a useful approach to understanding and helping such workers is via the disaggregation of 'non-MV areas' proposed in Chapter 3, i above.

Fourth, apart from the 'average' fate of labour, *particular groups* of poor labourers (and in bad times many such groups) may gain or lose from processes involving MVs. This crucial impact has two aspects. In sections c to e below, we ask how MVs have changed the structure of labour use. How have the proportions of labour supplied, and demanded, altered, as between family, permanent, and casual workers; men, women and children; farm operations; and peaks and troughs? In sections f and g below, we ask about the effects on workers of possible longer-term responses to the extra labour needed by MVs: migration by labourers from non-MV areas; or labour displacement by means of machines, weedicides, etc. At each stage, we need to ask to what extent these processes of structural change in the labour force are linked to MVs.

(c) Structure of labour use

Hired vs. family

Numerous Indian studies concur that with MVs 'the employment of family labour increases [proportionately less than that] of hired labour' [Visaria, 1972, p. 184]. This is widely supported by studies from other countries [Barker and Herdt, 1984, p. 39; Kikuchi *et al.*, 1982; Smith and Gascon, 1979; Roumasset and Smith, 1981]. Quite often a post-MV rise in hired labour input per acre outweighs an actual decline in family labour input [IRRI, 1978, pp. 73–5, 91, 101, 126, 396], presumably caused by income-effect within family-farm households (pp. 178–9).

Is this good for the poor? In other words, are hired labourers likely to be poorer than small farmers? Clearly so in the main irrigated MV areas, where a little land confers a much reduced risk of poverty, so that labourers are significantly poorer than landholders. The only clear exception arises where, as sometimes in Java [Lluch and Mazumdar, 1981], landed households with slightly different seasonal peaks give one another preferential work on a more rewarding

MV crop, and employ the poorest – who are more or less landless – on other crops at lower incomes.

In many unirrigated areas, however, households dependent mostly on landless labour face very similar risks of poverty to those dependent mainly on smallholdings [Lipton, 1985b; compare Ercelawn, 1984]. In such cases, the anti-poverty advantage of 'hired labour bias' in extra work generated by MVs is less clear. Also, we must not conceptualize people as either labourers or operating farmers. Most of the rural poor are a bit of both, often concentrating on employed labour when young and on own-farm work after inheriting land.

Casual vs. long-term

Usually real annual wages have risen somewhat in MV areas, while real day-wages have stagnated [Leaf, 1983, p. 268; Blyn, 1983, p. 711]. This is because (at any given set of rewards) the *supply* preferences of labourers are shifting towards day-labour and away from longer contracts, and because *demand* for labour is shifting from casual to longer-term. (1) Supply of labour to the market comes from pure landless labourers (A) and labourers with other income sources also, mainly family farming (B). Both requirements for, and returns to, family farming are raised by MVs. Hence the supply preferences of (B), and therefore of aggregate labour supply, i.e. of (A) + (B), shift from longer to shorter contracts – raising the reward to the former, relative to the latter. (2) As for demand for labour, the greater amount and lower seasonality of MV labour requirements, plus the reduced participation rates of families enriched by MVs (pp. 178–9), increase the gains to employers from longer contracts.

There are, however, remaining seasonal peaks. These would suggest rising demand for male casuals at harvesting and threshing time. Such demand in India is strongly correlated with the proportion of rice in MVs [Agarwal, 1984a].

Although there is little direct evidence, partial 'decasualiza-tion' of labourers is almost certainly an effect of MVs. If labourers are the poorest people, is it desirable? Certainly, it is for those labourers obtaining job security; they (and, through knock-on effects, their families) also benefit from the direct interest of the employer in feeding well those persons he or she expects to employ in the longer term, so as to raise their productivity. However, especially in the slack season when work is hard to find, shifts in labour demand towards permanent labourers reduce the prospects of the remaining casuals, who are presumably less strong or able. Partial post-MV decasualization, therefore, might well reduce the *numbers* in poverty, by pulling the stronger of the impoverished workers from casual into permanent labour, and thus, perhaps, above the poverty line. However, this process would presumably increase the *severity* of poverty for the remaining casuals, who are the weaker (and probably initially even poorer) workers; having been 'screened' out of the shift to decasualization, they face a lower demand for their labour.

However, even if MVs do promote decasualization, they are fighting opposed forces. In the ICRISAT study villages – where there has been rather limited spread of MVs – the trend is towards 'casualization'; the number of permanent servants is falling and the length of their contracts is shortening [Walker, pers. comm.]. This is probably true of many regions in India's semi-arid areas, where employment is becoming harder to get, except where MVs have spread substantially. Annual labour contracts in the rural Third World are 'Janus-faced': partly a relic of pre-capitalist exploitation [Mundle, 1979], weakened by MVs in the process of commercialization; partly a forerunner of settled agriculture with a permanent, but free, labour force, and as such accelerated by MVs.

Men vs. women

There are some village-level data suggesting that MVs reduce women's share in income, partly via the switch from family to wage labour [Ahmed, 1983, for Nigeria; D., V. and R. Singh, 1981, for Eastern Uttar Pradesh, India]. Post-harvest labour displacement, associated with MVs, appears to have selectively affected women in Bangladesh and Indonesia [Greeley and Begum, 1983; Jiggins, 1986, p. 55]. It is even probable that the extra demand for male labour, due to MVs, has increased women's burden of unpaid family labour in some cases [*ibid.*,

p. 25]. The only systematic survey, however, shows that total female labour use is positively related to proportion of area in rice MVs in all three Indian States reviewed, and significantly at 5 per cent in two [Agarwal; 1984; 1984a]. The results for wheat, if available, would probably be less favourable, as the greater cropping intensity of MVs creates more work that is 'traditionally female' in rice than in wheat.

In this case, it is not clear what, as 'friends of the poor', we want from MVs here, nor what specific research IARCs should prefer. Evidence from several Asian countries suggests that female-headed households – and households with large proportions of females – are little, if at all, over-represented among the poorest income groups (which is not to deny that women suffer from serious discrimination) [Lipton, 1983a, pp. 48–53, citing Visaria and others]. Moreover, poor women presumably want to be freed from excessive work if they are mainly farmers, but to be provided with access to more paid work if their earnings come mainly from hired labour. Also, how exactly are IARCs to aim at outputs providing incomes to specific groups, such as (poor) women? The same germ plasm from an IARC is used or adapted by many different national research systems. The resulting MVs may tend to reach wealthy men in one country, poor women in another, etc.

However, it would be a bit evasive to say that IARCs should leave entirely to national systems the task of focusing on poor groups, on women farm labourers, etc. Some physical characteristics and timings, which vary among crops and varieties, are almost everywhere especially helpful to participation or benefit by such groups. Furthermore, by careful disaggregation of the likely effects of their proposed innovations upon poor men as against poor women, agricultural researchers can avoid unwanted discriminatory effects, e.g. where men and women earn and retain separate incomes and pay each other for services within a household [Jiggins, 1986, pp. 26–7, 48]; or where particular MV options place quite separate demands on male and female peak-season labour [*ibid.*, p. 75].

A further problem, specific to the sex balance of benefits from MVs, concerns the work requirements to be aimed at. Some tasks, such as weeding and rice transplanting, are more usually female than others. Should MVs aim at more or at less work in such tasks? As indicated, the proportions of women, and of female-headed households, with very low household income per person are seldom significantly above the proportions of men, and of male-headed households, respectively. It is often claimed that women are likelier than men to pass extra income on as food to children at nutritional risk; but this view requires much more firm evidence.

In any event, it is not at all clear whether higher or lower proportions of work, casual or family, are an effective means to improve poor women's status, power, or retained income. Slack-season risk of undernutrition of children is apparently reduced by maternal earnings only if they are obtained from self-employment, e.g. with MVs on the owned farm, not in hired employment that may involve leaving children at home [Kumar, 1977].

(d) Labour use in specific operations with MVs

Various operations with MVs tend to increase and stabilize the demand for labour. This is a major service to the rural poor from the IARCs. Although much of their work underpins this excellent result, unfortunately some does the opposite.

At *sowing*, if MVs are to approach the economically feasible yield, higher densities are usually required [*CIMMYT Review*, 1981, pp. 31–2]. This raises demand for sowing labour, and in the case of rice for (mostly female) transplanting labour. However, this has led IRRI to research and develop (i) the mechanical direct seeding of rice, and (ii) a multi-row rice transplanter. These may well 'save transplanting labour', but is this a proper goal for researchers paid for out of foreign aid? Possibly, if poor consumers gain more, from extra rice not otherwise cultivable, than poor transplanting labourers lose. However, the constraint on adequate nutrition in most poor countries is effective demand, not supply. Appropriate research priorities could surely avoid undermining the favourable impact of MVs on poor people's income from employment in transplanting.

Since MVs increase the returns to *fertilizer*, extra labour inputs are also required to place it. But many small farmers cannot at first get credit or afford fertilizers, although small farmers, eventually, often come to use no less fertilizer per acre than big [Herdt and Capule, 1983]. Moreover, most farmers, being risk-averse, use less than profit-maximizing levels of fertilizers. IRRI has led the way in researching mudball techniques, deep placement, sulphur-coating and other forms of slow-release, increasing the incentive to use fertilizers and ancillary labour. Indeed, deep-placement of urea may be constrained in the Philippines by labour scarcity [Flinn and O'Brien, 1982]. But does this not make such methods especially suitable for the deficit family farmer, able to switch labour from hired work to his or her own enterprise? Such use, however, may depend on keeping the fertilizer (e.g. pellets or slow-release) not too expensive – not just in terms of nutrient equivalent, but for the smallest purchasable unit, which may be all that a deficit farmer can afford.

Similar issues arise for *weeding*. Dense planting and erect leaves somewhat reduce MVs' requirements for weeding, but are usually more than offset by the stimulus to weed growth from higher fertilizer levels. Much IARC research appears to be directed towards developing MVs, timing advice, etc., that reduce weeding requirements [CIMMYT, 1983, pp. 89–91; 1984, pp. 77–9]. This would be a desirable feature of MVs if it were costless, and we argue in Chapter 2 that weeds are grossly under-researched. However, research time and land, used to raise yields or stability, should if possible seek uses that increase (rather than reduce) the employment that a farmer can profitably offer. Some IRRI work [IRRI, 1983, *passim*] appears to go far in the opposite direction, by assuming that aid-financed research resources should be used to test, select, improve, or develop recommendations for, commercial weedicides. Even where there is a case for this – e.g. with perennial sedge in dwarf rice where hand weeding leaves weed tubers in the ground [*IRRI Reporter*, 2/1973, p. 1]; or with barley, where ICARDA research suggests that, by the time handweeding is feasible, the damage is already done – should not IARC resources aim to shift timings rather than to 'save the labour' of poor rural women?

Since MVs increase yields, they increase the requirements for *harvesting* labour, given the technology. A natural economic response is to seek a harvest technology that saves labour, for example by the switch from ear-by-ear harvesting to the sickle in Java [Hayami and Hafid, 1979]. Clearly, if migration can be encouraged or varieties selected so that harvest time shifts to a less labour-constrained period, that is better than 'unemploying' labour via reaper-binders. The harvest can also be 'pushed away' from the next ploughing season by varietal selection, again spreading the labour peak instead of cutting it off by labour-displacing mechanization; thus in Sri Lanka, at some cost in yields, shorter-duration *yala* (second-season) rice MVs can reduce the need for a quick turnaround, and thus for labour-displacing tractorization. Institutional innovations, too, can spread the peak and thereby reduce employers' incentive to save harvest labour; thus fragmentation inhibits early harvesting of MVs if the remote plots must be reached through fields of standing crop [Pal, 1978], so that consolidation of plots can permit a more 'labour-spreading' and hence employment-creating approach to the MV harvest.

In *post-harvest operations*, we have suggested that some MVs or related IARC research might displace labour. The attempt to develop an 'IRRI thresher' [Jayasuriya and Shand, 1986] is surely an inappropriate activity. Threshers were much the most clearly labour-displacing piece of mechanization in Ludhiana District [Oberai and Ahmed, 1981] and in the Philippines [Roumasset and Smith, 1981]. Even if a cheaper thresher is the only way to permit double-cropping in a few places in South-East Asian economies [Bray, 1986, p. 123], if successful it will spread to other areas, and constitute an aid-financed subsidy towards reducing the cost of labour displacement there, e.g. in Bangladesh. As for rice milling [B. Harriss, 1978], some MVs are structured physically in ways that tilt the balance towards displacing labour from traditional mortar-and-pestle, via 'intermediate' Engelberg huller systems, towards modern rice mills. Finally, since MVs are bred above all for a high grain weight relative to other dry matter, and therefore often have thin husks, care needs to be taken to screen varieties for storage characteristics.

Altogether, MVs tend to increase post-harvest labour requirements, which (except for drying) can mostly be deferred to a suitable time, and which are especially likely to employ women and the landless. It is important that misdirected 'labour-saving' research does not destroy these possibilities. Thus the modern rice mill in Bangladesh is highly profitable, but only because it 'saves' labour costs. It reduces labour requirements from about 270 to about 5 hours per ton. Quite modest income distribution weights (social prices) suggest that the adverse distributional effects of milling are much larger than efficiency gains. The losers are wage-labouring women from landless households [Greeley, 1987].

All these warnings apply to environments where there are many poor rural people relying on employee income. They might apply much less to MV research for Zaire or Zambia, where the decision to use or reject a 'labour-saving' process is usually in the hands, not of an employer, but of a poor (but cultivating) beneficiary (often with access to empty cultivable land). Also, the warnings are not advice to prevent *efficient* labour-displacing innovations that raise food availability. The advice, rather, is to avoid using aid-funded research to ease the path, in land-scarce countries, towards innovations that unemploy the rural poor.

(e) Seasonality and stability

The poorer people are, the more likely is it that they depend on casual labour for large parts of their livelihoods, and that they are (for reasons of health, nutrition, age, sex, language, or simple unfamiliarity or prejudice) relatively unattractive to employers. Hence we find that, the poorer a person or household, the higher the risk of unemployment – and the greater the fluctuation of unemployment; it is the poor whose labour is discarded in slack seasons and bad years (also, in part, because family farmers then lack cash to pay wages, and have enough labour to do the reduced levels of work on their own). Since unemployment creates expectations that the prospects of job search are bad, poor people respond, in bad seasons and years, to the greater risks of unemployment by slightly reducing their participation in work. Since in hard times the fall in demand for labour (i.e. the rise in unemployment) is much greater than the fall in supply of labour (i.e. in participation), wage-rates also fall.

Thus the poorest rural groups are in a sixfold bind as a result of the seasonal and annual variations in crop production, and hence in demand for their labour. (1) Of all groups, their income is the most heavily dependent on hiring out of labour – and increasingly so. (2) Because poverty and casual work go together, their incidence of rural unemployment is the highest. (3) In bad times, their unemployment rates rise most. (4) Their income is further reduced because they respond by reducing participation. (5) Because poor people's unemployment rises much more than their participation falls, their wage-rates fall too in bad times. (6) Finally, actual or expected food scarcity raises the price of food; and poor labourers are usually net buyers of food.[10]

This sixfold bind is dramatically reflected in rising infant mortality rates among the poorest in hungry seasons[11] and bad years – most clearly in famines, but not only then. Since growing proportions of the poorest depend for food mainly on labour income – and since stabilizing such income, at least in semi-arid places, is a much more powerful way to stabilize poor people's purchasing power than stabilizing farm output alone [Walker *et al.*, 1983] – a life-and-death question about MVs' impact on the poor is: how have they affected the seasonal pattern of labour use? First, MVs have, on balance, tended to shorten the duration of cultivation; the effect in concentrating each season's labour peaks has certainly been outweighed by the effect of spreading labour requirements across the year by permitting double-cropping, which both increases and stabilizes year-round labour requirements.

Second, however, there may be an offsetting ratchet effect. MV-induced double-cropping often renders it a paying proposition for farmers to buy or hire equipment to overcome labour-constraints during the period when late operations on one crop overlap with ploughing for the next season. Such equipment, once purchased or available for hire, can then be cheaply used to displace labour at other times of year. This may explain much of the apparent fall in the gains to labour from MVs [Jayasuriya *et al.*, 1981] – for example the tendency, in the double-cropped wheat-rice system in Western Uttar Pradesh in India, for the fall in labour-inputs per acre due to mechanization to pull ahead of the rise due to MVs [Joshi *et al.*, 1981]. The research implication is to increase emphasis on very short-duration *second-season* crops, even at some cost in yield, if it can be established that such varieties are likelier to allow the economic continuation of labour-intensive harvesting, threshing and/or ploughing methods between the crop seasons.

Generally, however, MVs reduce fluctuations affecting seasonal labour, e.g. fluctuations of real wage-rates in the Indian Punjab [Dasgupta, 1977, p. 336]; of employment in the Philippines [Barlow *et al.*, 1983, p. 42]; and of earnings in Kanpur district, Uttar Pradesh [Singh and Kanwar, 1974, pp. 66, 84–5]. In Bangladesh, the growing importance of MV *boro* rice and wheat has plainly reduced seasonal fluctuations in labour use. This is the general pattern, with MV sorghum, mainly a kharif crop, an exception [Rao, 1982].

*   *   *

What do MVs do to *year-to-year* fluctuations in poverty among labourers? We have seen (Chapter 3, h) that MVs probably raise the coefficient of variation of *national*-level output of 'their' crops, but nevertheless normally raise absolute output levels for most adopting *farms and villages* even in a bad year. In such a year, therefore, farm employment is normally raised by MVs, even if by less than in a good year – unless labour-saving changes, introduced by farmers in response to MVs, bite hardest in bad years. This is unlikely, since the main such changes are in harvesting and threshing; these activities are less important in bad years, when they are heavily reduced irrespective of technology. Well before farm employers can realize that the year will be bad climatically, they will usually have taken on hired workers for most operations, so that the associated income for the poor has mostly been generated – in activities such as planting and water management, which are less affected by possible labour-saving pressures in the wake of MVs than are the post-harvest operations. The post-MV shift (pp. 189-90) towards longer-term contract workers, moreover, may somewhat reduce employers' scope for laying off harvest and post-harvest workers in bad years.

An interesting analysis from semi-arid South India [Walker *et al.*, 1983, p. 21] suggests that direct year-to-year stabilization of output (via MVs or otherwise) would be an inefficient way to reduce poverty, especially among labourers. Perfect stabilization of crop labour income, generated by crop output, over the five crop years analysed, would have reduced the average landless household's variability of total income by zero, 5 per cent and 5 per cent in the three villages, as compared to 34 per cent, 20 per cent and 55 per cent respectively for perfect stabilization of labour income. However, stabilization of income from crops – and thus of local spending by farmers – would have indirectly stabilized part of labour income too. It is thus not clear that we can conclude that 'emphasizing crop income stability for small farmers [and landless] in India is a misguided means to an end' [*ibid.*, p. 36]. Subsequent research by Walker [1986, pers. comm.] in India's semi-arid regions, however, suggests that the issue of yield stability may have been overrated.

(f) MVs and the labouring poor: mechanization

There is very little doubt that most mechanization – not only four-wheel tractors [Binswanger, 1978], two-wheelers [Jayasuriya *et al.*, 1982] and threshers, but also the shift to more mechanized irrigation technologies [Joshi *et al.*, 1981] – is on balance labour-displacing. Animal care and animal ploughing (or hand-hoeing), use much more hours per acre than tractor care and tractor ploughing respectively [Farrington and Abeyratne, 1982]. In some cases, these labour-displacing effects of the shift to tractors are offset because the reduced herd size permits former pasture or fodder-crop land to be sown to cereals – usually to MVs, which would raise labour use per acre; however, there is little scope for such a shift in much of Asia, where most beasts are now stall-fed, or grazed on stubble, roadsides, etc.[12] Claims that tractors and threshers allow higher cropping intensity usually collapse when the inputs of MVs, fertilizers, and water are controlled for

[Jayasuriya *et al.*, 1982; Agarwal, 1980, 1984b]. Indeed, the labour-displacing effect of tractors may be more in double-cropped than in single-cropped systems, because replacement of animals is more complete [Cordova *et al.*, 1981]. Where mechanization pays in land-scarce economies, the reason is usually that it is cheaper than labour (even at a subsistence wage) – generally because of subsidies to fuel or credit – and not that it directly raises farm output [Binswanger, 1978]. Most analysts concur, however, that 'labour-saving effects of mechanization have been more than neutralized by the labour-increasing effects' of MVs [Dasgupta, 1977, pp. 323–4].

However, there is major controversy over whether MVs and associated factors have caused not only these effects, but also the 'very high and significant correlation between tractor use' and MV adoption [*ibid.*, pp. 96–7]. Some analysts explicitly assert that 'there is no sign that tractor adoption was accelerated by the dramatic diffusion of MVs' [Hayami, 1984, pp. 393–4; compare Hayami, 1981, p. 174; Kikuchi and Hayami, 1983]. Others speak of MVs as embedded in 'essentially a "package" of technical improvements including. . . tractors' [Gibbons *et al.*, n.d.], or claim that the inducement to double-cropping, provided by MVs, intensifies the pressure to mechanize [Byres, 1981]. A balanced account [Barker and Herdt, 1984, pp. 85–7] reveals the environmental specificity of the tractor–MV relationship – strong in the Philippines and Malaysia, modest in India, Indonesia and Thailand, and with tractors arriving first in Pakistan.

How might MVs be linked to labour-displacing machines? Does IARC research help to forge that link, or to weaken it? In what follows, we concentrate on tractorization, but most of the argument applies *mutatis mutandis* for other forms of mechanization.

A very low-yielding farm cannot afford to tractorize. The hourly costs of tractors in many parts of Africa would exceed value-added on the land that is ploughed in an hour. Even at somewhat higher yields, fuel and hire costs at normal rates can exceed the value, to the farmer, of savings on ploughing labour. By greatly raising yields, an appropriate MV often renders tractorization and threshers financially *feasible*.

But why should it make them economically *profitable?* The main reason advanced is double-cropping: first, because it enables the machines to be used for twice as long each year; and second, because they can permit timely sowing of the second crop. A subsidiary reason is that machines – purchased, or more commonly hired, profitably for the peak period (harvesting and threshing the wet-season crop, planting the second-season crop) that MVs have sharpened or even created – are then available cheaply in other seasons, and undercut labour there also. However, such motives to mechanize would often not suffice unless buttressed by subsidies to labour-replacing equipment [Binswanger, 1978; David, 1982; Gill, 1981; Farrington and Abeyratne, 1982; Jayasuriya and Shand, 1986]. A hidden form of such subsidies is research to cut operating costs with machines, not by the firms which make them (that is plainly a legitimate commercial activity), but by public-sector research agencies. IARC work, e.g. at IRRI, has sought to reduce the cost to farmers of mechanized operations, and in particular to develop and test mechanized threshers and reapers. The reaper was judged to be 'a highly profitable investment', reducing harvesting labour by some 80 per cent [*ibid.*, citing Moran, 1982].

It is possible to specify some conditions in which such research, implicitly subsidizing mechanization, might contribute to GNP and/or the welfare of poor people in LDCs, sometimes even through higher employment income. Sometimes, tractors introduced alongside MVs clear new land, which would otherwise have stood idle and employed nobody; some such land now produces hybrid maize in Zimbabwe and Zambia.[13] Sometimes – probably far less often than claimed [Farrington and Abeyratne, 1982] – tractors or threshers can 'make time' to improve water management, which is highly complementary with MV use and with double-cropping. However, such effects are unusual. In general – for the increasing majority of LDC farming areas where expansion of cultivated areas is infeasible, uneconomic, or not best done with farming tractors (footnote 13) – research to reduce costs of mechanization will displace poor workers with few prospects of other work, and will not raise yields, farm output, or GNP. As enthusiastic supporters of the IARC system, we see

no useful role for such research in it. It has been su
[Mueller and von Oppen, 1985] that apparently justi
caution to avoid labour displacement has created *seasonal*
labour bottlenecks impeding new semi-arid farm systems that
would otherwise have boosted *year-round* employment;
however, recent work suggests [Walker, pers. comm.] that
such bottlenecks have not impeded the adoption, where
otherwise appropriate, of new vertisol farming systems.[14]

   In general, as person/land ratios increase, the case against
subsidized research that displaces labour becomes steadily
more overwhelming. Admittedly, sometimes 'effort-saving'
innovations might reduce drudgery or human energy-costs
without reducing employee time much. And sometimes fairly
egalitarian access to land rights is secure enough to guarantee
that farmers, if they hire or buy labour-displacing machinery,
save their own labour, not employees' – i.e., reveal a preference for rest over income, and do not destroy employee
livelihoods. Such things are possible, but rare: increasingly so
in view of current advances in population, individualization of
land tenure, and 'Northern' cost-saving research on machinery. It is almost certain that, for example, the IRRI threshers,
reapers, and 'IRRI-PAK Mechanization Programmes' harm
poor workers (though of course far less than IRRI's new
varieties have helped them). Just as IARCs already breed and
time MVs to 'avoid' periods of disease or moisture stress, so
they now need to breed and time MVs, and (as at ICRISAT
and IITA) farming systems, to avoid creating incentives for
labour-displacing mechanization. The growing proportion of
the world's poorest rural people who derive their precarious
livelihoods mainly from hired labour, together with the spread
of MVs to new areas and crops, adds urgency to the need for
this reorientation.

(g) Migration and MVs

A major alternative to mechanization for easing seasonal
labour peaks is seasonal migration, especially in countries with
good transport between rural areas that have distinct peak
seasons. Usually workers readily return to their homes after

. different peaks there. On the other
. if acquired mainly to ease genuine
.our and thus to permit double-cropping
.n hand in slack seasons, and is used at
.ıs displacing already underemployed labour.
.ant labour responded to MV options? Can
.se the likelihood that farm employers will meet
.our shortages by recruiting migrants, rather than
.ing labour-displacing machinery?

.ı. .e is much quantitative evidence of seasonal migration
to MV areas [Dasgupta, 1977, p. 326; Kikuchi and Hayami,
1983]. Even where this evidence is less quantitative, it is made
more credible by being often given by commentators who
would rather not see what they have to report – e.g. by
commentators who fear that outsiders in these jobs are easier
for employers to manipulate [van Schendel, 1981]. In fact such
migration is a powerful way to spread the gain from a new
variety to the neglected regions. This is not only because their
workers obtain extra income away from home. Also, MVs in
advanced regions strengthen workers in backward regions
unaffected by the MVs – by reducing labour supply there,
when migrants from backward regions move to MV areas
instead of to other backward regions with different seasonal
peaks [*ibid.*, pp. 230, 142–3].

In assessing MVs' capacity to generate migrant incomes for
poor workers we must recall that it is unequal, progressive, and
food-surplus villages, whether in MV areas or not, that are the
most likely to produce two sorts of migrant: the moderately
well-off, 'pulled' towards known urban work or education; and
the poor, usually 'pushed' out of land or employment, and
searching for seasonal farm work [Connell *et al.*, 1976]. This
applies whether such villages are in an MV area or not, but
they are likely to be. Thus MVs impinge on villages that
initially have high propensities to emigrate. A study in a
Punjab village suggests that MV-based development increases
emigration for the better-off 'pull migrants' (presumably by
enriching them, so they can buy urban contacts and be pulled
out), but reduces it for the poorer potential migrants who are
rendered less likely to be pushed out [Leaf, 1983, p. 268],
presumably thanks to better post-MV employment prospects

in the home village. Such reduced emigration from MV villages by poor workers would reduce the prospects for seasonal immigration to them by workers from other, poorer regions. However, in a lowland rice village in Laguna, Philippines, the spread of rice MVs in 1961–80 was linked to substantially increased immigration by landless workers from poorer, non-MV upland areas – though also, as in the Punjabi village studied by Leaf, to a switch towards net emigration, on a substantial scale, by farmers and 'white-collar workers' [Kikuchi and Hayami, 1983, p. 4]. The Punjab's MVs have attracted many seasonal immigrants from remote, poor areas of Bihar and Eastern Uttar Pradesh, although not enough to prevent a considerable rise in *seasonal peak* real wage rates in the early 1970s, which may have accelerated labour-displacing mechanization later.

Mechanization in the wake of MVs, indeed, can induce net emigration by the poor. In the lead district of the Punjab, Ludhiana, emigration has exceeded immigration plus return migration, and the proportion of net emigrants comprising scheduled-caste and low-caste people has gradually increased [Oberai and Singh, 1980]. In lowland Laguna, despite 'a new contractual arrangement [*gama* that] reduced risks in finding employment. . . thereby facilitating intra-rural labour flow', the growing use of threshers (and herbicides) meant that wet-season use of labour-per-hectare – after rising from 86 to 112 days in 1966–75, as MV adopters rose from zero to all farmers – fell back to 93 days by 1981. This effect, by which mechanization outweighed migration, also reduced the proportion of all labour that was hired – also following a rise from 60 per cent in 1966 to 80 per cent in 1975 – back to 72 per cent in 1981 [Kikuchi and Hayami, 1983, pp. 3, 6–7]. These trends must threaten to reverse the rising capacity, in 1971–80, of the area to support net labour immigrants.

IARCs need to know more about the impact of alternative crops, varieties, linked inputs, and farm systems on poor people's migration, and about its impact, in turn, on wage-rates, incentives to mechanize, and subsequent employment and labour-income. If there is no effective land reform, an MV system may spread income *lastingly* to labour only if it generates sufficiently long, spread-out employment peaks to attract enough immigration to prevent a large rise in real wage-rates seasonally, and thus to maintain the incentive to employ. Paradoxically, in view of hostilities between immigrant and local labourers, the latter can probably gain longer-run employment in threshing and ploughing seasons from MVs only if the former arrive in sufficient numbers to moderate real wage-push in those seasons.

(h) MVs and labour: conclusions and omissions

Year-round, MVs raise labour demand, employment and real wage-bills. However, the real wage-rate in most MV areas shows little long-term uptrend, and the wage share in farm income falls, because there is plentiful labour but scarce land. Therefore, while MVs raise the demand for labour as well as land, it is only the latter whose price is substantially bid up: rent/wage ratios rise. So does the part of farm income that goes to producers of fertilizer and other purchased inputs – again, at the expense of labour's share more than of land's.

MVs spread labour better across the *seasons*, by permitting double-cropping. However, end-of-season peaking of extra MV labour requirements, especially the peak from main-crop harvest through second-crop weeding time, can have two effects. It can pull in seasonal migrants, sharing the benefits, and moderating even peak-season increases in the real wage-rate. Or it can first pull up local wage-rates in the peak season, but thereby stimulate the hire or purchase of machines: tractors, threshers, mills, and transplanters. These machines not only flatten the seasonal labour peak. Less happily, they are also available to displace very poor slack-season workers, at low marginal cost to employers.

The *year-round* rise in income that MVs bring to poor rural employees, some of them migrants from backward areas, is partly to the IARCs' credit, although the rise has been hidden by other factors, such as population growth and cost-cutting 'Northern' research on labour-displacing equipment and inputs. Unfortunately, several IARCs have lost some of that credit, and may have set back labour income. Directly, they have sometimes pursued research policies, especially in agricultural engineering but also to some extent on weed control, that encourage the labour-displacing rather than the migrant-employing response to *seasonal* MV labour peaks. Indirectly, the IARCs have not sufficiently researched the crop options that might make it more attractive, even in peaks, for farmers to employ rather than to mechanize. Also, research into the types and directions of migration, as it is affected by MVs, as between rich and poor rural areas (and between both and towns), should have some place in the IARC's economics research agenda. We understand that IRRI is starting a major project of this type.

Finally, three points need emphasis. First, despite MVs' possible link with *ultimately* labour-displacing inputs in some places and times, MVs help labour much more on balance, by *initially* raising demand for it (hence the link!), and by making food cheaper. These initial positive effects are still affecting new areas – under hybrid maize in Africa and hybrid sorghum in India, for example.

Second, although 70 per cent of Asians and 80 per cent of Africans live mainly from agriculture, labour displacement is a 'problem' only because land rights are unequal, so that many agriculturalists must get most of their living as employees of others. If a big farmer, or a collective of owners, displaces assetless workers with machines or weedicides to break a seasonal bottleneck,[15] then *the poorest compulsorily lose their income source* and must look, in hope, for others. If family farms or co-operatives/collectives (embracing all village cultivators, so that none must live mainly off hired farm work) decide to hire a tractor or buy weedicide, they *voluntarily take their own welfare as leisure instead of labour income*. In such a case the labour-displacing input – if fairly priced, and not subsidized at the expense of poor people's taxes – is a solution, not a problem.

Third, we have discussed mainly the effects on farm labour of MVs – and chiefly in areas where MVs have spread. The greater the prospects of off-farm employment, the less important are such effects. Even in traditional villages, about a quarter of working time and about a third of income are non-agricultural [Chuta and Liedholm, 1979]. Yet there is almost no research into the effect on such working time and income of MVs or of farmers' reactions to them. Even more seriously –

except for a handful of studies of migration – there is no work on the effects of the spread of MVs upon labourers in areas where they cannot be grown. Perhaps the competition, from lower-cost crop production in MV areas, leads farmers elsewhere to lay off employees from cereals production for sale. On the other side of the ledger, food gets cheaper for these poor net food buyers; and a shift in non-MV areas to crops abandoned by MV growers, or new MVs in other cereals or roots, may create employment in Type IV or III regions respectively (Chapter 3, i).

Labour displacement is a problem to the extent that there are population pressures, undiversified farm employment, and unequal land (no land reform). Into these three realities the IARCs' research, and the MVs, must fit. Such fitting is a matter partly of tailoring MV research, releases, and systems to help labourers and small farmers within those realities. Partly, however, it is a matter of researching how *different* land distributions, work diversifications [Bell *et al.*, 1982; Hazell and Roell, 1983], and person/land ratios could interact with the pay-offs from MVs to the labouring poor.

Ultimately, this could involve IARCs in some difficult tasks of 'speaking truth to power' about land reform and population policy. However, the first task is to uncover that 'truth'. At present IARCs and other researchers know little about how MVs affect, or are affected by, (i) rural population growth or (ii) distribution of land. These interactions will affect poverty mainly via their impact on landless and near-landless rural workers; for these workers comprise a rapidly rising proportion of the world's poor. As the next chapter will show, more food alone – even if it restrains retail prices – has rather limited power to help such people. More income, either from asset redistribution or (usually more plausibly) from extra paid employment, will be required as well.

Notes and references

1 Not necessarily the same farmers, or the same foods for any given farmer; the IARCs' mandate was and is certainly not mainly (indeed was and is not sufficiently) addressed to farmers' production for on-farm, household consumption. However, the poverty-

and-hunger problem was perceived as mainly one of inadequate output by, and therefore inadequate income and consumption for, 'small farm' households.

2 Although this chapter concentrates on poor people's employment incomes in the wake of MVs, it is possible that the most cost-effective policies – or research – to help the poor, even if they increasingly depend for income mainly on hired labour, could still involve raising their holdings of land, other assets, or skills, and/or the productivity of their remaining tiny holdings. This would require redirection of IARC or other MV research towards farming units so small that they accounted for below half household income; towards the impact of labourers' skills on their own and fellow-labourers' incomes; and towards production effects, alongside MVs, of alternative distributions of land, skills, and other assets.

3 First, on a given land area sown to the MV crop, the MV normally raises output per acre more than it raises output per person-hour in each season, and also tends to increase cropping intensity – the proportion of that land area on which (double-cropping) pays. Second, the MV also makes it pay for farmers to shift land towards the MV crop, typically from less labour-using crops. Third, if before MVs it almost paid to bring new land into cultivation (or, via irrigation, into double-cropping), MVs will probably make some of this activity pay.

4 In fact, the work is about the response of family farms to higher incomes due to better farm-gate prices [Barnum and Squire, 1979; Singh, Squire and Strauss, 1986], but the conclusion applies just as well if the income rise has another origin.

5 Real rural unskilled wage rates, except seasonally in the short run, tend to be little affected by either the labour-using effects or the offsets because of the high long-run elasticity of labour supply, and its growth alongside population. These wage rates tend to stay close to subsistence levels until rural population starts to fall, well on into the development process.

6 Abundant evidence is available; see, for instance, Berry and Cline [1979].

7 On the assumption that employment is first sought by those who need it most, and first offered to those who do it best, leaving others to provide extra work as MVs increase demand for labour.

8 By this we mean that the real wage rate is free to vary in a *subsistence band*, around a sort of 'family historical and moral subsistence equivalent' close to the often-measured norm: a 'subsistence wage' of roughly 3 kgs. of grain per man-day [Braudel, 1981, p. 134], between, say, 2¹/₂ and 3¹/₂ kg. as shown in the following diagram.

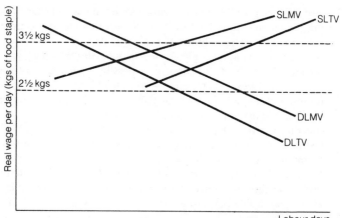

MVs require more labour per acre. So they raise the demand for
hired labour, from D^L_{TV} towards D^L_{MV1}, as adoption spreads. As
farmers divert labour supply from hired-labour markets towards
their own farms (also to meet the higher labour requirements),
they are willing to supply less labour to the market at any given
wage. So the supply of labour also shifts, from S^L_{TV} towards S^L_{MV}.
Real wages therefore increase within the 'subsistence band' (move
from bottom to top of the diamond within the dotted lines).

9　Assuming that the share of labour in net income from purchased
farm inputs is much less than in net farm income; this is
reasonable, since (i) most agrochemicals are capital-intensive, (ii) a
significant proportion, by value, is in most LDCs imported from
developed countries.

10　Evidence for the propositions in the last two paragraphs is
summarized in Lipton, 1984a.

11　To be precise, infants undergoing the transition between passive
(maternal) immunity and active immunity – i.e. aged 4 to 12
months – are especially likely to show increased death-rates
during hungry times [Schofield, 1974].

12　Draught animals, even if displaced by tractors for ploughing, are
often retained for other operations (water-buffaloes for puddling
paddies and weed control in Sri Lanka; oxen for farm and market
freight in parts of the Indian Punjab). In these cases, the labour-
displacing effects of the reduced demand for animal care are
reduced – but so are any employment-creating effects of a shift
from pasture or fodder crops to cereals.

13　Note, however, that it is often quite infeasible, and usually
economically sub-optimal, to use the same sort of machinery for
land clearance as for ploughing in subsequent seasons.

14 Seasonal labour bottlenecks would at worst comprise only one of the reasons why such systems were not readily adopted. Seasonality in the improved systems did not greatly exceed that in traditional systems in the three main test locations [Ghodake and Kshirsagar, 1983, Fig. 2]. Earlier estimates of seasonality were exaggerated because analysts assumed that all farmers would plant the same improved cropping systems [Walker, 1986, pers. comm.].

15 Examples are the adoption of combine harvesters alongside MV-linked double-cropping: by large private farms (in the wake of double-cropping with wheat and teff MVs) in Chilalo, Ethiopia; and by co-operative/collective farmers in Chile in 1971–3. In both cases, the landless poorest lost work and income.

5. Modern Varieties and the Poor: Consumption and Nutrition

(a) Malthusian optimism, entitlements, revisionism, and MVs

The poorest 10–15 per cent of households in low-income countries are so undernourished that their risks of death and disease are substantially increased. Perhaps another 15–25 per cent are quite often painfully hungry; some of them may be impeded by undernutrition (among other things) from full economic functioning[1] [Reutlinger, 1986; Lipton, 1983]. These poor people are at nutritional risk, yet they use 65–80 per cent of their income[2] to obtain food, and similar proportions of working time to produce and prepare it. It *looks* self-evident that MVs help the poor mainly because better cereal seeds grow more food. This should mean more food produced per hour worked – and more food to buy with each hour's earnings. If the hungry do not lose their jobs in the process, this seems clearly good for their command over food. Apart from these static appearances, neo-Malthusian dynamics suggests that growing populations can be fed, from constant (or, at the margin, increasing but qualitatively deteriorating)[3] land areas, only if new technologies such as MVs improve the efficiency of land use.[4]

The history of international agricultural research since the 1950s suggests that this half-true argument for 'more food from MVs' as the cure for inadequate food consumption is deeply entrenched. If we feel the need to modify it, we must understand that history; for the population argument was, from the start, at the centre of the IARCs' efforts (Chapter 7, c).[5] The celebrated Ford Foundation report [Ford Foundation, 1959] saw 'India's foodgrain crisis' as one of continued population expansion confronting the limits of land expansion, and creating a crisis of food consumption; 'steps to meet it' were to be concentrated on raising food production, largely through accelerated technical progress by 'progressive farmers' – not the poorest, as a rule – in lead districts [Brown, 1971].

This view of the problem and its solution extended far beyond India. It was the 'neo-Malthusian optimism' [Sen, 1986] behind the foundation and expansion of IRRI and CIMMYT. For their backers in the Ford and Rockefeller Foundations, and later in the aid donor nations, the statics of poverty were mainly about inadequate food output per person (in countries with limited capacity to import food);[6] the dynamics of poverty were mainly about Malthusian worsening in person/land ratios; and the medium-term cure for poverty lay in redressing to some extent the bias of policy against rural areas (a radical proposition in the 1960s), but mainly by technical means: MV technology, in irrigated lead areas, was to make much more home-grown food available to improve local consumption and nutrition. This was to provide a medium-term breathing-space, during which the demographic transition could be largely completed.[7]

The strategy did produce the extra output to 'keep ahead of population growth' in a South Asia that, after about 1963, had more or less run out of good spare arable land. Yet the prevalence of undernutrition and rural poverty hardly lessened. We nevertheless describe the arguments of the previous paragraphs as 'half true', for reasons that will now be clarified. Many would call them plain false. Most famines (and probably most cases of chronic hunger too) are caused, not by declining local availability of food, but by the exposure of at-risk groups to 'failures of food entitlements': declining availability of paid work; increasing food prices; reduced non-market claims on food;[8] or some combination of these [Sen, 1981, 1986; Ravallion, 1987]. Most chronic undernutrition is probably caused in large part by insufficient 'entitlements'; conversely, the rise in Indian food availability per person in the wake of MVs in 1965–80 was not accompanied by any substantial reduction in the incidence or severity of under-nutrition or poverty (i.e. in poor people's incapacity to afford, or claim, adequate food) – although the Government accumulated some 20–30 million tons of grain stocks.[9]

So there seems to be a major objection to the neo-Malthusian optimism that originally animated the 'technology first' strategy of the IARCs. The objection is that growing more food need not put 'entitlements' to it in the hands of the needy – nor, therefore, more food in their mouths. Initially, the IARC ethos did downplay the entitlements problem, with the early emphasis on 'progressive' (normally, not small) farmers in reliably irrigated lead districts [ADB, 1969]. Yet a case can be made for this. Extra food, given the political difficulties of effective redistribution, *was* needed to reduce hunger, though it was not enough to do so alone. And any new MV of tropical wheat and rice MVs, in the experimental phase, was perhaps best tried out in conditions where the chance of success was best – and on farmers who, if it failed, would not be destroyed by the losses.

By the early 1970s – partly due to the critics of the early MVs – most IARCs were gradually revising their approach. MV technology did generate more food production in many poor countries; this was still seen by IARCs as *necessary* to attack inadequate consumption, undernutrition and hunger; but, in order to be *sufficient*, neo-Malthusian technological optimism was seen by the IARCs' revisionists to require two amendments, to address the entitlements problem. First, more emphasis was placed on generating robust MVs, attractive to risk-averse small farmers. Second, IARCs increasingly stressed MVs' role (through the means of increasing locally available food output) in restraining food prices.

The revisionist claims that MVs in fact do these things are, in retrospect, broadly accurate. Unfortunately, these achievements have not produced as much improvement, in the entitlements of poor net food buyers, as expected. We shall argue that the neo-Malthusian thrust towards technical progress *is* half the truth, not a plain error; that the IARC revisionists' two entitlements-orientated amendments do perhaps turn it into 60 per cent of the truth; but that, to make a major dent in undernutrition and low consumption among the poor, the incomes or other claims of landless and near-landless labour must be addressed directly.

The first success of MVs, that should apparently have greatly improved poor people's consumption and nutrition, is their undoubted penetration to small farmers (Chapter 3). However – and only in small part due to MVs – 'small farmers in areas suitable for MVs' show less and less overlap with 'poor people at nutritional risk'. These (i) reside mainly in places little affected by MVs – including most of Africa, and/or (ii) derive their income (and hence most of their entitlements to food)[10] mainly from labour, not from managing 'small farms' that could directly gain from MVs. This increasing labour-dependence of poor people's food consumption is itself linked to population growth;[11] its threat to poor people thus not only transcends the crudely Malthusian (so that it cannot be met *only* by measures of 'neo-Malthusian optimism' [Sen, 1986] that raise food production per unit of land), but is increasingly a threat to non-operators of land (so that it does not even suffice to direct 'neo-Malthusian optimism' towards small farmers). Less and less are poor people reliant for consumption and nutrition directly, as 'small farmers', on land in areas transformable by MVs.

Moreover, evidence accumulates that undernutrition that endangers health – as opposed to extremely unpleasant and undesirable hunger – is almost entirely confined to 'ultra-poor' households, receiving below about 80 per cent of dietary energy 'requirements' although devoting about 80 per cent of income to food [Lipton, 1983]. These households comprise mainly landless or near-landless rural workers in South and East Asia; mainly low-income employees and the 'informal sector' in Latin America; and mainly small farmers only in parts of Africa and perhaps some semi-arid parts of India. These 'parts' became smaller and smaller as population presses on land, and forces farmers' younger sons and their wives to rely for income increasingly on hired employment. 'Food for small farmers', as against 'entitlements for labourers', therefore addresses a small (though still significant) and dwindling part of the problem of inadequate food consumption.

What of the other revisionist amendment of the IARCs' pro-poor perspectives, to modify neo-Malthusian optimism to allow for the entitlements problem: the argument that extra supply of cereals based on MV output, even if not mainly controlled by the poor, helps them by restraining the increase in food prices? This seems to respond directly to the evidence that most famine (and hunger) is caused mainly by failures of entitlements: if extra MV-based food supplies restrain food prices, thereby permitting a given real income to buy more food, surely poor people's entitlements rise automatically?[12] If so, the poorest should gain most from food price restraint. They spend a larger part of income than the rich; of that spending, a larger proportion is used to buy food; more of that food spending must be devoted to cheap carbohydrates, i.e. to the cereals and starchy roots affected by MVs; and, within such items, the poorest are the people likeliest to select products based on the usually coarser MVs, and therefore standing at a further price discount.

Once again, though, we come up against the problem that poor people's incomes (and hence entitlements) depend increasingly on hiring out their unskilled labour. This, except in the very short run, is in very 'real-wage-elastic supply'. A deceleration of food prices (e.g. due to extra supply of food from MVs), by raising the value of labour income, enables or induces many more poor and unskilled workers to come forward and compete for employment. Hence real wage-rates are not automatically improved by restraint of food prices; employers can correspondingly restrain the money value of the wage they pay (whether cash or kind), and can find as many workers willing to work, at about the same real wage-rate as before. We call this effect responsive money-wage deceleration.

Notice that, once again, it is population-linked effects – here as they affect the supply of workers, which is increasing at 2–3$\frac{1}{2}$ per cent a year in the rural Third World – that, ironically, destroy neo-Malthusian optimism even as amended by revisionist entitlements considerations. Just as population growth has helped to make growing proportions of the poor dependent on labour incomes rather than on MV farms direct, so it has helped to create a 'reserve army' of under-employed adults, Malthusian rather than Marxian. The former population effect limits poor people's consumption gains from extra income in cultivating MVs even on small farms. The latter effect, the growing reserve army – by permitting 'responsive money-wage deceleration' – reduces the prospects that extra MV-based food availability can help the poor by holding food prices in check. Only MV technology directly available to the poor – either because unavoidably labour-intensive (yet profitable), or because concentrated on crops (or areas or assets) that remain in the control of the poor – is, in our judgement, likely to lastingly overcome the 'population threats' to poor people's food entitlements.

(b) More food, lower prices, and labour's claims

For poor people not mainly dependent on either food sales or employment income – or facing employers who cannot responsively decelerate money-wage rates – the effect of MVs in moderating, and (by building up stocks) in stabilizing, food prices is important, both to improve nutrition and to free incomes for non-food consumption. Poor people need this price effect even more because several non-MV factors drive up food prices, both absolutely and relatively to other prices and to poor people's incomes, in developing countries; and several things – transport costs, protection, food tastes – partly de-link their domestic food prices from world prices. For example, in most countries of Africa in the 1970s and early 1980s, despite real falls in *world* food prices, national factors meant that *domestic* food prices to consumers outpaced other elements of the cost of living [Ghai and Smith, 1987]. What were these factors?

First, population growth, at $2\frac{1}{2}$–$3\frac{1}{2}$ per cent yearly, raises demand for food and thus its price. Second, so does growth of income per person. Third, the changing patterns of food demand, as income grows among the better-off, encourage farmers to divert land towards 'richer people's foods', especially meat and dairy products, which need five to seven times as much land per person as cereals; hence the average cost of dietary calories comes to include more land, and therefore their price is pushed up, while 'poor people's foods' become relatively scarcer.

Fourth, population pressure renders land scarcer, and costlier to bring into production. In much of South and East Asia, and in increasingly many parts of Africa, there is hardly any 'plausibly cultivable' land left unfarmed. Rising costs of marginal land help drive food prices up.

But why do these four factors throw the onus of Third World price restraint onto domestic cost reductions in agriculture, for example by MVs? With world real food prices falling, couldn't LDCs rely on imports rather than MVs to restrain food prices for poor consumers? Alas, the famous food surpluses of Europe and North America are of limited use to poor LDC consumers. The surpluses have always been unreliable, and are increasingly used for feed in producer countries. They dwindle in times of greatest need, such as 1972–4. Often LDCs, facing foreign-exchange crisis, cannot afford commercial food imports, yet find that long-term dependence on food-aid imports undermines domestic incentives, probably to farmers, certainly to agricultural policymakers. Also, national transport and marketing structures in many African countries can be so weak that stockpiles in the capital city are no guarantee of food availability for the poor in remote areas. Finally, the poor must have the incomes to purchase that food: availability alone is no guarantee of adequate consumption.

Since 1973, moreover, a fifth factor has driven up food prices in LDCs, and has made food and fertilizer imports especially costly. Despite recent oil price falls, each barrel in 1987 cost LDC importers over three times as much of their typical export products as in 1972. The costs of both transport and fertilizer include large components of fossil fuels. The price increases in such fuels, therefore, greatly raise the cost to LDCs both of shipping foodgrains from the West, and of fertilizing them at home. Cereals have high weight per unit of value, and the major cereals exporters in North America and Australia are thousands of miles from most poor food-importing countries; both facts raise import costs per calorie from grain, and further reduce the scope for imported cereals to restrain food prices for poor consumers at nutritional risk.

Clearly, the huge post-war expansion of food output in North America and Europe – largely due to artificial subsidies to cereal, dairy and sugar farmers (and to the research induced by such subsidies) – has dragged relative world food prices down. But this probably, on balance, impoverishes rural people in poor countries. Nor, as indicated, can they rely on cheap imported food. It thus remains crucial, to poor and nutritionally vulnerable households in an LDC, whether and how much MVs have moderated *domestic* food prices – and whether any such moderation does, or does not, induce a corresponding slowdown of growth in poor people's money incomes. This depends on (i) what MVs have contributed to output in LDCs, (ii) whether that extra output has raised food availability and entitlements for consumers (as opposed, say, to reducing imports), (iii) the effect of the extra food availability, if any, in restraining food prices, and (iv) whether any such effects have been offset by countervailing restraints in money-wages, or (v) via policy or market effects – through trade, health, regional balance, or research management itself – affecting MVs' impact on consumption and nutrition among the rural poor. We look at these five issues below.

(i) The contribution of MVs *alone* to annual outputs of rice in Asian developing countries in the early 1980s has been estimated at 10–27 million metric tons,[13] and to wheat output in all LDCs at 7–21 million[14] [Pinstrup-Andersen and Hazell, 1984]. Rice MVs (with other inputs made profitable by them) increased rice production in Latin America in 1981 by 2.5–3 m.m.t. or about 20 per cent, valued at $854 million [CIAT, 1984]. Sorghum, millet and hybrid maize must add at least another 3–5 m.m.t. of grain. Extra output due to extra fertilizers and to other inputs induced by MVs probably raises these figures by at least 50 per cent.

(ii) Yet in India, despite extra food *output* due to MVs estimated at 5–7 m.m.t. of grain in 1970–1 [Rao, 1975, pp. 6–9], and surely over 12 million tons today, food *availability* in terms of dietary energy (calories) has barely outpaced population growth [FAO, 1984]. This is because almost all the extra MVs have been used to replace imports or to build up stocks. However, the entitlements of poor people to food usually

matter more than its availability in determining the incidence
and severity of famine as well as of chronic undernutrition
[Sen, 1981, 1986; Reutlinger, 1986]. Unfortunately, alongside
stagnant food availability per person, poorer people
(especially in non-MV rural areas) fell even further behind
India's growing average real income per person, and may not
have got much better off between the sample surveys of the
early 1960s and that of 1983.[15] Hence in India – and this is
even clearer in many developing countries with much less
explicit public-sector programmes against poverty – the rise in
MV-based food output has left poor people's consumption
and nutrition almost unchanged. Effective demands for food –
i.e. income-based entitlements – have not expanded to absorb
the extra output-per-person due to MVs; so it has been
allowed by governments to build stocks and to replace imports
instead, in order to avoid catastrophic collapses in farm
incentives.

(iii) Some economies have operated trade policies 'fixing'
net food imports (instead of allowing them to fall as domestic
output rose, as in India). There, the effect of 'extra MV
output' on domestic food prices and hence on consumption,
overall and for the poor, can be isolated. In Colombia,
households with below $600 in 1970 appear to have gained
12.8 per cent of income, because rice MVs grown in Colombia
restrained food prices [Scobie and Posada, 1978], while better-
off households gained proportionately much less. (This analysis assumes no responsive money-wage deceleration.) If the
extra rice due to MVs in South and East Asia had been added
to market supplies (as in Colombia) – instead of being used
largely to reduce net imports or to build up stocks – the new
IRRI varieties would have enriched South and East Asian
consumers by about $160 million in each year in 1972–5, plus a
further $400 m. yearly from new national rice varieties
[Evenson and Flores, 1978]. Assumptions about just how open
the economy is, and how much trade would have occurred
without MVs, are crucial in calculating such numbers [Munchik
de Rubinstein, 1984, p. 20].

Of course, producers initially lose some part of what
consumers gain from these price effects. That part is estimated
in the above studies at 50–60 per cent. However, this loss is

offset by the various factors increasing the demand for food, and by producers' ability to switch into other crops if prices of MV crops fall very fast relative to production costs. *Poor* producers may also 'internalize' many of the consumption gains by eating more of their own produce as MVs boost it, thereby reducing extra marketed supplies and shifting some rural benefits from pure consumers to poor growers [Hayami and Herdt, 1977]. Benefits to poor urban consumers as well as rural semi-subsistence producer-consumers are even greater when the MV affects 'inferior goods' like cassava meal in Brazil [Lynam and Pachico, 1982]. Even for costlier staples, however, a major benefit is possible. In the Philippines in 1982, a 7–8 per cent real rice price fall would have allowed the lowest income group to escape undernutrition [Gonzales and Regaldo, 1983]; since a 30% real fall was achieved in 1975–80 (because local MVs, although used largely to replace imports rather than to increase supply, reduced transport and storage costs), this seems feasible. Unfortunately, there are limits to this story of major consumer gains via the 'price effect' of MVs.

(iv) We have argued that, if extra food supply due to MVs restrains the price of food to the workforce, employers can respond by restraining the growth in the money value of the wage-rate – indeed, if their competitors can so respond, they must do so too – leaving the real wage-rate (and therefore food consumption) much less improved: responsive money-wage deceleration. Papanek [1986] has estimated that, in India, 'over a two-year period, nominal wages. . . fully adjust to the price changes resulting in constant real wages', despite price restraint if extra grain supply (from MVs) restrains prices [de Janvry and Subbarao, 1987, p. 1003]. We judge that this *total* fade-out, or transitoriness, of consumption benefits to the employed poor (except via higher employment at the constant real wage), is implausible; other evidence has linked food price restraint in a region, to a decelerating, not to a falling, money wage-rate [Agarwal, 1984].[16] The extent of this responsive money-wage deceleration – and hence of the erosion of poor employees' consumption gains, when MVs raise food supply and thus cut prices – depends not only on the wage-elasticity of labour supply, but also on the nature of the wage. (1) Fixed food wages, as part or all of a day's pay, contribute *less* (relative to cash wages) to poor people's consumption as MVs drive down food prices, especially if MV cereals (usually 10–20 per cent cheaper) are paid instead of the costlier TVs. (2) Wage shares of the harvest, still common in many areas (e.g. one sheaf out of twenty for each of several labourers), are worth *more* than fixed wages, as MVs increase harvests.

However, 'the nature of the wage' – while often rigid in the short run, until more workers move into the MV area to compete for new work – is in the longer run flexible. To the extent that workers are paid with food *and* cash, as food prices drop and new workers immigrate, employers switch from cash to food as a wage. Whole wage systems can change, in effect reducing a 'rigid' harvest share indirectly by employers' insistence on provision of weeding labour by harvest workers [Kikuchi and Hayami, 1983] – again, if new workers come into an area (or if established workers offer more labour-time).

Whether or not cheaper food is a lasting benefit to the labouring poor (i.e. is *not* offset by wage-restraint) – even if the outcome is mediated by superficially rigid payments systems – thus depends on the wage elasticity of labour supply. In the short run, this may not be high [Berry and Sabot, 1981, p. 153]. In the long run, however – at least in those LDCs where populations are growing rapidly, and where there is increasing unemployment among unskilled (poor) labourers already – it is, [Lipton, 1984a] so that the longer-run benefits from MVs in restraining food prices are mostly passed on from producers not *to* poor consumers, but *via* them to employers. None of the cited estimates of 'poor people's gains from MV price effects' allows for this crucial possibility.

(v) It is not only such 'responsive money-wage deceleration' that erodes consumers' gains from food price restraint as MVs raise food supply. How will governments respond if prices seem likely to fall? In the early years of MV-based farm growth, early big-farm innovators and input suppliers get most extra income, and spend little of it on cereals. They do shift demand from grain to costlier animal products as their income grows; and some extra early MV incomes do go to rural labourers, who spend it mainly on extra cereals. However, on the whole, in early MV-based growth, the poor gain too little money income to create enough extra demand for food to absorb the extra MV-based cereal supplies without downward pressure on domestic prices. How, then, can a government retain incentives for farmers to maintain delivery townwards of the major urban staples, so as to provide wage-goods for industrialization? It will probably prefer that – where the purchasing power of poor people is not increasing fast enough to mop up big farmers' food surpluses without price declines – the prices *to larger, surplus farmers* for *critical urban staples* shall be maintained, often by import restraint or even export promotion. Such policies have been used to maintain surplus farmers' incentives (and urban supplies) for wheat in India, maize in Zimbabwe, and rice in Nigeria, while incentives for crops largely grown and eaten by the rural poor – notably sorghum, millet, and cassava – have been allowed to fall. If, as in India, the price of the favoured crop (in this case wheat) is maintained by building stocks for emergencies and for schemes of nutritional support – food-for-work, infant feeding, etc. – the net impact on nutrition may be good;[17] otherwise, the price gains to poor consumers of the MV crops are largely destroyed by such policies on foreign trade in grains.[18]

A second source of erosion of poor consumers' gains from MVs is that – although 'small farmers' in MV areas may at last be sharing significantly in MV gains – very poor farmers and labourers elsewhere, while benefiting if they are net food buyers and MVs induce food price restraint, lose consumption power as work opportunities are reduced because their home areas cannot compete against the lower average costs of MV production expansion. Such effects are perhaps not so impor-tant in middle-income countries, for example in Colombia, where there are many more poor urban consumers, who gain when food prices fall, than poor upland rice farmers, who seldom get MVs yet suffer from the price decline as other farmers sell MV outputs [Scobie and Posada, 1978]. However, in low-income countries such as India, such lost consumption power by poor cereal growers and their employees in non-MV areas looms much larger, and there are relatively fewer urban poor to enjoy offsetting gains. Millions of not very mobile rural poor in Rajasthan and Madhya Pradesh are selling small wheat surpluses (often, probably, to buy cheaper cereals) at prices 'undermined' by the burgeoning MV wheat surpluses of the Punjab. IARCs should help to fill the vacuum in empirical research on MVs' impact on poor people in non-MV rural areas, new crops in the 'crop remixing' Type IV regions (Chapter 3, i) may compensate for any decline in wheat and rice income.[19]

Third, in the MV areas, small farmers' and labourers' extra incomes – and hence gains to consumption and nutrition from MVs – can be offset by associated health costs. If MVs stimulate the spread of gravity-flow surface irrigation systems, there are dangers to health, mainly from schistosomiasis in Africa and from malaria in Asia [Goldsmith and Hildyard, 1984]. The latter – and other insect-borne diseases – can also be worsened by the perceived need to use, on sensitive varieties of cereals (and especially if the same crop is farmed two or three times a year), amounts and types of insecticide that steadily increase as pest resistance builds up; this 'pesticide treadmill' often causes mosquitoes and other insect vectors of disease to 'select' resistant strains [Bull, 1982]. With proper research planning, such side-effects of MVs are avoidable, indeed reversible: appropriately pest-resistant MVs, especially if screened for storage pest resistance,[20] can *reduce* pesticide requirements; standard guidelines exist to design irrigation schemes that reduce, instead of increasing, health hazards [Lipton and de Kadt, 1987]. Unfortunately, in default of proper integration of policies for agriculture and for health, neglect of such options can remove some or all of the health-nutrition gains from MVs to poor rural consumers, especially if extra dietary energy for under-fives is diverted to fight increased numbers of parasites.

Fourth, MV nutrition research strategy has unduly concentrated on protein and amino-acid balance and on food 'quality' (Chapter 5, h). As we shall see, this too has eroded the central advantage of MVs to the poor consumer: the provision of his or her greatest need, cheaper calories.

(c) The nutritional background

The development of MVs has nevertheless been the main means of moderating food prices – helping the poor as food *consumers*, since responsive money-wage deceleration (pp. 219–20) was not complete, not immediate, and did not apply to all the incomes of all the poor. Also, the negative effects on *farmers* (and hence on their *employees*) of MV-induced restraint in the price of their food output, and of the 'cost-pull' of MVs on the prices of fertilizers, and other inputs, have in MV areas usually been outweighed by MV-induced rises in the conversion efficiency of inputs into outputs (Chapter 4, a). Therefore, there also usually have been net gains to adopting poor producers in MV regions.

Both as members of the farm economy and as food buyers, therefore, poor people in MV areas of LDCs have usually been helped by MVs to improve their incomes, consumption and nutrition. But how many people gain how much, nutritionally, from MVs? This has to depend on the nutritional features of MVs, and of their users producers and consumers. In order to adapt the features to the feeders, improved nutrition of persons at risk is not only a desirable by-product of MV research; it should be the central objective.

How to reach this objective must depend on *who* is vulnerable to undernutrition, where, when, by how much, and with what trends. It is reasonable to conclude, from the evidence so far in this book, that MVs can reduce the vulnerability of these families, and of the most vulnerable people within families. But what choices of research strategy are likely to do best at this? What characteristics of MVs should researchers seek, in order to do most for the undernourished at a given cost? Will some sorts of MVs, in helping some undernourished people, increase the vulnerability of others? The above 'price effects' suggest that MVs have helped the nutrition of the poor. However, as we shall show, this is *despite* most nutritional components in MV research so far. The land, talent and cash devoted to these components is being re-allocated to alternative approaches – but much too slowly, in our judgement.

Recent research has clearly identified the main problems of those who suffer or die from nutritional problems. By far the most prevalent and harmful nutritional deficit is energy, not protein[21] (except possibly in some of the areas where the staple is yam, cassava or plantain). The major vulnerable group is the very poor, especially under-fives and pregnant and lactating women. They eat mainly coarse grains, root crops and cheap varieties of wheat or rice. Vulnerability is most acute in specific seasons, and in bad years [Schofield, 1974; Chambers *et al.*, 1981; Longhurst and Payne, 1979; Lipton, 1983].

The 'ultra-poor' at nutritional risk, among whom the under-fives are heavily over-represented, are found almost entirely in one of six groups: the landless in irrigated areas of Asia; the landless and very small farmers in unirrigated Asia; small farmers and increasingly the landless in Africa; and the urban poorest. Within these economic groups, children aged under five have higher energy requirements per unit of body weight, because their brains and bodies are growing; because, in the transition from passive to active immunity, they are more exposed to infections and need energy to fight them; and because play is needed for psychomotor development in children, and uses much energy. All this suggests that MVs cheap enough to reach the poorest, and readily absorbed via breastmilk or weaning foods, will do most nutritional good. It is also important for the extra dietary energy to reach very poor pregnant women in the first trimester, when undernutrition increases the risk of dangerously low infant birthweights. Otherwise, however, the extra energy requirements of pregnancy and lactation appear much less than was once thought [Whitehead *et al.*, 1986].

These facts suggest MV research targets in terms of production, consumption, regional and commodity mix, and varietal priorities. The choice of targets can either strengthen or weaken the benefits from MVs for poor consumers, discussed in Sections 5 a–b, via restraining food prices. Poor consumers' nutrition requires more, cheaper, more robust sources of dietary energy – 'calories' – in forms likeliest to reach small children in hard times. Instead, MV nutrition research has concentrated excessively on protein content and quality and consumer acceptability (palatability and cooking and aesthetic characteristics). Such efforts (i) produce costlier MVs, (ii) divert research resources away from the goals of dietary energy and robustness, (iii) often face a direct physical trade-off (plant nutrients, water or sunlight, if used to produce high protein levels, are normally diverted from raising yields), and (iv) delay the release of suitable MVs, both because more characteristics must be selected for, and because it is often obscure to what extent high protein levels, or acceptability characteristics, in a variety are heritable in general, or sustainable in particular environments.

Such research would have been better directed to other topics, and indeed may have raised the costs of energy intake for the poor. MVs' impact on the nutrition of at-risk people, i.e. (mostly) small children in very poor households, has been good despite some agricultural scientists' diversion to the quest for such things as high-protein maize. The benefits have been achieved almost entirely because MVs have meant more, more local, and more stable dietary energy, and hence have restrained and stabilized its price; and because the production process has generated employment income, which the poor can spend on food.

(d) Commodity choice, production strategies, and consumption

The large increases in rice and wheat production have prevented major deterioration in consumption for the poor in Asia. Public distribution programmes, and much greater and earlier reductions in average production costs than for rice, have since the late 1960s made wheat into a staple for very poor people, even in places like Calcutta that traditionally have eaten very little of this crop. Recently, this has happened also to less-preferred MV rices (and to rice 'brokens') in some places.

Yet in most of Africa, and parts of Asia and Latin America, wheat and rice remain the foods of consumers well-off enough not to be at nutritional risk. Thus in Colombia and Brazil extra output (and lower prices) for maize and cassava do much more for poor people's consumption and nutrition than comparable efforts for wheat and rice [Pinstrup-Andersen, 1977; Pachico, 1984]. The latter crops, moreover, are not readily or safely expanded into the marginal soil and water environments of poor growers – and, indeed, of entire countries in Africa. Some of these may be getting increasingly and riskily dependent on MV-based crops that they must import, or grow – usually on big farms – at high disease risk (tropical wheats in

Zambia) or forbidding cost (irrigated rice in Ghana). 'Food security' for many of the poor – in Africa, for the large majority of them – continues to depend on sorghum and millet [Jodha and Singh, 1982], maize, or cassava. What tropical wheat development as it is pursued by CIMMYT might do for the poor is discussed more fully later in this section. What have MVs to offer for poor people's nutrition through improving more typical 'poor people's crops'?

The poor in Kenya, Zambia, Zimbabwe and much of Latin America have benefited from *maize hybrids*. However, the main breakthroughs were in the 1950s and early 1960s, though there was some further spread in parts of Zimbabwe and Zambia in 1983–5; more generally, progress has recently been slow. Despite major growth of *sorghum* and *finger millet* MVs in some regions (and seasons) in India [Rajpurohit, 1983], much slower progress has been made in farmers' fields with these crops in Africa; attempts to spread hybrid sorghums based on ICRISAT germ plasm have been largely abandoned in West Africa (but have achieved much success in Zimbabwe). There is little real progress with millets and root crops on small farms in most LDCs. In India the diffusion of *pearl millet* hybrids has been even faster than that of the sorghum hybrids – the all-India figures for diffusion of MVs in 1983–4 were 43 per cent of area for pearl millet and 29 per cent for sorghum [Walker, pers. comm.] – but the impact on yield has been much less, and less reliable [Nadkarni, 1986]. Large claims for *cassava* MVs in West Africa have not yet, in our judgement, been properly validated in farmers' fields. Overall, poor consumers have gained less from MVs in rainfed (or 'rainparched') countries of Asia and Africa than elsewhere.[22]

*   *   *

There is reason for concern that MV research – nationally rather than in IARCs – has slighted the needs of poor consumers by overplaying rice and wheat relative to coarser, cheaper foods.[23] However, there is less reason to worry about the undoubted fact that the progress in wheat MVs has intensified a pre-existing trend [Grewal and Bhullar, 1982] towards replacement of pulses (grain legumes) [Ryan and Asokan, 1977]. This switch to wheat has been deplored by some, but must be analysed in the framework suggested in section c above. Land will yield more and cheaper dietary energy in wheat than in pulses, with major nutritional gain if the poor get the same proportion of food as before, and with little detectable nutritional harm from the small protein loss.[24] Since 90–5 per cent of undernourished people lack energy, but do not lack protein – or do so only because energy shortage compels them to divert protein foods to energy uses – the switch means a clear gain in nutrition. Apart from this, MV cereals often provide more protein per acre than the pulses they displace.[25]

The protein-calorie debate aside, much 'nutrition research' has sought to improve the aesthetic qualities (texture, colour, etc.) of MV crops, especially wheat. Certainly, the rich can afford to care about such things. However, MV wheats are currently 10–15 per cent cheaper than traditional wheats. 'Success' in aesthetic breeding would erode the advantages of wheat MVs in providing cheap energy sources for the poor.[26] Similarly, 'success' in breeding crops for use in breadmaking diverts them away from very poor people, who can seldom afford bread (rather than, say, chapatis or tortillas). And 'success' in breeding rice varieties suitable for milling and high polishing, while attractive to better-off urban consumers, can harm the nutrition of the poor in two ways: unusually, a protein problem can arise as the protein-bearing outer parts of the grain are polished away [Bray, 1986, p. 13]; and modern mills can destroy jobs.

Understandably, the IARCs are not able to provide precise information on how their yield-increasing programmes affect human nutrition [Ryan, 1984]. Inferences have to be drawn from data on MV adoption and resulting extra production, combined with estimates of demand parameters. It can be inferred that, despite the distortions discussed above, urban, and (in MV areas) landless rural, *consumers* have clearly gained nutritionally on balance,[27] as MVs have restrained the price of food. Even though such effects have been due mainly to increases in yields of rice and wheat in irrigated areas, some of the consequent price restraint has affected both 'poor people's crops' and unirrigated areas. Small *growers* and their employees have lost from these price restraints, but have been compensated in MV areas by higher consumption on family farms and out of wages in kind, and lower non-labour unit costs. Many poor farmers in unirrigated areas may be eating worse due to MVs; which have increased their yields only a little, but have substantially restrained the prices they can obtain for their output, owing to the large increases in supply (at lower unit cost of production) from adjacent irrigated areas. As we point out elsewhere, these non-MV areas may contain 'new' MV-induced vulnerable groups. The extent of their vulnerability as families will depend partly on how far they have benefited from MVs of millet and sorghum, and on their success in gaining income from those crops such as groundnuts which MV adopters have dropped (Chapter 3, i).

(e) Some specific crop developments and issues

Wheat consumption by the poor has increased in India following the introduction of MVs, but partly at the cost of other foods; in some years MV wheat is cheaper for urban consumers than classical 'poor people's foods' such as pearl millet. At least until the hybrid sorghum advances in the mid-1970s, the Green Revolution was, in T. N. Srinivasan's words, for India a wheat revolution; and in that period National Sample Survey and National Institute of Nutrition work shows no net reduction in incidence or severity of undernutrition [George, 1980]. The 1983 NSS (and parallel NIN) data appear to show some improvement in energy consumption in India, but this remains controversial [Ramachandran, 1987]. Certainly wheat MVs are partly responsible for restraining Indian urban wheat prices, and for a clear reduction in undernutrition in the Punjab [Chadha, 1983].

A recent development at CIMMYT is the encouragement of tropical wheats [CIMMYT, 1985, p. 16]. This is a CIMMYT priority for Thailand, the Philippines and Sri Lanka, and the Cervados region of Brazil; many would add Zambia and other African countries. Tropical wheats require breeding for heat tolerance, resistance to *Helminthosporium* and leaf rust. Such developments should be assessed along the lines indicated in section c of this chapter. Tropical wheats are not necessarily a nutritionally sensible solution merely because acceptable yields can be obtained under tropical conditions, and/or because we observe an increase in (imported) wheat consumption in developing countries. Rather, tropical wheat development needs to be demonstrated as an effective strategy for meeting consumption needs of undernourished people.

It is not clear if such a programme is based on an identification of the victims or causes of malnutrition in these areas, nor that tropical wheat development is thought to be the most cost-effective means of reducing it. A breeding programme such as this has costs, in terms of work foregone on other crops, locally long adapted but often largely without efforts at scientific improvement. Also, if tropical wheat breeding is a 'frontier technology' – and if (as is clearly the case) IARC crop varieties are usually successful and safe only to the extent that an adopting LDC can develop and retain its own adaptive research capacities, to breed or screen, as new pathotypes develop – the approach may have serious risks, especially if the poor come to rely on tropical wheats for food. Yet the argument that the world's poor – and food processors – are increasingly hooked on wheat (and rice), in countries at risk of lacking access to imports, cannot simply be ignored.

*   *   *

High-yielding, mosaic-resistant and mite-resistant IITA *cassava* varieties, as and when they get into farmers' fields on a large scale, must improve self-consumption among the rural poor in Africa. However, weanling children dependent on cassava may well feature protein deficiency even when they have enough dietary energy (calories); whether this is so, and if it is whether high-protein cassava MVs are a sensible route to improvement, are not self-evident, but require research. (Recognition of analogous issues saved much wasted effort for millet and sorghum at ICRISAT; lack of recognition of them has probably caused such waste in the case of maize: section h). In areas of high rainfall where cassava is a staple, the wide range of crops – especially legumes – consumed suggests that there will not be widespread protein deficiency, although small children may require to supplement cassava with a more energy-dense staple.[28]

Caution in imposing specific, allegedly nutritional, criteria (which might delay higher-yielding cassavas) is needed, because poor people can gain greatly from higher cassava yields. For Brazil, a CIAT study indicates that the calorie consumption of the poorest 25 per cent of the population could be increased by 45 calories per day by improved cassava production technology. In the rural North-East, 20 per cent of the calorie shortfall in the diets of the poorest 25 per cent would be alleviated [Pachico, 1984; Lynam and Pachico, 1982]. The distributional gains from cassava MVs are confirmed by expenditure elasticities for fresh cassava in Java: positive and moderately large for rural consumers, small and negative for urban consumers [Dixon, 1984]. Increases in cassava production are likely to be consumed by poor rural people [Okigbo, 1980].

Cassava is used as food and feed, consumed domestically and exported. In India and Indonesia virtually all cassava production is used locally; one suspects the same is true of Africa. In Thailand about 75 per cent is exported to Europe; elsewhere in Asia two-thirds is used as human food. In Latin America it is divided between the two uses with 70 per cent being marketed, none exported [CIAT 1984].

Cassava can help poor people's nutrition in three ways: directly, by increasing their consumption of cassava they themselves grow, or of cheap purchased cassava products; indirectly, especially in exporting countries, by providing cash incomes to workers; and elsewhere, also indirectly, by replacing grain imports for feed (especially for poultry and pigs), rendering it more readily feasible to divert such grain imports to very poor people, perhaps with some subsidy. The cases probably require distinct research strategies to assist poor people's nutrition. In cassava-exporting LDCs, ways to increase the labour-intensity (and reduce the scale) of production processes with MVs may be required. Where cassava is to be purchased by poor people used to flours of wheat or maize, some screening of cassava MVs for suitability in flour-mixes is indicated. However, in the majority of cases where poor people grow and eat cassava, higher yield and greater disease resistance are the priorities; tailoring cassava production to particular sorts of farmer – or breeding for protein or palatability, or absorption – can unduly divert scarce research resources away from yield and robustness.

*   *   *

Apart from a Colombian urban study [Pinstrup-Andersen, 1977], the work on the impact of *maize* and *rice* MVs upon nutrition remains largely hypothetical. Even in that study, the observed responses of consumers to changes in the prices of crops, given pre-MV consumption bundles, are used to infer what apparently must have happened when MVs changed prices. For rice overall, IRRI concludes that MVs must have increased consumption more among malnourished households, because of their higher income and price elasticities of demand for rice [Flinn and Unnevehr, 1984]. Such arguments seem logically sound – unless rice MVs added a rather new commodity – and our own are certainly no better or less hypothetical. However, post-MV consumption patterns need to be followed up with panel data, tracking affected households over several years. Otherwise, the effects on nutrition of money-wage adjustments to MV-based price restraint in food prices of changes in imports, and of much else, remain speculative.

Potato is sometimes consumed as a low-cost staple. A study from CIP [van der Zaag and Horton, 1983] shows that consumers are highly responsive to potato prices. Breeding MV potatoes for high quality rather than for calories, therefore, will (as it raises prices) *harm* the poorest if they are found mainly among potato eaters, but *help* them if they are found mainly among potato growers or their employees.

*   ··   *

Research on grain *legumes* (pulses) – chickpea at ICRISAT, and lentils, chickpeas and faba beans at ICARDA – is often cited as favourable to the nutrition of the poor. But this depends on their spending patterns and on the type of their nutritional deficiency. In the usual case where deficiencies are not primarily protein – and even where they are, but pulses are costlier than cereals as a protein source – legume MVs can do little for nutrition of at-risk groups, unless (i) initially undernourished growers sell pulses to better-off buyers *and* (ii) demand is price-elastic, so that a reduction in unit costs and hence prices (due to MVs) brings a more than proportionate rise in sales of pulses by poor growers, whose net farm income (and hence nutrition) therefore improve.

Protein per unit of land or labour input may in any case be higher with cereals than with pulses. In India, replacement of pulse area by MV wheat taken in isolation led to more and cheaper protein as well as calories for the poor. This was so even between 1961–5 and 1971–3, although changes in the crop-mix were then offset by other factors so that poor people's total calorie intake per person stagnated [Namboodiri and Choksi, 1977, p. 33].

Pulses are usually less affected by MS than cereals. Pulse MVs, if they increase the proportion of land under pulses, can therefore make diets – and output and income from a piece of land – more or less vulnerable to drought. If readily absorbable and cheap, pulses can also assist vulnerable groups, such as weanling children and their mothers. However, the nutritional role is often quite limited. It appears to be necessary that ICARDA identify target groups, whose nutritional needs can be shown to be most cost-effectively helped by its research activities [Somel, pers. comm.].

For all crops, however, the procedure for defining nutritional priorities in MV research is similar, as outlined in section c and expanded here in the context of MVs. First, researchers need to identify the people at nutritional risk who are (or may become) reliant on the crop, *either* as their main staple (or other important food source) *or* as a main source of farm or employment income. Second, the problems with such people's nutrition – usually, energy deficiency among small children – need specification. Third, the 'solution role' – if any – of MVs of the crop, in regard to those problems, needs to be specified: to provide income (and hence food) for poor workers, calories for farmer-consumers, lower-cost flour-mixes for urban consumers, readily absorbable protein for weanlings, or what? Finally, the cost, feasibility and scope of MV research to fill the specified 'solution role' have to be assessed, and compared with other approaches to the problem. Then, the money can be spent – or not spent – on the proposed nutritional research into that crop.

(f) Variability

Since very poor households contain infants near the margin of survival even in normal times, a great nutritional threat is plainly posed by 'bad times', when food is costly and harvest employment hard to come by. These times are related to, though not the same as, times of low output or supply of crops (dear crops, dearth). Surprisingly, MVs may have somewhat increased the year-to-year variability of national cereal *production* (Chapter 3, h). Variability, however, is merely a convenient measure of year-to-year fluctuation – namely the 'coefficient of variation' (Chapter 3, fn. 30) or CV – not of the risk of dearth as such. Moreover, most CVs have increased only at national level. This rise is mainly due to increased covariance among areas producing crops heavily, namely MV areas. It does not mean that in 'bad' years the nation – let alone particular small producing areas, and thus the people who live there – will obtain lower output (let alone lower consumption) after MVs than would have occurred without MVs (Chapter 3, h). In fact, recent, robust MVs usually do better, at the level of the individual farm, in a 'bad' year for rainfall or pests than the TVs they replace.

Even if this were not so – if MVs made cereal output at national level unequivocally riskier – they could well permit the accumulation of larger grain stocks, as in India. Then consumption becomes less risky, as shown by India's successful use of stocks to defuse a potential consumption crisis during the 1987 drought. However, at national level – unless compensated by changes in stocks or net imports – even slight increases in the variability of *production* substantially increase the riskiness of *consumption* by vulnerable groups in towns and in non-MV areas [Murty, 1983]. Since small growers, normally in surplus after MVs, try to meet family food needs first, they cut net sales in bad years[29] more than in proportion to output. Therefore fluctuations in marketings – and hence yearly cereal price instability – increase more than might be expected from the post-MV increase in yearly output fluctuations. This effect is made worse for the growing proportion of poor rural consumers who depend mainly on labour incomes, because these too are cut back when the harvest is bad – the small growers, many of whom become net hirers of labour only after

MVs, are in years of bad harvest both willing and able to bring it in with more family labour, and much less hired labour, than in a normal year.[30] All this can leave the poor even more vulnerable, as consumers, than before – especially if the cash value of wages adapts to the long-run restraint in food prices after MVs, but fails to respond swiftly to short-run rises in food prices in bad years, as seems to be the case [Parthasarathy, 1977].

Rises in net imports, or falls in stocks, can modify or remove these price risks to consumers. But LDC governments often cannot afford the imports in bad years. Stocks, on the other hand, carry high costs. The 20–30 million tons of grain typically stocked by the Indian government, through the Food Corporation, tie up, as working capital alone, some 3–4 months' worth of India's total net investment. Moreover, central stocks and international trade seldom deal much in root crops or coarse grains for human food. These are sold in remote areas, and have high ratios of weight to value, and therefore of marketing costs to value. Thus, for such 'poor people's crops', markets are 'thin', i.e. release a small proportion of normal output (most of which is consumed by growers). In such cases, as with sorghum in India – which absorbs 10 per cent of consumer budgets, much more for nutritionally vulnerable poor groups [Murty, 1983] – moderate rises in variability of outputs, due to concentration of MVs in a few covariant areas, can greatly increase risks to undernourished groups. This is not only because (as with other food crops) slightly below-average harvests mean substantial falls in marketings, and hence substantial price rises. Also, these falls in output – seldom made good out of stocks or imports – comprise a very large proportion of the rather small amount of coarser foods that gets marketed under normal circumstances. The result is that bad years bring very expensive coarse foods for very poor consumers [Walker, 1984].

Shifts in the crop-mix modify this effect, for better or for worse. In India, cheaper MV wheat – with a higher *output floor* in bad years, due to the growing role of irrigation, and despite the fact that the output ceiling in good years is raised even faster, so that *variability* also is higher – modifies the harm done by price instability for coarse grains. In much of Africa, however, the success of maize MVs (mainly hybrids) has caused them to displace more stable but less dynamic crops – millet and sorghum – and that, often, in regions or countries with only one peak per year of seasonal output. Instability of supply, and hence prices, of marketed crops has thus probably become even more serious as maize MVs have spread.[31]

Though MVs may in some cases have worsened, or raised the cost of avoiding, *year-to-year* instability of output – and therefore of prices, and of consumption by vulnerable groups – they have probably reduced *seasonal* instability. MVs usually raise output (and employment income) proportionately more in the subsidiary season, owing to controlled irrigation, than in the less certain conditions of the main rainy season, when most grain is produced.

However, the widespread famines of 1982–5 have underlined the nutritional risk, to poor consumers, of yearly price and output instability in Africa. This has not been helped by the spread of maize hybrids. This suggests a higher nutritional priority for stabilizing MVs' output, especially as they spread to climatically riskier environments. In much of semi-arid Africa, an alternative (or complement) is urgently needed to the high-yielding but drought-prone and consumption-destabilizing maize hybrids. Possibly, the alternative will be found in maize populations (or hybrids) better able to resist MS, but the problems (pp. 41–2) render it likelier that the answer lies in appropriate drought-resistant millet and sorghum MVs.

(g) Vulnerable groups, children, and women

Much of this book has discussed the impact of MVs on the poor treated as groups or classes of households: 'small farmers', rural labourers, urban workers, etc. However, *within* poor households, some people are at much the greatest risk of lasting damage from undernutrition: above all infants (aged 0–1) but to a lesser extent pregnant and lactating women and other pre-school children.[32] Their particular vulnerability can be ascribed partly to special physiological need. However, it is still legitimate to ask what MVs can do for them. That is one of the two linked issues in this section.

The other issue is the constraints placed, upon the capacity of MVs (by raising poor households' income or food output) to improve nutrition among 'vulnerable groups', by the low status of women – including mothers – within many households. The household or family cannot be assumed to be a democratic decision-making unit with equitable distribution of resources or control over product. Much evidence suggests that women, and small girls especially, do not, in *some* societies at nutritional risk – notably in North India and Bangladesh – receive a fair share of resources: health care, leisure, or (in rather special circumstances) food. Therefore we have to ask (i) if the impact of MVs varies between men and women; (ii) if so, how this affects the impact of MVs upon consumption and nutrition (including labour requirements and stresses) for those in need; and (iii) whether this has any implication for MV research. We have examined some of these issues already in Chapter 4 (with regard to MVs' effects on work among women and men). Their importance for MV-linked food consumption and nutrition is based on evidence that – although adult women, in general, receive much the same proportions of dietary needs as adult men [Schofield, 1979; Harriss, 1986; Lipton, 1983] – income flows to, and time allocation by, rural women are important determinants of the nutritional status of the under-fives. In particular, we need to ask if MVs would do more to improve the nutrition of sucklings and weanlings if MV research were explicitly redesigned to favour women or girls, over and above its goals of increasing entitlements to food among poor households.

Pre-school children are heavily over-represented among the poor [Lipton, 1983a] because of very high birth-rates and infant and child mortality rates in the two lowest deciles of households by income per person. It has therefore seemed common sense for planners of nutritional interventions to assume that vulnerable groups – especially under-fives, but also pregnant and lactating women – benefit more or less automatically if poor families can grow or buy more or better food. This approach, however, overlooks the fact that the proportion of extra income and food, allocated by a household to women and children, depends on their status, prospects, and power, all seen in the specific socio-economic context,

very senior, often foreign, researchers of either sex can substitute for research involvement by farmer-clients, especially women. We must return to this issue (Chapter 6, i), but here we revert to the issues affecting poorer women's access to food (and power over it), that research institutions – however participatory, or however top-down – may need to address in developing MVs.

The most important issue is to design MV-based systems raising forms of employment, for households at nutritional risk, that provide women with control over more income – without unacceptable increases in energy stress on mothers, especially in peak seasons, that may conflict with child care (Chapter 6, i). The second issue is the utilization of non-grain, non-tuber biomass of MVs, e.g., in the case of rice, straw for thatching and mat-making, fodder for livestock, bran for fish ponds and husks for fuel. MVs have not been selected, nor often even screened, for their ability to meet these end-uses, and to do so in ways that employ women rather than displacing them. Yet women figure disproportionately among those who depend on such products, because of their assignment in households to manage, as well as because of their need for, 'fuel, home-based income opportunities and effort-reducing convenience food' [*ibid.*, p. 15]. MV research could well enhance women's status and income – and perhaps thereby improve food consumption and nutrition among vulnerable groups – by more emphasis on such 'women-orientated', often non-food, end-uses and income sources from MV crops.

A third issue concerns whether MVs can be selected, or at least screened, for their capacity to release mothers from some tasks of food acquisition or preparation, leaving more time for care of children's health, nutrition, and general development. For example, if MVs of cassava and legumes are screened for the palatability and digestibility (and ease of cooking) of their leaves, mothers are freed from securing and preparing sources of vitamins and proteins in time-consuming ways. In general, the 'intrinsic links between varietal characteristics and domestic food processing, preservation and preparation technologies must be investigated and considered at an early stage in the research process' [*ibid.*, p. 85]. However, there is an important question of cost-effectiveness in research here.

There are high fixed costs in developing a little-researched crop – and these crops cover smaller areas than, and are usually secondary to, the main staples [Longhurst and Lipton, 1987]. In general, nutrition among vulnerable groups will often gain if MVs (main-staple or secondary) are *screened* for characteristics other than yield and robustness,[36] but seldom if research planners defer or dilute these two objectives by *breeding* for others, especially where the crop is not a main staple for many consumers.

That is especially the case because increased levels of women's income have never been *proved* to do more for vulnerable children's nutrition than a similar rise in men's incomes. However, the hypothesis is plausible and deserves testing. Even if it is false – and although adult women seldom suffer worse food intakes, relative to requirements, than adult men – women's generally depressed, discriminated status certainly deserves special remedial action, and some of this could appropriately come from agricultural research. Yet MVs, by raising the proportion of rural activity devoted to crops for sale, may have raised men's share in household income [Ahmed, 1983].[37] Moreover, the shifts to hired labour and to shorter-duration crops – alongside MVs' association with reduced (off-season) post-harvest work for women (Chapter 4, c) – may have intensified the seasonality of labour demands on women's dietary energy. Both factors would reduce the amount of time and 'calories to spare' that women have for their children. Recent evidence suggests that the time allocated to children by their mothers is an important determinant of nutritional status [Tripp, 1981; Popkin, 1978; Wolfe and Behrmann, 1982].

We do not, of course, suggest that the research goal of larger farm output – almost inevitably leading to a higher ratio of sales to self-consumption – or of shorter-duration crops (as a major cause of that larger output) should be abandoned. These are part of development, not just goals of MV research. We suggest only that harmful side-effects from such goals on women – a somewhat poorer (and much lower-status) group already in many LDCs, and a group directly responsible for vulnerable small children's consumption and nutrition – need to be anticipated, and where possible prevented or reversed, in

MV research planning. For example, in many cases, there are special environments in which women grow MV crops, especially in home gardens and in mixed stands. If such environments receive more emphasis in MV research, perhaps child nutrition will be improved indirectly – as women, including pregnant and lactating women, get more income and power – without being filtered through the family unit. Unless the crops are grown in or near the home gardens, however, this may have to be balanced against the negative impact on mothers' provision of child care. Pregnant and lactating women earning outside the home compound may fare worse, and feed children worse, than women with similar incomes but able to work nearer home [Schofield, 1979; S. Kumar, 1977; Lipton, 1983]. The neglect of staple crops, and to some extent of legumes, in home-garden circumstances by IARCs – complementing the familiar [e.g. Bond, 1974] neglect of women farmers by extension workers – is thus especially unfortunate. Putting this right could be one nutritional benefit from greater concentration of new research inputs – and of extension – on poor women; this is important because *extra* nutrients, from new sources such as MVs, are often maldistributed within households [Carloni, 1981; Longhurst, 1984], even though such maldistribution is much rarer for average or long-standing nutrient availabilities [Lipton, 1983, pp. 50–4].

The role of women, and agricultural research strategy towards it, is one of two key areas in which MVs may be 'tuneable' to help vulnerable groups' nutrition. Another is through research to improve MVs' direct impact on infants in poor households who consume them, or breastmilk substantially derived from them. Although it is almost certain that more income (or cheaper food) for poor households due to MVs usually helps child nutrition, we know very little about how various types and timings of MVs contribute to small children's special nutritional needs: high energy and nutrient density, more frequent ingestion, enhanced nutrient absorbability, nutrient complementarity, nutritional availability in terms of ease of preparation, and favourable interactions with infection. It is actually rather amazing that – given the 'anti-hunger' motives behind much support for MV research – almost nothing is known about such matters!

Despite some welcome recent shifts in a few institutions, IARC research has not been planned explicitly enough to discover how MV outputs or work inputs affect the nutrition of these vulnerable groups – whether indirectly through changes in causal sequences such as the relative power of women in households, or directly through changes in the ecology of infant nutrition, absorption and infection. Even recent, innovative research has generally been confined to the proximate problem of 'getting more food to infants and women', and has not often enquired how MV *options* might actually reach vulnerable groups in the prevailing family and social structures. Indeed, most 'nutrition research' into MVs has concentrated on issues such as their protein quality, which are quite unrelated to this (Chapter 5, h).

We do not here pretend to determine how MVs might best help vulnerable groups. We merely raise the questions, which should perhaps have been raised in the IARCs before they began nutritional research.

(i) In the total food-work-infection context facing major groups of vulnerable households, what is the impact of consuming different varieties of a main staple – or of a research-induced shift, from a main staple without a successful MV to one with – on a pregnant woman and the foetus she is bearing?

(ii) Does the balance of consumption by a lactating mother (or by a weanling), as between different crops (e.g. cassava vs. millet) or varieties of a crop – and the different work requirements for mothers, or even for growing children, posed by each – significantly affect at-risk children's nutrition?

(iii) What is the differential impact of crops and MVs on the volume, absorbability and quality of breast milk?

(iv) Can the nutrition of sucklings be influenced in this manner, especially during infections?

(v) As some IARCs work on the energy density, fibre content, and absorbability of staples used as weaning food, do they assess the impact in the context of vulnerable households' total diets, intra-family food allocation and distribution procedures, and food processing and cooking arrangements, including cleanliness, cost, and wastage?

(vi) Is such research avoiding the pitfalls of past, costly and nutritionally not very relevant, efforts to breed for amino-acid content?

(vii) Analogously, are there nutritionally important differences among MVs and TVs in their effects on bulk, absorbability, and quality of at-risk weanlings' total intake?

(viii) Do crop or varietal nutrients, and work inputs, interact, for vulnerable groups, with the type and timing of infection and the building or weakening of mechanisms of immunity? Varietal and crop-mix priorities, including the seasonal timing of food flows, *may* not make much difference, but they should be further investigated.

There are differences among crops and varieties – in isolation, and in the context of vulnerable groups' total diets and needs – in energy density, fibre content and anti-nutritive factors.

(ix) Are these differences significant, and are costs of improvement justified by benefits to target groups?

(x) Conversely, might some MVs increase the work (or travel) required of women, perhaps at times when they are already hard pressed to muster enough dietary energy (or time) for child care?

(xi) Is research important that reduces preparation time and firewood costs or increases energy density of food eaten by vulnerable groups?

(xii) Might research that improves palatability raise the attractiveness of the commodity to non-poor buyers, and hence harm the poor (unless sellers of the crop) by raising the price of calories from it to them?

(h) Nutrient quality and palatability: the wrong menu

Most research on how MVs might best be selected to improve nutrition, by IARCs and others, has not addressed such issues. Instead, it has sought to 'improve' nutrient quality and palatability. This is the *inappropriate menu* of conventional MV nutrition research. Of this, the improvement of maize amino-acids via the opaque-2 genes has taken most resources. Yet extra dietary energy, not extra amounts of protein or of a specific amino-acid, is the overriding need for almost all vulnerable *humans*. They do not live on maize alone, and get their balanced amino acids by supplementing it with beans and other foods. (Storage pests do live almost entirely from the grain stored, so that a balanced-protein MV, with all amino-acids well represented, while normally of little value to humans, does wonders for storage pests: Sriramulu, 1973; Rahman, 1984; Podoler and Appelbaum, 1971). There is an enormous scientific literature on protein improvement, however, and attempts to justify it continue, albeit with increasing unease and defensiveness [Valverde *et al.*, 1983].

Many CIMMYT researchers state that appropriate varieties of high-lysine opaque-2 maize give yields equal to ordinary MVs, store and cook as well, are available in acceptable non-floury form, and improve the nutritional status of children under 2 years of age (for discussions see Ryan, 1984 and Tripp, 1984). But these claimed properties, and the last is questionable, were made possible at the huge cost of diverting land and researchers from yield improvement and stability towards amino-acid enrichment. That cost included calories and even proteins foregone. Many poor children are calorie-deficient, while few are protein-deficient.[38] Of those few, most would have enough protein if they were not forced to burn it up for want of calories. Only where root crops or bananas are main staples, with very few pulses added, is protein research likely to do much for *human* nutrition.

Protein quality analysis and breeding have been carried out to a lesser extent with barley at CIMMYT (now abandoned); with potatoes (to obtain amino-acids) at CIP, in Peru [Valle-Riestra, 1984]; with chickpeas at ICARDA; with cassava and beans at CIAT; and, on a small scale, with coarse grains and pulses at ICRISAT and rice at IRRI. Research efforts at CIMMYT on improving protein content of bread wheat were reduced in 1983 [CIMMYT, 1985, p. 104]. It was found that protein content was generally no greater than in conventional varieties, presumably because the high-protein characteristics, formerly identified in 'MVs', had proved after all not to be stably heritable. More generally, these protein emphases have dwindled as researchers have come to perceive trade-offs between protein content, yields and stability – at least, if research sought to *maximize* one of these goals, others would suffer.

But there is a deeper problem. Should breeders be seeking MVs with high protein, let alone high lysine or tryptophan, at all? If rats eat only millets or sorghum, they are likely to die of lysine deficiency. Yet, right from the start (though accompanied by long debate), ICRISAT rejected the improper inference that it should breed high-lysine cereals; for in semi-arid areas people (i) mix millet or sorghum with lentils, chickpeas, pigeonpeas or beans, (ii) hardly ever suffer protein deficiency unless they have energy deficiency too, (iii) normally cease to suffer from the former once the latter is removed, (iv) if, exceptionally, they are genuinely protein-deficient, do not usually suffer especially for want of lysine. Confirming its wise decision to seek MVs robustly providing plentiful energy rather than MVs with high protein (let alone high lysine), ICRISAT has found that the major dietary deficiencies in their sampled villages were in energy, plus items probably best treated by non-MV-based interventions (calcium, and vitamins A, B-complex and C) – not in overall protein or specific amino-acids [Ryan, 1984]. Nevertheless, the world's leading sorghum breeder advocates research to improve absorbability of sorghum protein [Doggett, in ICRISAT, 1982]. That view must be respected, resting perhaps on a perception that quick and inexpensive research prospects exist (which could justify such work even if only very few vulnerable children benefited). However, in general, only weak arguments can be advanced for protein research into MVs, either by national institutions or by IARCs.

For example, at IITA, the farming systems and grain legume programmes have been studying soybean production and utilization to help farm families who 'cannot afford to buy expensive protein to meet nutritional needs'. The justification is that at mealtimes children come to eat after adult males, when 'there may not be enough soup that contains fish, meat or another protein source, and the children get only the starchy foods' [IITA, 1986, p. 33]. Admittedly, the risks of protein deficiency are higher in root-based diets than in grain-based diets. However, the form of IITA's observation suggests that no analysis has been made of whether there is a binding amino-acid constraint, or even an overall protein constraint, upon the nutrition of vulnerable groups (mainly sucklings and wean-lings) in poor households – let alone of whether research into soybean MVs could plausibly be a cost-effective means of relieving any such constraint. This, indeed, seems unlikely, given post-harvest problems with soybeans, its frequent con-centration on wealthier cash-crop farms, and the neglect of research into alternative traditional pulses. The major con-centration on soybeans in much recent research – not solely or mainly in IITA – indeed suggests a belief in 'protein gaps', as major components of the food problem, that is at least twenty years out of date. If it is supposed to be directed towards helping vulnerable groups, IITA's protein research – given the centre's largely African mandate – further overlooks the evidence that nutrient deprivation, compared to require-ments, is hardly at all concentrated on children (or on females) within households in most African circumstances [Svedberg, work-in-progress, pers. comm.; Schofield, 1979], in marked contrast to North India and Bangladesh. The main need in West Africa, certainly where farming systems are based on cereals more than on root crops, is more and safer MVs, to provide income or output to increase dietary energy for poor, largely farming but also sometimes labouring, households – not more protein for children, especially from little-farmed yet much-researched crops.

*   *   *

Apart from protein, research continues at IARCs on consumer acceptance – palatability and cooking characteristics. This has been summarized [Ryan, 1984] as improvement of: (i) potatoes at CIP, by selecting for increased specific gravity to improve transportability, shape, colour, size, eye depth, and culinary and processing characteristics; (ii) chickpeas at ICARDA, with respect to taste and cooking time; (iii) cassava at CIAT, with respect to storage characteristics; (iv) beans at CIAT, for seed size, colour, thickness and cooking time. There is also (v) evaluation of rice breeding materials at IRRI for milling percentage, grain size, shape and appearance.

IITA has concluded that, in view of the low elasticity of demand for roots and tubers, processing improvements, i.e. mainly MVs with better processing characteristics, must accompany production increases if consumption levels are to be maintained as incomes rise [Okigbo and Ay, 1984]. This appears to assume that vulnerable groups growing cassava need better chances to sell it. Even if that were so, work in Java suggests quite high income-elasticity of urban demand for some forms of processed cassava [Falcon *et al.*, 1984]. However, market demand – and hence improved processing characteristics – for cassava matter to the nutrition of Africa's *poor* only if they live off cassava, not as a food, but as a source of income from sales. We doubt whether this is often the case. Similarly, ICRISAT has vigorously advocated its programme of consumer preference studies [Doggett, 1982], although millet and sorghum are surely grown and eaten substantially by poor people whose prime need is for more energy rather than for subtler things, and who do not sell much to those able to afford more elaborate preference structures.

More understandably, most IARCs have paid attention in their breeding programmes to screening out anti-nutritional factors such as tannins and trypsin-inhibitors. Also sensibly – given the importance of custom hulling to poor landless women, and the threat to their income (and hence food entitlements) from mechanical milling – ICRISAT is carrying out research on the impact of mortar-and-pestle dehulling on food quality [ICRISAT, Annual Report 1985]. The varieties ranged in recovery from 90 per cent of dehulled grains to only 62–5 per cent in the case of varieties with soft endosperms. At first sight, screening MVs for high recovery could, in such circumstances, greatly improve the impact on consumption in poor groups; however, it may be that the 'non-recovered' food matter is in fact retrieved or scavenged by some of the very poorest people.[39]

Overall, the allocation of scarce research funding to these activities must be viewed in the context of what causes malnutrition and how far they lift constraints on its improvement. Where the poorest people, at most nutritional risk, are found among small farmers who sell a higher-value food crop to rich people (and thereby acquire income to buy cheaper calories), it makes sense to improve that crop's market value rather than its caloric value, since such farmers get more command over calories that way. But most MVs are *either* grown and eaten by the poor or sold to the poor by farmers at little nutritional risk themselves. In these much more usual circumstances, the major nutritional advantages of most MVs to the poor consumer are (i) their greater and nearer supply (restraining local prices), (ii) their greater robustness, and (iii) as against TVs, their 10–15 per cent price discount. Breeding for stability and quantity maintains this discount, concentrates on the other two advantages, and does most for poor consumers. Breeding for 'quality', palatability and gourmetry harms the poorest, by removing the price discount, by delaying the release of higher-yielding varieties,[40] and in both ways by raising prices. Both such breeding and protein emphases divert scarce IARC resources from their primary functions of providing poor people, especially in hitherto neglected areas, with high-yielding and stable crops that they can grow and/or consume cheaply.

However, the stability, quantity and quality interactions – both trade-offs and complementarities – and the price implications are complex. Generalizations are hazardous, whether across crops or among situations with very different nutritionally-vulnerable groups (e.g. self-consuming farmers; farmers selling to other poor people; farmers selling to the rich; farmworkers; or the urban semi-unemployed). Farmers allocate area not only between low-quality MVs and high-quality TVs, but also between high-quality MVs and high-quality TVs. Moreover, one cannot assume 'linearity' in the implied trade-offs between extra research resources spent on quantity as against on quality characteristics; if 100 per cent of research is on quantity and robustness, it may well pay to divert 1–2 per cent of resources into appropriate 'quality' characteristics, if the scientific judgement is that a high return is likely, *and* if socio-economic research reveals that poor farmers (or poor employees) will benefit from resulting higher offtakes and/or prices from better-off buyers. But the returns to 'quality' research, like those to any other research, may well diminish sharply, once its proportion of research resources rises above a certain limit.

Also, 'quality characteristics' cover a multitude of virtues and (possibly) of sins. (i) An MV that can be cooked quickly and easily can be very important to desperately poor women, and to the nutrition of their inevitably undersupervised children. (ii) Palatability is a different matter: does it absorb research resources at the expense of yield and stability? (iii) Higher-quality grains are often bought at a premium by the rich. Suppose a rice-growing area is divided between three crops – a standard TV, a high-quality TV (such as basmati rice), and an inferior MV – with research resource allocation to be made between the latter two. Normally, the poor would be helped by improving the inferior MV rather than the high-quality TV. Usually – not always – basmati is grown by the rich as well as for the rich.

In a few cases, however, investing in consumer acceptance has advanced the nutrition of vulnerable groups. Sometimes, lack of acceptance has put off earlier, better-off innovators, thus constraining diffusion of MVs to later, poorer 'follower' further along the logistic curve of adoption rates against time. For example, some hybrid sorghums are rejected in parts of Andhra Pradesh in India yet are accepted in Maharashtra; if the medium-to-large lead innovators in Andhra could be persuaded, by a more palatable hybrid, to adopt, the very poor subsistence farmers could gain from diffusion later on. Similarly, it may be necessary to adapt an MV to a cooking practice that has been found inexpensive and attractive by the very poor; for example, maize hybrids in El Salvador have been bred for tortilla quality. The desirable level of investment in such forms of 'palatability research' hinges on the level of adoption. Once a robust and high-yielding MV is widely adopted, the payoff to the poor from such research will be negligible; elsewhere, if it is established that palatability is constraining adoption (and that adoption would benefit the poor), the returns could be high.

Notes and references

1 The latter proposition is much more controversial. Clear damage from 'mild to moderate undernutrition' has not been demonstrated, and may not exist except in the presence of infections (often easier to prevent than hunger). Those alleged to suffer from it *may* often be 'small but healthy': extra income and welfare leads to larger body size, but not necessarily vice versa [Lipton, 1983]. However, unappeased hunger is an evil, even if there is no nutritional harm; certainly this 15–25 per cent often suffers this evil.

2 Including (i) earned cash income; plus (ii) the market value of (a) earned income in kind, (b) items produced on the family enterprise and used at home (e.g. eaten, from the family farm) or sold, *minus* costs of purchased inputs and rents.

3 Because the best land is farmed first, so that population growth pushes out the margin of cultivation into less fruitful areas (i.e. raises average production costs).

4 Boserup [1965] argues that population growth actually *causes* such technical progress.

5 Indeed, the notion that the hunger problem is one of 'population-food balance' remains the reason why these efforts concentrate entirely on *tropical food* crops in LDCs – although many of the world's poor and hungry people rely on export income from *cash crops* such as tea or cotton (whether grown on smallholdings or plantations) to import their food from *temperate* developed countries.

6 Because import-intensive industrialization – then widely assumed to be necessary for development, and to contribute substantially to efficient (i.e. labour-intensive) growth and hence to employment – was to absorb most extra imports.

7 The evidence then [e.g. Thompson, 1959; Davis, 1951] suggested a 25-year transition between falling death-rates and responsively falling birth-rates in South Asia. However, the 1961 Censuses revealed population growth in the Third World well above earlier assumptions, and accelerated the thrusts *both* towards MVs *and* towards family planning. Eventual applicability of MVs (and of population pressures requiring them) in Africa – then still largely colonial – was widely assumed in 1959–63.

8 These claims can be rights to various forms of relief, or of fixed food payments, from 'bonded' employers or other patrons in pre-capitalist labour systems; and/or rights to food relief, free school meals, food-for-work, etc., from the state, charities, or religious bodies.

9 See folio 48 To the extent that the Indian Government, or other governments, dispose of such stocks in support of public-sector entitlements programmes – such as employment guarantee schemes or other food-for-work, school lunches, etc. – undernutrition is reduced, the more so as the schemes reach groups at risk. This, however, merely underlines the fact that 'food availability' is not sufficient to reduce undernutrition.

10 Why does their private income increasingly determine poor people's claim on food? Because population growth both contributes to the privatization of claims formerly based on access to common property resources such as grazing land [Jodha, 1983; Lipton, 1985c], and accelerates the disappearance of 'feudal' claims by labourers upon employers and landlords as a group, since that group can more cheaply hire labourers who are in growing surplus upon 'capitalist' labour markets. For the contrast, compare 'feudal' Wangala and 'capitalist' Dalena [Epstein, 1973; the over-simple epithets are ours, not hers].

11 Partly because, with several children, parents increasingly resist extreme subdivision of land, pushing younger sons and their wives on to the labour market; partly because of the reasons in fn. 10.

12 Such restraint would also appear to enable a fixed wage to command more non-food, since a smaller part of it is pre-empted to buy food needs; but this shift would, by raising demand for non-foods, bid up their prices.

13 The minimum estimate of 10 m.m.t. was for South and East Asia in the late 1970s. The 27 m.m.t. estimate (medium to high range) was for the MV-induced increase in rice production in eight Asian countries which produced 85 per cent of Asia's rice in 1980.

14 Estimates vary according to different assumptions about yield: from 7 m. tons in 1982/83 worth $1200 m. to 21 m. tons worth $2500 m.

15 The Seventh Five-Year Plan devotes unprecedented sums and efforts to schemes to get assets and work – and hence food entitlements – to the rural poor, and this follows a consistent Central policy priority, supplemented by major non-market or semi-market entitlements via State schemes of school food provision (Tamil Nadu) or employment guarantee (Maharashtra). Yet even the gains shown in the unpublished 1983 Round of the massive National Sample Survey show the incidence and severity of Indian poverty not significantly different from the levels of the early 1960s, though somewhat redistributed away from the MV lead areas.

16　In the longer term – as labour supply increases in the MV area, and as employers respond to the *earlier* labour shortages by adopting weedicides or tractors – this can even, on occasion, reduce the real wage-rate.

17　Will this not destroy commercial demand for food, reviving the disincentives problem? Far less than one might imagine, because extra free food for the very poor frees up their extra spending power (that otherwise would have had to be used to buy that food) – and experience suggests that most of this spending is used to increase food consumption further. However, the fiscal costs of carrying large stocks readily become prohibitive.

18　Extra MV-based tropical output does, however, help the poor who eat partly imported grain, by restraining food prices at *world* level somewhat [Flinn and Unnevehr, 1984], although (except possibly for rice) the effect has been much smaller than that of growth in European and North American cereals output. Such world price restraint helped to slow down inflation for poor people in Indonesia, the Philippines and India, while such countries are – or were – net rice importers [Siamwalla and Haykin, 1983].

19　This can combine with 'Type III' second-generation MV crop effects (Chapter 3, i). For example, producers in Madhya Pradesh are now reaping the benefits of the rapid diffusion of modern soybean varieties on land that was fallowed in kharif. Technical change in soybeans has partially redressed some of the differential impact of technical change in wheat.

20　This is especially important because MVs, being selected for high ratios of grain to dry matter, tend to have thin husks, prone to penetration by insects. For major field pests and vectors, this effect is more than offset – indeed, is sometimes itself avoided – by breeding plants for resistance, tolerance, or (temporal) avoidance of the insect. Probably rightly, IARCs have judged that it is seldom worth *breeding* for resistance, etc., against the 1–3 per cent losses [Boxall *et al.*, 1978] caused by insect pests in storage. However, pesticides to defeat them *are* worth the farmer's while – and carry cumulative health risks. This probably helps justify *screening* for storage-pest resistance at IARCs.

21　Barely 5–6 per cent of people with insufficient dietary energy – and a similar proportion of those with insufficient protein intake – would not get enough protein if they received sufficient extra energy; most protein deficiency (and there is *much* less than was believed in 1960–75) is a side-effect of calorie deficiency, as the victims of the latter burn up high-protein foods for energy uses. Some micro-nutrient deficiencies – especially in vitamin A and iodine – are very grave, but can usually be met much more cost-effectively by fortifying an appropriate carrier (e.g. salt), than by other strategies such as MV breeding or screening. For discussion and references, see Lipton [1983].

22 Transfer of 'Indian' MV sorghums to West Africa has been unsuccessful, but there were signs in 1986–7 that – suitably crossed with local varieties – they might fare better in Eastern and Southern Africa.

23 Of course, it needs to be proved that MV research in coarse crops, and/or for less-favoured areas, can succeed. The poorest *as consumers*, it might be thought, would benefit more from concentration on the most promising crops and areas. However, that is not so if such concentration denies the poor the extra purchasing-power (entitlements) over the food. Also, extra agricultural research into the Punjabs, etc., must bring decreasing returns – and neglect of backward areas must conceal opportunities.

24 We have argued that there are grounds to expect, in the wake of MVs, a shift from workers to landowners in the share (though not absolute amount) of income command over food. However, the associated shift from pulse production to wheat production – with its much lower cost per calorie, and probably only slightly lower employment per acre – will reduce this undesirable real-income shift away from the poor.

25 Wheat typically contains 8–12 per cent protein, and legumes 15–25 per cent. If MV wheat produces twice as much weight per acre as legumes – or more – it can outyield them in protein *per acre*.

26 This is not to deny the usefulness of reduced cooking time, cost, or loss to the poor – though whether such features are sufficiently attainable and variety-specific to be worth *breeding* for (as opposed to *screening* varieties for) is doubtful.

27 This need not mean that incidence and severity of undernutrition have declined (as they have in East Asia, a few MV lead areas of South Asia, and much of Latin America, but probably not elsewhere) – only that they would be worse without MVs.

28 It has been argued that, with bulky crops like cassava and yams, children cannot get enough nutrients; they are full before they are properly fed. Hence the stress in some quarters on breeding MVs for 'energy density', (oilseeds?). More research into *actual* gruels and other components of weanlings' diets, however, is needed before one makes possibly costly attempts to re-jig (or add to) breeders' criteria for such reasons. Do children really get ill, even die, because parents persist in serving energy-diffuse meals although they can afford to do otherwise? We doubt it.

29 Some such families are forced into larger-than-usual distress sales (gross) immediately post-harvest; such families, having sold more than usual (gross) from a smaller-than-usual harvest, must then strive – by borrowing or otherwise – to buy back more than usual (at higher prices) later in the year. *Net* sales fall more than gross.

30 This fall in labour incomes, and hence in demand for food, in bad years does moderate (but not normally remove) the rise in food-price inflation; the two effects reduce poor people's consumption more together than either separately, but not by as much as the sum of the two effects.

31 Many of these African countries are, or have become, food-staples deficit countries in typical years, despite the maize hybrids; these, therefore, have not enabled governments to hold large inter-annual stocks. Nor are the families at nutritional risk likely to be able to do so. Hence the argument on p. 233 that, even if MVs had raised output instability, they had increased stocks and hence lowered consumption instability – is probably inapplicable in these cases, though the argument does apply to Zimbabwe, a grain surplus country in normal years.

32 Death-rates in the first year of life (typically below 10 per 1,000 in developed countries) are at least 150–200 among the poorest 10–20 per cent of households in LDCs; undernutrition is the main, or a major contributory, cause of at least one-third of these deaths. In other age-groups, differences among death-rates are both much smaller and much less related to nutrition. Pregnant and lactating women appear, on recent evidence, to be much less exposed to risk from undernutrition (to themselves or their unborn children) – i.e. to have greater capacity to adapt upwards the efficiency of energy conversion – than was once believed [Whitehead *et al.*, 1986].

33 Special care is needed for post-harvest activities. These are usually particularly important for women's income – yet particularly prone to displacement, partly assisted by MVs with thin husks that are not amenable to non-mechanical pounding. Such effects can harm nutrition (via poor women's income entitlements to food) much more than the extra 'food availability' due to the MVs helps nutrition [Greeley and Begum, 1983].

34 We reject the denial, by Behrman and Wolfe [1984], that low incomes are by far the main cause of inadequate household access to dietary energy. Many of their variables, other than income, which appear to account for differences in intake among households, are themselves largely due to income differences.

35 This requires a special explanation even in the context of the generally lamentable (and worsening) state of the statistics on

smallholder food production in Africa. Work in progress by J. von Braun at IFPRI, however, suggests that, in The Gambia, household adjustments to improved earning opportunities from 'men's' vis-à-vis 'women's' crops are complex and often counter-intuitive; and that, after such adjustments, household income gain from MVs in 'women's crops' may prove *less* helpful to women and girls than similar gains from 'men's crops'.

36 Robustness, of course, requires many 'characteristics'. Most of these are capacities (often polygenic) to cope with a wide variety of biotic, water-related, and other risks.

37 This would not, presumably, be the case in the substantial parts of Nigeria, Papua New Guinea, and other LDCs where marketing is largely a female occupation.

38 Even fewer face a lysine constraint on their utilization of protein. It is not the binding amino-acid in a maize-only diet, but the inadequacies in the actual diets of vulnerable families (especially infants) on diets including a maize staple, that are important.

39 Similarly, research to find MVs that avoid 'waste', by reducing grains left in the field after harvest (or the proportion of rice brokens), may damage nutrition among the poorest gleaners (or gatherers) who had relied on such crumbs from the tables of the less-poor.

40 Even this is an optimistic view; pessimists claim, for many crops convincingly, that there is a trade-off between yield potential and quality characteristics, especially protein content, however long the breeding programme continues.

6. Putting Together the MV-Poverty Mystery

(a) Adding up to a problem: holistic solutions?

The problem is: why have MVs, apparently good for the poor, not improved their lot much more? We showed in Chapter 2 that the *biological features* of MVs were good for the poor as farmers, workers, and consumers, probably increasingly so as IARC and other research responds to their problems of pest and water risk, and to their need for high conversion efficiency at low input cost. Chapter 3 showed that in most MV areas *'small farmers'* (often after a time-lag) adopted no less widely, intensively, or productively than others. Chapter 4 showed that MVs increased *labour use* per acre-year, especially via hired employment, albeit less so recently than in the late 1960s – raising the real wage-bill, and thus the total returns to the working poor (though real wage-rates rose little, while labour's share in income usually fell). Chapter 5 showed that poor people's consumption and nutrition were better and cheaper with MVs than without them.

At each stage, however, we have had to make major qualifications and reservations, both about the findings and about the poverty-orientation of some agricultural research.

(i) Threats to crop diversity, and hence dangers from diseases and pests, require more strategic concern from IARCs. Otherwise, the poorest farmers, who cannot afford back-up chemical protection, are at greatest risk (section 2, k).

(ii) 'Small farmers' are not the same as 'farm households of poor people'. Research on the good performance of the first group in MV areas leaves big gaps in our knowledge about the welfare of the second group there (section 3 a).

(iii) In some areas without MVs, poor farmers have probably lost out because of competition from lower-cost cereals from MV lead areas. The non-MV 'Type II' areas (Chapter 3, i) include most of Africa, and much of the semi-arid and upland-rice zones of Asia and Latin America. Their populations are very large and often very poor. The impact of MVs in 'adopting' areas on small farmers elsewhere has been neglected by researchers.

(iv) A growing proportion of the world's poor do not receive income mainly as own-account farmers, but as rural labourers; yet most research on MVs' poverty impact concerns 'small farmers'. MVs' success in creating work (and labour-income) cannot always be separated from a less happy side effect. MV-linked crop intensification may later encourage labour displacement, first at new seasonal peaks but later year-round, via tractors, threshers, weedicides, etc. Research has sometimes supported the wrong way of meeting MV-related peak labour demands; there is more IARC work on ways to cut the unit cost of farming by using weedicides or tractors (and hence to replace labour) than on ways to increase the efficient use of hoes or hand-weeders, let alone on how to relieve labour bottlenecks via migration from non-MV areas, or from places with different seasonal peaks.

(v) There must have been gains from MVs to poor people as consumers. However, the supporting calculations are largely hypothetical. We do not know what farmers or governments would have done about availability to the poor of domestic, or imported, cereals in the absence of MVs. Nor – even if MVs do restrain food prices – do we know to what extent employers can capture the gains by holding back wage increases. We do know of major cases, such as India, where the incidence of undernutrition has failed to decline, despite big MV-induced rises in food output. There, such rises have displaced food imports and raised stocks, but have not substantially increased food availability per person, especially among the poor.

(vi) Consumer gains, moreover, have not been helped by most IARC nutrition research. It has diverted plant-breeding resources away from increasing the yield and stability of output of cheap calories. Instead, it has emphasized increasing plants' proteins, amino-acid balance, and palatability. Such issues are at best of secondary importance; often they are unreal; at worst, their pursuit makes staple foods dearer.

(vii) Benefits to the poor are reduced by problems of timing. Poor farmers usually adopt MVs late, after better-off early innovators have raised output supply and input demand; hence poorer, later innovators pay more for inputs, and get less for outputs. Poor employees still usually gain work in a switch from TVs to MVs, but less so than in the early 1970s; real wage-rates for the unskilled, too, rise much less in the wake of MVs in the longer term, as labour moves to MV areas.

*   *   *

Despite these major reservations, the balance of advantage to a typical 'poor person' in the Third World, from MVs, appears large, if we 'add up' their various effects on such a person as small farmer, hired worker, and consumer (typically, she or he is all three). Yet, in spite of such all-round gains to poor people (as well as others), few countries of sub-Saharan Africa contain many farmers or national research systems that have worked systematically with MVs; and since the late 1960s Africa's poor have become poorer [Ghai and Radwan, 1983]. In South and even some East Asian countries or sub-regions, massive spread of MVs – although accompanied by clear, significant, and fairly steady growth of real average income per person – has nevertheless brought no clear uptrend in unskilled labour incomes, nor in real income per person in the poorest two household deciles [ADB, 1977, p. 63; Griffin and Khan (eds.), 1977; Ahluwalia, 1978, 1985; Lipton, 1983].[1] How is this possible?

Part of the answer is that the MVs have had to contend with population-linked factors weakening the poor: not just with absolute rises in the number of persons requiring food (and hence in its local price), but with induced rises in the supply of labour – although in many places land cultivated could not be significantly increased – and, increasingly, with more unequal access to land, as it became scarcer [Hayami, 1984; Hayami and Kikuchi, 1981]. But this is not a complete explanation. Consider the Indian Punjab. There, MV-induced real income growth far exceeded the growth of workforce or population. Moreover, MVs did not much affect the distribution of land ownership – most, though not quite all, the adverse redistribution of owned land had taken place before MVs arrived, and the resulting maldistribution of owned and occupied holdings was by Asian standards modest [Bhalla and Chadha, 1983]. Yet in the mid-1970s, over a decade after the arrival of MVs, the proportion of Punjabis who were in absolute poverty appeared greater than in the early 1960s [Rajaraman, 1975]. Also, the real value of annual wage income per agricultural labour household in Punjab-Haryana in 1974–5 was apparently 11 per cent below the 1964–5 level [Bardhan, 1984, p. 190]; modest rises in real wage-rates [Sheila Bhalla, 1979], and in employment, had been outweighed by rises in the population (including immigrants to this leading MV area) that relied mainly on agricultural labour for employment.[2]

These apparent deteriorations, admittedly, were offset (to an unquantifiable extent) by extra real income from Punjab's MVs for people even poorer than Punjabi labourers: migrant workers from then non-MV areas in Eastern Uttar Pradesh and Bihar. Also, the deterioration was reversed as MVs spread, and as wheat (and later rice) performance improved further. Yet only in the early 1980s did it become certain [Bhalla and Chadha, 1983; Chadha, 1983] that the poor of the Indian Punjab have become significantly better off than before MVs arrived. How is it possible that, with all the 'bit-by-bit' logic and empirical work pointing to gains from MVs by the poor, it is such a hard and slow process, in the (not especially unequal) conditions of the Indian Punjab – probably the Third World's leading area for MVs – for the poor to establish clear gains?

We suggest three linked explanations, from different approaches to social analysis: general-equilibrium economics, political economy, and comparative history. Although these approaches are practised by different, often mutually hostile, specialists, all three suggest 'holistic' methods of analysing how MVs affect the poor within 'whole' social units; and all three concur that, because a national or village society or economy is a complete and interacting set of parts, the *adding-up approach* implicit in almost all the analyses of how MVs affect the poor, including Chapters 3–5 above, is at best seriously incomplete. In other words, it is never enough – and can be very misleading – to take the various effects of MVs at the level of the individual poor farmer, worker or consumer separately, add them up, and infer a total effect on 'the poor'.

These three holistic critiques should not be overstated. They do not create serious doubt that, without MVs, most of the world's poor would today be poorer still. However, the critiques contain enough force to suggest important changes in both agrotechnical and socio-economic research priorities.

(b) 'General equilibrium' in economics and the MVs

Very few economists in the standard Western tradition have tried to assess the total effect of MVs on poverty. These attempts have mostly used the adding-up approach [for outstanding examples see Hayami, 1984; Barker and Herdt, 1984]. Yet, even without any 'holistic' alternatives, we have to be uneasy about that approach. It 'adds up' effects on poor people as farmers, workers and consumers, but does not reconcile these effects. And it tends not to look beyond first-round impact.

The problems of *adding up* unreconciled first-round effects are exemplified by the various calculations of the gains to consumers from greater supply of post-MV cereals, and hence lower food prices in economies with restricted imports [Scobie and Posada, 1978; 1984], or in large regions with little net grain trade with other areas [Evenson and Flores, 1978]. Such calculations often show those gains well above the GNP gains from MVs on most assumptions. This implies net losses to producers of MV-affected crops. Usually, such net losses are much too big to be contained in non-MV areas, or even in larger farms in MV areas. Hence some net losses go to adopting, surplus, but still poor MV growers. How do we reconcile this finding from the consumption studies of MV effects, with the many production studies (Chapter 2) claiming that such farmers gain?

An example of the inadequacy of *first-round* analysis can be drawn from the consumption side alone, and has been discussed already. The main single benefit claimed for MVs (Chapter 5, b) is that, by making the food staple cheaper, they enrich its consumers – above all the poor, because they spend the largest part of their incomes on food staples in developing countries (50–80 per cent, as against 20–40 per cent for better-off groups). However, we know that (partly due to population growth) poor, unskilled labour is in very wage-elastic long-run supply.[3] Therefore, when extra MV-based food (by restraining food prices) raises the real wage, many more person-hours of unskilled labour are offered for sale. These compete for work, driving the real wage back down again. 'Responsive real-wage deceleration' operates (Chapter 5, b.).[4] Thus cheaper food due to MVs appears to benefit employers more than poor workers, once second-round and subsequent effects are allowed for.

To transcend one-round adding-up, standard 'Western economics' offers three sets of mathematical methods called general equilibrium (GE). Neo-classical GE analysis assumes that each competing firm (including farms) seek to maximize profit through its choice of products to make and sell, and through its choice of ways to make those products – e.g., by varying its mixture of purchases of types of labour, equipment, and inputs such as seeds; and that each household seeks to maximize its own welfare by varying its choice of purchases out of a given income, in response to their relative prices and to their 'utilities' (contributions to household welfare). Neo-classical GE – by looking at the sequence of changes in prices, outputs, and consumption that is implied by transactions between each 'maximizing' firm and each household – finds the point at which the economy settles down, i.e., at which people's actions do not change the set of outputs that it produces each year. Walrasian, or neo-classical GE, then, is analysed as a consequence mainly of *transactions between firms and households*, each responding to, but none manipulating, the variations in prices of transacted items – commodities, labour, or other inputs – induced by changing supply and demand.

The other two forms of GE analysis concentrate not on transactions but on flows. Leontief, or input-output, GE is analysed as the end-result of a given set of demands for commodities by households. It enquires what *flows between firms and firms* are needed to produce those commodities: flows of inputs; of labour and inputs to make those inputs; of labour and inputs to make 'inputs to make inputs'; and so on. Keynes-Goodwin, or matrix multiplier, GE examines the spending pattern of households out of a given amount of extra income; their extra purchases provide incomes for a new group of households, who in turn spend some of their gains on a further set of commodities, enriching a third group of households whose members make those goods; and so on. This GE is thus the result mainly of successive *flows of spending between households and households*, mediated by firms.

Readers are politely requested to defer their questioning about the evasions in the above condensed account. Economists will be unhappy about the fudged assumptions on markets and prices, notably in our use of the word 'mainly'. Biologists, and students of agricultural development, will wonder how these three processes, with their strange assumptions, bear on the realities of rural life, especially on responses to MVs by poor farmers, workers and input suppliers – many, in each group, integrating their activities as households and firms. Sections c–f should reduce some of this unease.

What these sections cannot do is reconcile these conflicting versions of general economic equilibrium. They conflict, in assumptions and approach, with each other. They conflict, too, with the 'softer', less computable, but perhaps more important, holistic effects of MVs via social systems (sections g–k). They also conflict – but desirably so – with some of the puzzling results and inconsistencies of the partial-equilibrium, first-round, adding-up approaches to the effects of MVs. In so doing, the GE methods may provide new insights into why MVs' impact on the poor has been less favourable than the partial approaches suggest.

(c) Neo-classical GE

The first of the three GE approaches was originated by Walras [1902]; an accessible account of later developments, by their leading exponent, is Arrow [1968]. The approach allows us to analyse how changes in demand or supply lead to new sets of price incentives. These are acted upon by 'profit-maximizing' producers and by 'utility-maximizing' consumers respectively. Producers respond to changed demand patterns and demand prices by adjusting the use of their land, efforts, machines, and buildings, so as to change the composition of what they supply to the market; consumers respond to changed cost structures and supply prices, by adjusting their purchases. These responses again change prices and costs. A long series of price changes, responsive adjustments in the structure of supplies and purchases, further price changes, etc., was shown to 'converge' eventually – to new, stable equilibrium levels of prices, wages and outputs [*ibid.*; Arrow and Debreu, 1954]. Such Walrasian GE analysis is able to make its strong predictions only on rather strong assumptions: that – up to the level of use where further efforts do not cover extra cost – land, labour [Binswanger and Quizon, 1986, p. 5] and equipment are fully employed; that no producer (or consumer) is big enough to influence prices, e.g. by selecting a different level of sales for purchases; and that labour and other non-land current inputs are perfectly mobile in search of higher incomes, while land can shift uses freely but not, of course, location.[5]

These assumptions are used to analyse the equity effects in GE of different forms of technical progress in agriculture (such as the labour-using, land-saving and fertilizer-using MVs) in two notable papers. Binswanger [1980; for the above assumptions, see pp. 195, 210] simplifies the problem by examining the distribution of extra income, due to technical progress, between labour, land and capital in one 'agricultural' and one 'non-agricultural' sector. Quizon and Binswanger [1983, especially p. 526] examine distribution on a different set of simplifications, namely the existence of an agricultural sector only, but with internal differentiation not only between sellers of labour, land and capital, but also between regions. These models are developed for economies with and without foreign trade; if and only if the Government allows variable levels of foreign trade 'and the country or region is [too] small [to affect world] commodity prices, [they] are given from the outside' [Binswanger, 1980, p. 195], and are not affected by technical change in agriculture. MVs can still shift wage/rent ratios, but not food prices. The whole class of Walrasian 'computable GE' models has been used to generate predictions of the effect of alternative MV policies on various poor groups, in India and elsewhere [de Janvry and Sadoulet, 1987].

The flavour of just how Walrasian GE differs from the adding-up approach to the effects of MVs is best given by citation. Adding up 'neglects GE effects such as the effect of [MVs] on the demand for output (and hence inputs) of other sectors via price [e.g. MVs bid up demand for fertilizers. It also neglects] 'income effects' – not only the effects of price changes on income, but also such effects as that, when MVs make them richer, landowners buy more clothing, and thus provide incomes to workers and employers who make it.

Neglecting GE. . . is unimportant if a sector or region is very small, but it may become unsatisfactory when we consider very large sectors. . . In [adding-up approaches] all factors in one sector or one region either gain or all lose, whereas in GE one factor will always lose and another one will always gain [Binswanger, 1980. pp. 195, 210].

That is because only by GE can we allow for feedback effects of labour migration, on its regions of origin as well as on the destinations.

Walrasian GE analysis of the effects of MVs is at an early stage. Results still tend to change rather drastically as models are refined.

Most troublesome [is prediction of effects of technical change, etc., on] income distribution [between] landowners and labourers. A complicated interplay across markets for different commodities, land and labour prevent[s] easy generalizations except that in all cases labour-saving technical change adversely affects labour [Binswanger and Ryan, 1977, p. 230].

It would follow, conversely, that 'in all cases' *technical change raising demand for labour* per unit of land, *such as MVs, would benefit labour.*[6] Yet once land-labour-capital interactions and the special features of agriculture are allowed for more fully, the result evaporates [Quizon and Binswanger, 1983, p. 532]:

'When *[technical changes that* save or *use labour] occur at the expense of inelastic land.* . . no definite signs can be proved', *so that the effects on real or money wage-rates or shares can go either way.* 'Intuitively [this is because an innovation that uses more labour per unit of land usually also uses more] capital relative to land, but it [saves] capital relative to labour. . . the net effects are unclear'.[7]

It has indeed been found that labour's share falls in most areas following the introduction of MVs. This is inconsistent with the 1977 and 1980 GE predictions, but consistent with the 1983 GE doubts. The non-GE adding-up approach has a simple explanation: with MVs either 'neutral' (i.e. raising by the same proportion the demand of farm employers for labour and for land) or labour-using and land-saving, there would be 'large price rises for [labour] in relatively inelastic supply' in the short run [Binswanger and Ryan, 1977, p. 228; cf. Anderson and Pandey, 1985, p. 9]; but then employers enthusiastically seek to replace labour with threshers or weedicides, reducing the labour share for a time. Perhaps, indeed, since full employment and mobile labour and capital are assumed in GE, 'where unemployment is large and [mobility slow], the [adding-up] models will do better at predicting distributional outcomes for. . . 5 to 10 years' [Binswanger, 1980, p. 211]. However, the following claim that Walras-style 'GE forces. . . will tend to dominate in the long run', determining the ultimate impact of, say, MVs upon labour-land distribution, depends on at least a *tendency* towards full employment. In LDCs, however, especially with rapidly growing populations, the tendency is in the opposite direction [Lipton, 1984a]. Also, any sort of GE effect can dominate, in predictable fashion, only if the technical change does not alter the structure of production and distribution too greatly (Chapter 6, j).

Nevertheless, work to develop computable Walrasian GE models could be very useful, ultimately – at least for countries such as India, Pakistan, China, Bangladesh, Indonesia and the Philippines. These countries are big enough that internal, rather than world, food and food-transport prices[8] are of great importance in the medium term. Also these countries' data base permits some estimation of outputs, inputs, incomes, prices, and elasticities for major sectors, products, regions and income-groups. For this small – but very populous and poor – group of developing countries, such models, if greatly improved, could help to provide much more reliable guidance on the impacts of options in agricultural research and diffusion, e.g. upon labourers or backward regions, than is now available [Quizon and Binswanger, 1983, pp. 533–6].

However, the numerical prediction of equity impact from such models, to be useful, will require substantial work to develop them. An extremely useful comparison of computable Walrasian GE models for six countries – India, Peru, South Korea, Mexico, Egypt and Sri Lanka – shows that the predicted outcomes of extra output due *purely* to MVs (or other exogenous factors) for three poor groups – landless workers, small farmers, and urban workers – depend critically on assumptions about how wage-rates are determined; about whether extra output cuts food prices, or alters net imports; and (in the long run) about how savings plans become consistent with investment plans [de Janvry and Sadoulet, 1987].[9] Work to incorporate appropriate assumptions, and to make the models more robust, could usefully be done in association with specifications of the inputs and outputs expected of a pending MV-based technology, as partly attempted in Ghodake [1983] and Ghodake and Kshirsagar [1983a]. The work will need to disaggregate farm products at least into MV-foodcrop, other foods, and other sectors; to specify key products and factors of the non-farm sector and their role for agriculture; to introduce the possibility of responsive changes in investment; and at least to explore the role, in the model, of the assumptions of perfect competition, notably full employment and instantly mobile non-land inputs. Also such 'Walrasian' work, on GE in factor and product markets with flexible prices and mobile non-land factors, needs to be integrated with two equally valid GE approaches to the total effects of MV options on the poor: the approaches implicit in the work of Keynes and Leontief.

(d) 'Keynesian' GE: multiplier equilibrium

From the work of Keynes, or rather from his fully acknowledged borrowing from Kahn [1931], has been extracted just one 'Keynesian' tool, the multiplier, for the task of examining GE effects of the introduction of new agricultural methods. The multiplier is a useful contribution to this task, and a valuable corrective to the Walrasian analysis, in which flexible prices (including wages) and full employment are assumed[10] – not very plausibly in the short run. However, the central assumption of multiplier analysis – that extra resources are available to respond to every rise in demand (e.g. for industrial products, from farmers enriched by big yield rises with MVs) without rises in prices or average cost – is also not plausible in the short run. In real life, extra demand calls forth some extra supply, generates some extra work and real income, but takes time to do so, and (especially in the short run) pushes up costs and prices somewhat.

But let us see how the multiplier assumption works. We shall describe how multiplier-based GE is used to analyse the impact of MVs on the poor. Then, however, we argue for a much more thoroughgoing incorporation of Keynes's work into that analysis: for a transition from 'Keynesian' to Keynesian GE. Throughout, we shall assume no changes in imports or exports – this assumption keeps the argument simple, though it could readily be dropped.

Suppose that output of all product lines in India could be increased (without big rises in average cost) if extra demand made it pay the producers to do so. Then suppose there is an initial boost to farmers' income, due to extra MV yield and sales;[11] and that, of any extra income, one-quarter is saved, and the other three-quarters is spent, all on Indian products. Then total extra income in rupees (call it R) equals the initial boost to farmers' income – call it B; *plus* the proportion of B spent (namely $3/4$), which enriches the Indian workers and businessmen who make the extra goods the farmers buy; *plus* a further $3/4$ of that proportion, as these workers and businessmen spend their extra income; and so on. So

$$R = B (1 + 3/4 + (3/4)^2 + ...);$$

after numerous rounds of spending, this process tends to the result

$$R = B \frac{1}{(1-\sqrt[3]{4})}$$

or 4B. In this process the proportion of initial extra income consumed (here $\sqrt[3]{4}$) is called the *marginal propensity to consume* or MPC; the amount by which initial extra income must be multiplied to obtain total or final extra income –

i.e. $\frac{1}{1-MPC}$, here $\frac{1}{-0.75 \text{ or } 4}$ -

is called the *multiplier*.[12]

How is this useful for analysing the effects of MVs? For a start, we need to split up the workings of the MPC. The gainers from MVs are spending particular sorts of extra income, on particular sorts of purchases. To trace such effects, instead of a single MPC, we can use survey materials to estimate the proportions of extra income (received by workers as wages, and by employers as profits) from each product line – e.g. rice, wheat, shirts, bicycles – that will be spent on each of those product lines in turn; then divided between wages and profits there; then partly spent again; and so on. This information can readily be written as a square table of 'marginal propensities to consume', and used to calculate a whole set of 'multipliers', called a matrix multiplier. This shows how *initial* rises in income for particular groups, such as farmers adopting MVs, after successive rounds of spending (and assuming that, within each product line, there is a roughly constant ratio: of wages to profits; of extra spending on each product to an extra unit of wage income (and similarly for extra profit income); and of labour per unit of output), will eventually lead to a new set of *final* incomes for workers and employers making each main commodity [Goodwin, 1949].

Since Keynes [1936] was analysing the conditions of Britain's 'Great Depression' – with widespread spare capacity, fairly free competition, and mobile unemployed labour – he assumed, quite naturally, that this increased demand can generate extra output from domestic suppliers, further extra domestic supplies to meet the new consumer demand of those suppliers, etc., with rather little extra imports (or diversion from exports), or cost-price inflation for either products or factors. In most circumstances, however, some of the extra demand may suck in imports, divert exports, or bid up prices of domestic inputs or outputs whose supply cannot readily be increased. In so far as this happens, the extra domestic money income does not correspond to extra domestic real output. If we cannot predict how the effects of extra rounds of spending will be divided between higher real output, higher prices, and higher imports, the 'Keynesian' multiplier GE approach is rather less useful.

Therefore, in applying this approach to predicting the income-distribution effects of MVs and associated changes,[13] economists have usually analysed *consumption linkages* in rather small areas – a decision also dictated by the formidable data requirements. The areas studied have all been largely irrigated rice regions: North Arcot in Tamil Nadu, India [Hazell *et al.*, 1987]; and the Muda and Funtua irrigated scheme areas, in Malaysia and Nigeria respectively [Hazell and Roell, 1983; Bell, Hazell and Slade, 1982]. Within a small area, it is reasonable to assume that local income rises for farmers enriched by MVs will not, via demand, much affect *local* prices of consumer goods bought by farmers;[14] and one can make a decent shot at identifying which of such extra items can be produced locally.

Obviously this process of successive rounds of demand, converging on a 'Keynesian' GE, is a very important part of the MVs' effects, especially on non-farm income and employment. It is not captured by the adding-up, first-round approaches of Chapters 2–5, nor by Walrasian GE. Unfortunately, the matrix multiplier poses heavy data requirements for computable estimation. One needs to establish the spending behaviour of many different groups of rich, average and poor farmers, and other producers, in response to income changes.

The fullest estimate so far published follows out the effects on local incomes, via local demands, of the income expansion initially due to irrigation, fertilizer, and MVs in Muda, Malaysia [Bell, Hazell and Slade, 1982]. It shows that each \$1 of extra initial (first-round) income generated for food staples producers in the MV context, when it is again spent, generates a further 80 c. of extra local incomes. The authors show how this is divided among local households in various income-groups. The spending of the 80 c. is not broken down between real-income and inflation effects, but locally the latter cannot have been large (due to competition from other areas). However, even this is only a 'second-round' effect. A later study, in press, looks at multiple-round effects – i.e. the full matrix multiplier system – in North Arcot, India [Hazell *et al.*, 1987]. This study argues that local experience supports the assumptions that extra demand for local non-farm products in fact calls forth local output and employment, and that these do not respond to extra local demand for farm goods, once the extra MVs have been grown and the first-round income created.

Similar assumptions are made in an attempt to use the Muda results, and less detailed parallel work on the Funtua project in Northern Nigeria, to estimate the poverty effect of alternative policies for the allocation of MVs [Hazell and Roell, 1983]. This estimate relates (i) income per person in the group of households – call it G – who *initially* obtain income-enhancing innovations such as MVs to (ii) the effects of G's extra spending upon income of the poorest, second-poorest, etc. deciles of households in the project area. It is assumed that, once the MV-linked innovations are in place, extra local demand for farm products (e.g. by those who gain income from the MVs) calls forth no *further* extra local farm output (only 'imports') – but that extra local demand for non-farm goods and services generates extra local output and incomes, without higher costs, prices, or 'imports'. Surprisingly, $100 of spending out of extra MV income appears in both Muda and Funtua to generate more demand for products from the poorest local households – and hence more income for these – if G comprises big farmers rather than small ones. Big farmers seem likelier to direct their extra outlay towards locally-made non-farm goods and services, which are *assumed* to be more readily produced – labour-intensively, without inflation, and in ways helpful to the poor – than farm goods and/or 'imports' from other areas.

But one cannot infer [Anderson and Pandey, 1985, p. 10] that the distributional effects of allowing MVs to benefit bigger farmers first are better when GE effects are considered than they appear to be from the adding-up model. The above approach looks only at the *first round* of *local, consumer* spending on *non-farm* outputs, as a first step towards a '*Keynesian*' *GE*. (i) First-round spending is likely to be much more different in impact, as between various initial gainers from MVs, than the total of all the successive rounds of spending from different hands (the North Arcot study will tackle this by looking at successive rounds). (ii) Since non-locals do not gain direct project benefits (or many gains from MVs), it seems implausible that local spending of such benefits – depriving non-locals once again – will be better for income distribution than non-local spending.[15] (iii) When big farmers get extra income from MVs, even if they devote higher proportions of their extra *consumption to local products* of poorer households (often, to direct labour services like domestic service), the gains to the local poor are reduced because big farmers may well save more – and thus devote a much smaller proportion of extra MV *income to consumption* (local or other) than do poor farmers. (iv) It is not quite right, especially in the long run, to assume that only non-farm spending helps the local poor: that extra farm products cannot be produced (or worked on) by the local poor, in response to extra local demand, once the MV gains have been exploited. This assumption appears to see a direct loss, for the poor, in small farmers' propensity to disburse a bigger part of income gains from MVs upon food than do big farmers. But food is made highly labour-intensively; is not all staples; and may be supply-responsive, to extra demand, even in regard to local MV products after full adoption. (v) Further, even if rich farmers' consumption behaviour is more pro-poor (not just more pro-local-nonfarm-firstround-poor!) than that of small farmers, their production behaviour – which is not considered in these 'Keynesian' GE analyses – probably is not; it often involves lower ratios of outlay on labour hire to outlay on weedicides, tractors, etc.

Furthermore, these models of 'consumption linkages' tend to blur three sorts of choice: between policies tending to allocate MVs (with supportive extension, fertilizer, credit, etc.), to (i) big or small farms, (ii) households with big or small total income, (iii) households with big or small income per person. The overlap between these choices is known to be extremely imperfect. For example, 'How do the smallest 10

per cent of farms use extra income?' is quite a different
question from 'How do the poorest 10 per cent of rural people
use their extra income?' (Chapter 3, a). MV policy choices, if
they are to benefit the poor, mainly relate to the first question;
currently available 'Keynesian' GE models (and the supporting
household surveys) mainly relate to the second.[16]

A multi-round tracing of consumer outlay by MV benefici-
aries, converging through the multiplier on a new GE, is
nevertheless an important part of the total picture of MV-
poverty relationships. So far, however, economists have taken
only a few steps in this direction. For example, estimates of
impact on welfare via self-consumed vs. sold MV-crop output
for each income-group, along lines indicated in Hayami and
Herdt [1977], would need to be included in any complete
model of 'Keynesian' GE distribution impacts. So would (i)
non-local consumption effects for rich and poor; (ii) allocation
of the effects of extra demand (e.g. by farmers initially gaining
from MVs) in raising prices, imports, and real domestic
incomes (requiring integration of the 'Keynesian' with the
Walrasian GE approach); and (iii) Leontief effects, dealt with
in section 5, f.

Yet the consequences, for the poor, of successive rounds of
spending out of income, when income changes due to MVs –
not only out of income gains to farmers, discussed here, but
also out of income gains to consumers (who can reassign some
income away from food as MVs make it cheaper), and perhaps
out of income losses to 'Type II regions' (Chapter 3, i) – may
greatly alter the conclusions of the adding-up approach. Some
'Keynesian' GE results therefore need to be used in forming a
reliable picture of what MVs have done to the poor. As with
the Walrasian GE, so with 'Keynesian' GE: the few studies done
so far get us from Square Zero to Square (or Round) One. That
is a long way, but as yet far from Square 64.

(e) Wider Keynesian equilibrium: relevant to MVs?

The multiplier method is obviously relevant to understanding
how MVs affect poor people in rural areas. But Keynesian
analysis and GE mean much more than deriving the final

incomes produced by successive spending rounds via the multiplier. Keynes's General Theory [1936], however, was designed for advanced Western economies. In Walrasian GE, long-run structural unemployment cannot exist; in advanced Western economies in the 1930s (and 1980s), it clearly did exist. Many ingenious evasions of the facts of involuntary unemployment have been advanced, but Keynes sought an explanation, not an evasion. How, in such economies, can households (each maximizing welfare) and firms (each maximizing profit) produce an *underemployment GE*? His overall argument – unlike the specific multiplier process – seems at first glance irrelevant to rural areas in poor countries.

To motivate readers in tolerating what will, at first, seem like a digression from the argument of this book, we recall Keynes's central theme: that markets for big aggregate items (like labour or cereals) were different from Walrasian micro-markets (for plumbers' mates, or for broken rice in a village) in one crucial way. A significant change in the price of a big aggregate would, as a rule, drastically change the real income of persons buying or selling it. This, in turn, would change their total demand, and its allocation among purchases, in ways that could offset, even reverse, the normal, Walrasian, equilibriating impact of the initial price change.

For example, in a Walrasian model, an unemployed worker could always get work by lowering his offer price, thereby persuading employers to hire more labour, so that unemployment could not persist; in a Keynesian model, if many workers do this and thereby bring down the real wage,[17] then aggregate demand for goods and services by labourers is likely to fall,[18] and this will provide a discouragement to employers that offsets – perhaps exceeds – the inducement to employ labour because it is cheaper. Similarly, if food gets much cheaper in the wake of MV innovations that cut costs and raise supplies, there are major *income effects* – both on its buyers and on its sellers – that change the whole level and pattern of demand for goods and services; the results probably cannot be adequately evaluated on a Walrasian GE model, although this can accommodate modest income effects,[19] and is an improvement on 'partial equilibrium' models.[20]

We first look at the sort of economy where Keynesian underemployment GE could not happen. Then, we ask why the anti-Keynesians (in the 1930s as in the 1980s) thought that it could not happen in real-life modern economies, either. Next, we summarize the Keynesian view of how underemployment GE may persist in developed economies, notwithstanding the apparent possibilities of reaching full employment via falls in wage-rates or interest-rates. Our exposition is deliberately crude, and makes little or no reference to monetary issues, or to the crucial issue of how expectations are formed [see Leijonhufvud, 1968, and Clower, 1968, for subtle and rigorous formulations]. Next, in the labour and capital markets in turn, we suggest analogues to the Keynesian approach that might help to explain the apparently disappointing aftermath of MVs for poor people in some developing countries. All this gives a sketch[21] of how a true Keynesian GE analysis of MVs' impact might look, and of how and why it might differ from a Walrasian or multiplier-'Keynesian' GE analysis. We assume a closed national economy throughout – no foreign trade, lending, borrowing or migration; to drop this assumption would greatly complicate the argument, without in essence changing it.

*   *   *

Suppose that all output is produced by family firms. They compete to sell consumer products – foods, cloth, carts – to one another. However, each depends entirely upon its own family workers for labour, and upon its own savings (i.e., in this case, family labour diverted from producing goods for sale or self-consumption) for investment in equipment or the improvement of land or works. In such an economy (i) planned savings equals planned investment, (ii) planned supply of labour equals planned demand for labour. This is not a pure subsistence economy, as there is trade in products; but there is no wage-rate to be adjusted, and no interest-rate on business borrowings (because there are no such borrowings). Departure from Walrasian GE to, or persistence in, a Keynesian underemployment GE seems impossible – at least, it cannot be caused by any problems in the working of aggregate supply and demand for labour, or of imbalance between savings and investment plans.

According to the position attacked by Keynes [1936] and attributed by him to Pigou,[22] even complex modern economies, in essence, tend to adjust in ways that remove structural unemployment, and in this respect to function like the above family-firm economy. As regards equilibrium in the capital market, everybody agrees that, in modern economies, the agents in a modern economy who save (mostly households) are different from the agents who invest, i.e. add to equipment (mostly firms). The standard neo-classical view is that these two agents' sets of plans, respectively to save out of income (or to lend) and to finance investment (or to borrow), are brought to equality by changes in the price of money-to-lend, i.e the interest rate. According to standard neo-classicals, this process does not require changes in real national income, and is always consistent with full employment of both equipment and workers.

Similarly as regards equilibrium in the labour market Keynesians and Walrasians concur that in modern economies most workers are hired by employers (instead of self-employed in family firms). However, Walrasians claim that employers' and workers' plans – respectively, to hire and to be hired – are equilibriated by movements in the price of labour, namely the wage-rate; and that this is always consistent with full employment, and requires no adjustment in real national income.

Thus unemployment cannot be due, for Walrasians, to inadequate demand for labour, or to inadequate planned investment. 'Unemployed' workers (or 'frustrated' business investors) are, on this view, simply people who are willing to work – and to incur the costs of seeking work – only at higher wage-rates (or to invest only if they can borrow at cheaper interest-rates) than the market will provide. Indeed, most Walrasians deny the possibility of involuntary long-term unemployment, and of frustrated demands to undertake business investment.[23]

From the depths of the Great Depression, Keynes challenged this view, on three grounds; and it is here that striking analogies arise to possible weaknesses in the view that MVs must 'add up' to benefits for the poor. First, he argued that there were lower limits below which wage-rates (or interest-rates) could not fall – but that such limits might well leave the rates still too high to persuade firms to buy enough labour (or to borrow enough for business expansion) to create full-employment, full-capacity levels of national production. Second, falls in *money* wage-rates (or perhaps money interest-rates), even if generated by excess supply of labour (or of savings), need not bring about sufficient falls in *real* rates to induce employers to absorb all the labour supply (or business-men to absorb the planned savings, given the national income). Third, even if there were falls in real wage-rates (or interest-rates), and even if there were no lower limits to these rates, those falls might not be able to induce an increase in demand for labour (or for money to borrow for business expansion) sufficient to revive the economy significantly – not, anyway, to do so enough to attain full employment.[24] The upshot of Keynes's case was that – because the markets for labour and/or savings, if affected by falling business demand, could in some circumstances not be equilibrated by downward adjustments in price (i.e. wage-rates or interest-rates) at a given level of national income and hence employment[25] – they could be equilibrated, in free markets, only by downward adjustments in that level.[26]

It is labour markets that provide the clearest analogies between the three Keynesian obstacles (to Walrasian price adjustments, i.e. wage cuts, as a path to full-employment GE) in developed economies, and the situation in rural areas of poor countries. How did those three obstacles allegedly operate in developed countries? Why did (and do) Keynesians deny that wage-cuts there could open the road to full-employment GE?

First, wage-rates often appeared to be set at a lower limit. Trade unions, or custom, or some notion of a 'family wage', or of a 'just wage' (corresponding, perhaps, to Marx's 'historical *and moral* subsistence costs' of producing and reproducing labour), kept rates above this limit – it was claimed – even if there was unemployment. Wages, said Keynes, were 'sticky downwards'.

Second, even where recession, via reduced demand for labour, led to lower money wage-rates, these often tended not to mean significantly lower real rates, because prices followed money wage-rates down. Competitive employers, faced with reduced unit labour costs, would judge that they must reduce product prices, because otherwise (i.e. if some of them tried to earn excess profits by keeping prices unchanged even after money wage-rates had fallen) their rivals would undercut them in product markets. So money wage-cuts might well not render labour cheaper (and hence more attractive) to hire, because wage-rates might well not fall relatively to the value of labour's product – or relatively to the cost of alternative ways of making that product, e.g. to labour-displacing capital (also probably cheaper, because also made by labourers whose wage had fallen).

Third, even if real wage-rates were to fall – e.g. if, as in Kalecki's model [Robinson, 1965], employers have enough market power, or collude sufficiently, to keep product prices up despite the cuts in money wages; or perhaps for other reasons [Keynes, 1939] – demand for labour need not rise, or at least not enough to end unemployment. Falling real wages per hour mean falling demand out of income, per hour worked, by labourers for consumer-goods. Any impact of cheaper labour in inducing employers to hire more workers (or more hours per worker) – unless, implausibly, it is large enough to raise the number of 'hours hired' by proportionately more than the fall in 'wage-rate per hour' – reduces the total demand, for consumer items, out of labour incomes. This reduction in workers' demand for the product of labour must create an incentive to *employ less* labour. That incentive offsets, and may exceed, the inducement to *employ more* labour because it is cheaper.

This account, despite its origins in the West of the 1930s, has many analogies that help us to grasp how a developing rural area today might stick at an 'underemployment GE', and thereby to understand persistent poverty in the wake of MVs. Before exploring these analogies, we must look at the 'capital market' – because Walrasians argue that incentives created for more investment, at times and places of high unemployment, can mop it up, even if (for Keynesian or other reasons) labour markets work badly (compare footnote. 24). Unfortunately, Keynesian arguments also suggest that simple price-response also often fails to bring about full-employment equilibrium via this capital- market route (i.e. by encouraging employers to

borrow more for investment, by making loans cheaper). In the
capital market, households' supply of savings (which they have
held back from income, and used to earn interest[27] instead of
to consume) is set against firms' demand to add to their stock of
productive buildings, machines, or land by means of investment.[28] In an economy with no foreign trade – or in an
economy with foreign trade in balance – actual savings must
always equal actual investment in any year: firms pay out
incomes (i.e. wages plus profits) equal to the value of making
consumer-goods plus investment-goods; households use those
incomes to buy consumer goods or to save; so *actual* savings
equals actual investment. However, given the rate of interest –
i.e. the return to saving, and the costs of borrowing[29] for
investment – the level of national income at which *planned*
savings equals planned investment may be too low to generate
full employment.[30]

A Walrasian would argue that this sad state of affairs cannot
last long, and would expect the following events to restore full
employment. First, as unemployment depresses demand and
hence reduces businessmen's wish to borrow, the rate of
interest falls. This discourages households from saving (i.e.
they divert income from savings to the purchase of more
consumer-goods, and thus employ more workers to produce
them), and encourages firms to borrow cash to pay for
investment-goods (also employing extra workers to produce
them). This raises the value of economic activity (income) until
– with planned saving equal to planned investment, at a higher
level – national income suffices to employ all who seek work at
the going wage-rate.

To this sequence by which interest-rates fall and thus
increase economic activity, a Keynesian has three objections,
closely parallel to the objections to the proposition that falling
wage-rates can price workers into jobs. First, interest rates
'cannot' fall below a certain level (in the West, during the
Depression of the 1930s, typically 2 per cent), because they
have to cover costs: transactions costs of lenders, and of
banking intermediaries; loan supervision costs; and the costs
of insuring or making good, to lenders, even a very small
incidence of defaults. Second, even if money interest-rates fall,
the real cost of borrowing[31] (i.e. the money interest-rate as

compared to the expected profit-rate on the product of new investment) may not, e.g. in a climate of decelerating inflation, or of worsening profit expectations. Third, even if there are falls in the real interest rate, they may fail to increase planned investment, and may also fail to reduce planned savings and thus increase consumption demand, at a given income level; lower interest-rates may thus fail to set up sufficient pressures to a new, higher equilibrium level of income, employment, and investment (equal to savings). Of the three 'motives for savings' discussed by Keynes [1936] – as a precaution; to prepare for future transactions; to speculate on rises in the value of the asset saved – not one is unambiguously likely to weaken in response to lower real rates of interest.[32]

Thus there exist Keynesian GE models with economies below full employment, and not tending towards it. The correction of such miseries via Walras-style adjustments of prices is deemed infeasible or ineffective. Money wage-rates cannot fall; or cannot affect real rates; or cannot, even if they do, stimulate employment much. Money interest-rates cannot fall; or cannot affect real rates; or cannot, even if they do, stimulate planned investment, and/or (by deterring planned savings) planned consumption, enough to raise employment much.

These models are controversial. To neo-Walrasian critics, they appear to rest shakily on the assumption that people's economic behaviour is irrational – on widespread failure to learn from experience. However, the models do give some account of a Western phenomenon widely perceived as 'involuntary unemployment'. The neo-Walrasian denials of this phenomenon, the claims that all the unemployed can price themselves into work, are frankly implausible: victim-blaming. Moreover, the period 1945–71 in the West was one of unprecedented growth and prosperity, apparently produced largely by quasi-Keynesian, counter-cyclical, public-works adjustments of 'planned investment' (plus its multiplier effects on incomes) to remove unemployment; the subsequent explosion of inflationary expectations was probably due, not to inherent faults in a quasi-Keynesian approach that had worked well for a quarter of a century, but to outside forces.[33] It is at least worth asking how this apparently plausible approach – and the implicit 'cures' for the alleged under-employment GE – mighı be applicable to the mystery of MVs' disappointing record in generating high-employment, low-poverty paths of progress in rural areas.[34]

*   *   *

Contrary to many current views, most rural labour markets in poor countries function rather smoothly, with quite rapid flows of knowledge, commuters, and longer-term migrants between tasks, and nearby areas, facing different patterns of seasonal, or other, requirements for work [Berry and Sabot, 1981]. Indeed, the case for radical change – political, institutional, or technical – in poor countries is not that 'markets fail', but that their relative *success* in swiftly and accurately matching buyers with sellers (given the real costs of transport and search) is consistent with persistent and appalling poverty. Hence much more is needed to remove such poverty than 'letting markets work' by improving transport and information, or by reducing public or private monopoly power.

Nevertheless, the smooth functioning of (and integration among) small-scale, local labour markets in poor rural areas is consistent with phenomena very like Keynes's three 'obstacles' in rich countries. Such phenomena prevent the *local* responses to specific price changes from adding up, in total national labour markets, to behaviour in response to changed wage-rates – directly or via changed food prices – likely to create, or maintain, full employment at a Walrasian GE. At local level, a year-round average proportion of person-days reported as involuntarily unemployed (on careful definitions and surveys) at 10–20 per cent is normal in developing rural areas – with the higher figures applying in slack seasons, to women, and to unskilled and poor casual workers [Lipton, 1984a].

Of Keynes's three obstacles to falls in wage-rates as a cure for unemployment, the first is the 'stickiness downwards' of wage-rates. When MVs reach a village, farmers might want to increase their labour hiring by, say, 30 per cent – rather than hiring threshing-machines or buying weedicides – if only wage-rates were 5 per cent less; yet, despite a lot of joblessness, wage-rates do not fall before – let alone after – MVs arrive. Rural wage stickiness in LDCs has little to do with union rates or prolonged negotiations, both very rare in tropical food farming. The reasons for wage stickiness are as follows.

First, wage-rates have to reflect subsistence costs.[35] Wages are in many poor rural areas barely enough to keep labouring families alive and working. Thus there is no scope for attracting employers to buy more labour via further reductions in wage-rates, unless big rises in working time, paid *in advance*, more than offset such reductions, *plus* energy costs of seeking, reaching, and undertaking extra work. Often this condition cannot be met, because the domestic and productive employment of poor families, especially in peak seasons when their health and nutrition is often worst, already approaches their physical limits [Lipton, 1984a].

Second, many rural areas appear to feature segmented markets for unskilled labour. In substantial parts of these markets, real wage-rates for one group of workers are kept above apparent equilibrium levels, and are very resistant to falls, despite widespread unemployment among another group. Such a 'privileged' market position has been noted for rice employees in Java, Indonesia [Hart, 1986; Lluch and Mazumdar, 1981]. In West Bengal, India, *some* villages have privileged arrangements between the farmer and local, often attached, employees. These receive year-round payment for work at customary wage-rates, even in slack times when little work is available, and cheaper unemployed labour is on offer from other villages. The villages with such local privilege appear to feature more than usually high and unpredictable fluctuations in farmers' needs for extra workers. Thus the above-market, 'sticky' wage-rates can be seen as insurance premiums, paid by farmer-employers in slack periods to ensure that their (year-round) employees feel an obligation – and have an economic incentive – to provide effort well over the norm during unexpected labour peaks [Bardhan and Rudra, 1980; 1981; Rudra, 1982].

Third, maintenance of 'above-market' wage-rates for a paternalistically privileged group of unskilled workers, while others – typically from particular ethnic, linguistic, or caste groups – remain discriminated against and/or unemployed, has also a 'non-economic' explanation, in terms of custom and reputation [for full exposition see Akerlof, 1984]. If a substantial proportion of workers (or employers) strongly believes that it is improper to accept (or offer) wage-rates below some norm or standard – and if the subjective losses, and perhaps objective sanctions, from being known by fellow-villagers to do so loom large, in the minds of transactors, relatively to the apparent economic gains from selling more (or hiring cheaper) labour – then 'above-market' rates for privileged workers, and non-Walrasian unemployment both for them and for non-privileged workers, can co-exist, even across generations.

Fourth, even if all potential workers are treated similarly, wage-rates in developing rural areas may be sticky downwards (at terribly low levels) because employers feel that lower wage-rates could reduce a specific employee's work quality. (i) The 'nutritional efficiency wage argument' points out that, if extra payments enable an employee to eat more, his or her strength and care at work may be increased, so that it can be profitable for an employer to pay a higher real wage than is necessary to purchase the labour. Such above-market payment benefits the employer, especially if the real wage bonus includes meals at work, and above all if the employee both is attached on long-term contract to one employer and has minimal family ties. Then, extra cash or food is not diverted, either to extra productivity for other employers or to extra food for relatives. But such a process of hiring fewer workers, in order to 'buy in' nutritional efficiency, leaves other workers – casual labourers, parents of big families – with significant rates of unemployment, while privileged full-time employees receive a a wage-rate 'too high' to clear the entire labour market. (ii) Above-market wage offers can also benefit an employer who is ignorant of the quality and application of casual applicants for work, yet who seeks to increase the proportion of good candidates, and thus to improve the chances of employing a really able worker. These 'nutritional' and 'screening' effects generate above-market wage-rates, sticky downwards despite unemployment, only under rather restrictive conditions [Binswanger and Rosenzweig, 1981], but do sometimes operate.

There is scattered but strong evidence that the above factors – alongside ill-health and hunger – do mean that we cannot expect labour markets to squeeze wage-rates (miserably low as they are) downwards and thus to employ poor workers fully.

- A standard '3kg. of grain per day', or the cash to buy it, appears to have been approximately the floor rural wage-rate in very many different periods, economies and societies, many apparently featuring involuntary unemployment [Clark and Haswell, 1970, p. 47; Braudel, 1981].

- Careful village enquiries [e.g. Drèze *et. al.*, 1986] suggest that attempts by a worker or employer to negotiate reduced wage-rates, in exchange for longer periods of employment, do exist but are very rare.

- A study of bonded labourers in four villages of South Bihar shows that they obtained only 68 per cent of the lowly wage-rate of free labourers, but 60 per cent more hours of work [Mundle, 1979, p. 114]. The loan received in return for the bond, plus the tiny plot often assigned to the employee (although unproductive, and typically only ⅕ acre) and the higher security of employment, appear partly to compensate for the lower bonded rate. The absence of these advantages to 'free' employees – and the employer's absence of certainty about their job performance and nutritional status – correspondingly fix a premium wage, above the market-clearing rate, for free labour.

- In a study in West Bengal, free labourers typically reported a wage expectation, before they would seek *extra* employment inside the village, 58–75 per cent above the typical current wage-rate (the higher differential being for the peak season); outside the village, the differential was 29–134 per cent; workers with higher wage expectations tended to have significantly lower employment. This again suggests that employment expansion is impeded by stickiness downwards of wage-rates, especially for unfamiliar, insecure or remote work [Bardhan, 1984, pp. 21–2].

In two ways, this stickiness helps to explain why the impact of MVs on poverty has been less than 'adding up' (Chapter 6, a) would predict. First, farm employers, collectively or individually, have little hope of successfully offering workers a deal in the wake of MVs: accept restraints (or small cuts) in real wage-rates per day, in return for a substantial increase in hirings –

say of weeding and threshing labour, rather than of weedicides and threshing machines. Second, in many places the *pre-MV* wage-rate – owing to such things as subsistence minima, screening, nutritional efficiency, and so forth – has been kept well above the equilibrium that would appear indicated by market supply and demand for labour; therefore, even if MVs induce a considerable increase in demand for labour (and some reduction in its supply),[36] this need not pull up the equilibrium wage above the actual pre-MV rate. So past, pre-MV 'stickiness downwards' helps explain why MVs, despite tightening labour markets, do little to raise wage-rates; and present, post-MV 'stickiness downwards' helps explain the rarity of individual or group bargains, between workers and farmers, for expanded employment at restricted wage-rates (instead of adoption of labour-displacing techniques) to meet the extra farming requirements per acre of MVs.

*   *   *

So Keynes's first obstacle to 'Western' labourers' scope for pricing themselves into full employment – money wage-rates' stickiness downwards – is clearly relevant to the disappointing impact of MVs on the poor in developing rural areas. What about Keynes's second obstacle: that, even if unemployed people can accept lower money wage-rates, there may be little or no response of real wage-rates – money-wages relative to product prices, or to prices of inputs alternative to labour – and hence of incentives to employ? There are straws in the wind which suggest that this, too, may be a problem in developing countries in the aftermath of MVs.

First, the extreme localization of the studies of consumption linkages (Chapter 6, d) is not due only to data difficulties. The authors of the studies presumably judged that – while changes in money wage-rates in a few restricted localities (with food and its buyers readily mobile in and out) could reasonably be assumed not to be eroded by changes in food prices – such an assumption would not be plausible at national level. Nationally, if money wage-rates fall, so probably do the prices of food, because it is labour's main product and main purchase. This erodes the incentive of food farmers to employ more labour when its wage-rate falls. Yet if (in the wake of extra demand for labour due to MVs) money wage-rates *rise*, the price of its products (and hence the cost of living) is also higher than it would otherwise have been. Wage-rate changes either way in the wake of MVs – because of this deficient conversion from money to real rates – might well do little to help poor workers. Higher wage-rates bring higher food prices and thus no more real labour income; lower wage-rates bring lower food prices and thus no incentive to employ more labour.

A second hint, that real wage-rates in rural LDCs may not change when money rates do, is the calculations for several developing countries of large consumer gains, due to the downward impact of MV-induced rises in food supply upon its price to consumers (Chapter 5, b). While these calculations look reasonable, how are they to be reconciled with the slowness, or even absence, of improvement in poor people's living standards in many countries with widespread adoption of MVs? Presumably, the answer is that the ample supply of labourers – and hence the ready response of potential labour supply to higher real wage-rates – enables employers to respond to cheaper food by slowing down money-wage increases.

Third, there are some studies [e.g. Parthasarathy, 1977; Papanek, 1986] showing that rural money wage-rates shift in the same direction (and to much the same extent) as food prices, though with a time-lag. Thus, although an individual worker and employer (or even many such, in a few villages or districts) can negotiate lower money-wages and reduce *local* unemployment – or can raise *local* labour-income through higher money wage-rates, if MVs make it pay the employer to demand more labour – at national level money wage-rates change with food prices. If food absorbs most income, and if most workers are hired to grow it, such changes do little to affect the inducement to hire (or to seek) work; higher demand for workers in an MV lead area will mop up some local unemployment, but the resulting rise in local money wage-rates will not raise national average real wage-rates much. Further, MV crop expansion is partly at the expense of other crops; when an MV crop expands it diverts labour from them to itself, leading to 'limited employment response' and hence to a 'very small response' in 'real wage rates' [Binswanger and Quizon, 1986, p. 25]. Thus both Keynes's first two obstacles, to full employment in a 'Western' GE, also appear to operate in the rural 'South', to erode the advantages that the poorest people – rural labourers – can obtain from the spread of MVs. (i) MVs do raise labour requirements per acre somewhat (Chapter 4); but the pre-MV forces had already maintained money wage-rates (while terribly low) somewhat above the levels that would persuade employers to absorb all the labour on offer, so that the extra MV demand for labour tends not to pull up these rates. (ii) Even when money wage-rates do change, prices change in the same direction. Real wage-rates thus neither rise sharply as labour becomes more attractive to employers, nor if they fall are there large gains to employment; indeed farmer-employers, and even self-employing farm families, pay over a substantial, rising share of the benefits of MVs to suppliers of inputs competing with labour (weedicides, etc.), although there remains some direct net gain to the poor as workers.

*   *   *

The disappointing – though positive – effect of MVs on the labouring poor is compounded by the third Keynesian obstacle to labour market adjustments: the rather low response of demand for labour to economic incentives, be they lower real wage-rates (as discussed by Keynes and his critics for developed economies) or higher conversion rates of labour into end-product (as achieved in the wake of MVs). In a Walrasian simulation for India, 'the total [demand] elasticity of rural. . . labour' was only -0.48; a 10 per cent fall in money wage-rates (or a 10 per cent rise, due to MVs, in the value added by an hour of labour-time) raises demand for employees by just below 5 per cent [*ibid.*, p. 25]. Thus, if workers combined to offer employers a swap of wage-restraint for labour-intensity after MVs arrived, they would normally lose by doing so.

Conversely, substantial rises in real wage-rates for unskilled labour – which MVs are unlikely to achieve – would help the poor, at least in the short run, since employers' demand for labour-time would fall less than proportionately. However, the

Punjab experience (higher real rates, gradual adoption of threshers to displace workers, fewer jobs) casts doubt even on this as a long-run strategy to link MVs to higher labour incomes. It can, however, work when – as in the Indian Punjab – it is accompanied by substantially rising demand for non-farm labour.

*   *   *

We shall not look in detail at Keynesian interactions in rural capital markets. These too are relevant to the prospects of fuller employment and higher labour income in the post-GE equilibrium. The three Keynesian obstacles, to interest-rate adjustments equating savings supply with investment demand at full employment (even if wage-rate adjustments prove insufficient), are certainly relevant.

First, rural rates of interest cannot simply fall until they encourage enough extra borrowing (and investment) to bring full employment. Risks of default, transactions costs on small loans, and perhaps local credit monopolies, all mean a high 'floor' to rural money rates of interest. This discourages employers from borrowing to finance new equipment (such as wells) – even if, due to high unemployment, plenty of workers could help to make it (and could later work with it).

Second, suppose that extra supplies of institutional credit, by competing against traditional moneylenders, cut the money rates of interest in the wake of MVs. That is possible, even though MVs also increase farmers' demands to borrow. However, hardly anything is known about the effect of a change in money rates upon real rates in developing rural areas. So we do not know much about the incentives, to borrowers or to savers, created by changes in money rates alone.

Third, even if we knew how real rates of interest changed as MVs spread (and credit expanded), what would the response to such incentives be? It is widely argued that credit subsidies destroy rural capital markets: that they do not stimulate investment, but encourage borrowing to consume, and discourage savings [Adams *et al.*, 1983]. If so, easier institutional credit in the wake of MVs would not reduce rural unemployment by stimulating investment, though it might by cutting savings (with inflationary side-effects). Also, the critique of interest-rate subsidies suggests that MVs' direct effects, in raising demand for credit and thus interest-rates, will not harm employment by choking back investment outside MV farms. However, the evidence does not clearly support the critique. In India, cheaper credit apparently encourages a lot of extra investment demand, does not greatly discourage the supply of savings [Iqbal, 1983], and is thus likely to stimulate non-inflationary growth and hence sustainable employment (provided the subsidies are not too large, threatening the viability of rural financial institutions). Conversely, if Iqbal is right, dearer credit after MVs – if extra demand for loans were not offset by extra, or subsidized, institutional supply – could severely harm the level of investment (and hence employment) in non-MV areas and activities.

*   *   *

The central Keynesian message is that markets in items that form a huge part of total supply or demand – national markets for labour, credit, or food – cannot be treated like local markets for refrigerator technicians, or for asparagus. That is because – while refrigerator technicians provide a negligible part of the demand for the service of refrigerator repair, and while a tiny proportion of the offtake of asparagus depends on the incomes of asparagus producers – demand in huge national markets feeds on itself. Responses to changes in money wage-rates or interest-rates or food prices, following the spread of MVs, are *not* in general likely to help the poor far towards the fuller employment and higher food consumption that MVs appear to offer – not, at least, in the increasing majority of developing countries where most of the poor depend on rural labour (rather than smallholder farming, or urban jobs) for most of their incomes.

Demand for labour depends partly on the level of output that employers find profitable to produce. That level depends largely on labour's purchases of its own products. Thus demand for labour depends substantially on the level of domestic spending (as against saving or importing) out of wages. Lower real wage-rates (if feasible) – although they encourage employment, by making workers cheaper to employ – also discourage employment, because workers can afford too few of its products . Cheaper *credit*, if feasible, can lead to heavy borrowing for investment, overproduction of its outputs, falling output prices, and hence reduced investment later on. As for the *food* market, if MVs raise cereals output by 5 per cent but thereby cut its prices by 10 per cent, lower income among poor growers (and their employees) can compel them to cut their demand for cereals; that will later cut prices again – and induce growers to produce less.

Keynesian GE tells us that price adjustments to clear markets, helpful to poor and underemployed sellers for individual outputs, *can* be useless – or worse – in aggregate markets as tools to help poor workers or growers to dispose of their saleable resources. Also, Keynesian GE reminds us that market integration, specialization, and trade (all in some senses at the heart of development) bring new problems. None of the above difficulties, discussed in this section, in steering potential gains of MVs to the poor, is nearly as serious if 'the poor' are mostly small farmers who largely provide their own labour, use part of it to 'finance' their own productive investment, and eat most of the extra food that their improved crops provide.

(f) Input–output GE

The matrix multiplier path to GE, on the expenditure side (section 6, d), has an exact analogue in the input–output (Leontief) path to GE, on the production side. They are two sides of the same question. The matrix multiplier tells us the following: after an *initial change* in incomes – e.g. as people are paid for a crop that is switched to MVs – what is the *final change*, after all the successive rounds of consumers' spending from extra incomes have worked themselves out? The input–output process tells us the following: after an initial change in outputs – e.g. as people grow more of a crop that is switched to MVs – what is the final change in output, after all the successive rounds of producers' purchases (to buy inputs to grow extra crops, inputs to make those extra inputs, etc.) have worked themselves out? Because the two processes are so similar, in logical structure and in implications for how we understand what MVs do to the poor, we summarize the matrix multiplier process on the income side, before examining its analogue on the output side.

In the former process (Chapter 6, d), the *initial change* involves several sorts of extra first-round incomes: in the Indian Punjab at least four, for labour and land used in growing rice and wheat, as these move into MVs. At the 'second round', each of the four extra incomes is used up in extra outlays on rice, wheat, clothes, bus rides, etc. – i.e., in creating labour incomes and other incomes for the producers of each item – or 'leaks' into extra savings or imports.[37] At the third round, recipients of these outlays use up their extra labour and non-labour income as further outlays, generating a new set of incomes; and so on. Because the 'leaks' remove a fairly steady proportion of income, each round of spending is less than the last, and the matrix multiplier converges to a new equilibrium, with a nation's incomes eventually increased by a *final change* $1\frac{1}{2}$–3 times as much as the initial income rise from the MVs. The 'consumption linkages' literature (Chapter 6, d) examines how various sorts of MV-related rises in initial income – e.g. for big or small farmers – pan out as rises in final income, as between well-off, poor, and very poor workers and others growing extra MV rice or wheat, making extra consumer goods for these people, etc.

Just as the matrix multiplier follows up the initial extra MV income via successive rounds of consumer spending, converging towards final income, so input–output analysis traces the extra *initial output* due to MVs – followed by successive rounds of producers' purchases of inputs, inputs to make inputs, etc. – as it converges to its result as extra *final output*. Suppose that, owing to MV wheats, producers demand to grow extra wheat. Each extra ton of wheat is associated with extra cash payments for inputs of various fertilizers, weedicides, irrigation water, draught power, etc., and of labour – 'intermediate demands'. Each extra unit of each of these inputs, in turn, needs extra cash payments for inputs of some or all of such items as steel, oil, electricity, etc., and labour. So, next, do the steel, oil, etc. At each stage, a fairly steady proportion of the extra 'intermediate demand' leaks into imports, so the process converges. Just as the matrix multiplier helps to assess the flows of income via consumption to rich and poor, so input–output seems designed to do the same for production flows; if we are dissatisfied with the Walrasian approach to GE via market clearing at full employment, both these other approaches seem desirable. Yet the Leontief approach has been little used in analysing MV impacts.

This may be because of doubts about whether, as in the process described above, an extra million tons of wheat (or of anything else) uses extra inputs (including labour and imports) in the same proportions as does existing output of that commodity – by value as well as by physical amounts. This cannot be correct unless labour and equipment to produce extra output are available, for all inputs at all rounds, at constant prices and unit costs. This assumption of widespread, readily available, idle resources[38] is rather doubtful, especially for big output expansions. So is the assumption that extra initial output of, say, wheat is made with the same, fixed balance of inputs as existing output. This looks especially implausible if the growth of output is due to a new technology, such as MVs; this usually requires different levels of inputs (per unit of rice, wheat, etc.) to the old technology. However, input–output analysis can be modified, to move away from these assumptions – and also from the related assumption, also unrealistic, that prices do not change during the process.

The real reasons why input–output analysis is little used to explore the impact of MVs on poverty groups are different. First, especially in small open economies, a large, variable, and often unpredictable part of any big rise in inputs, induced by MVs, will not increase incomes among domestic input producers – poor or not – but will correspond to increased imports of fertilizers, etc. Second, even if the extra inputs are made within the country (e.g. urea fertilizers in India), they tend to be made capital-intensively, so that the impact of their increased production upon poverty is small. However, these objections assume away a possibly available policy option, which input–output analysis can explore: the option of steering MVs towards crop-mixes, locations, etc. so as to increase the prospect that extra demands for inputs, inputs to make inputs, etc., can help the poor by being met (labour-intensively or by family enterprises, but also without undue price inflation[39]) in the country where the MVs are grown.[40]

The real impediments to the use of input–output analysis to explore poverty impact of successive input rounds, implicit in (alternative) initial MV-based output expansions, may be statistical. *First*, separate methods, organizations, timings, etc. are usually used to gather data on (i) inputs per unit of different sorts of outputs (Censuses of Production, etc.), and (ii) income, consumption, and poverty (household sample surveys, etc.).[41] It is therefore awkward and costly – often infeasible – to construct, or to collect freshly, data sets that 'map' the effect, upon various poor or non-poor groups, of increases in income for workers or enterprises corresponding to extra production of various inputs. For example, it is hardly ever known whether there will be a better effect on income among the poorest one-fifth or one-tenth of poor households if MVs expand demand for hydro-electric rather than for oil-based power; or for ground or surface irrigation; or for nitrogenous rather than phosphatic fertilizers. *Second*, the information is especially weak on the use, per unit of output (or of extra output due to MVs), of local, 'informal', often labour-intensive inputs, such as manure or traditional irrigation sources.

(g) 'Political economy' and paths to new outcomes: interacting groups

We began this chapter by pointing to a paradox. MVs seem to help the poor, via production, employment and consumption (Chapters 3–5 respectively), in 'partial equilibrium' – i.e. when we look at each effect separately, locally, and in respect of the affected crop alone. Yet the results of MVs for the poor, while positive, have been disappointing. In African countries, where most of the poorest are family farmers who would gain directly by growing and eating higher-yielding crops, their spread has been slow. In South Asia, where MVs have spread rapidly in several areas, the incidence and severity of poverty have declined surprisingly little.

The standard GE approaches of pure economics, discussed above, help resolve this paradox in some respects, but in others they sharpen it. For example, if the real wage-rate does not rise much following MVs (whether for Keynesian or for Walrasian reasons), their long-run employment effects should not be weakening, as they apparently are (Chapter 4, b). Another approach to the paradox is by looking at the economic behaviour of political agents – groups, States, or classes – and asking how the impact of MVs might affect that behaviour or its outcomes. Such 'political economy' traces MVs' overall impacts on poor people rather differently from the GE methods of sections c–f above.

(h) Group responses to changing food prices after MVs

Politically organized groups respond to changes in food prices or input prices by other methods than market decisions. These methods have been examined, partly in the context of the impact of technical change following MVs [Taylor, 1983; Adelman and Robinson, 1978].

First, one major power-group is the government itself. One consequence of changing food prices, following MVs, comes via budgetary policy. In many LDCs the government is a substantial buyer or seller of food staples on home or foreign markets. If progress with MVs leads to a large change in volume and perhaps price of these transactions, there is often a big change in the government's budgetary surplus or deficit. In the short run, there are effects on aggregate demand, and hence on prices, net imports, and/or domestic output. A little later, governments alter their tax and spending patterns to allow for the change in their revenues and outlays caused by the arrival of MVs. All this can have large effects on employment and on absolute and relative poverty. The models cited here usually assume that the 'first cause' of changes in government budgets is a movement in the farm–nonfarm terms of trade – e.g. because the price of imported food changes sharply. However, it is possible also to use such models to estimate what a major MV innovation would do, via farm prices, to government revenues and hence decisions on expenditures. This could well have significant impact on poverty groups. At the simplest level, the Indian Government would not have felt able to undertake the huge expansion of anti-poverty programmes in the current Seventh Plan – or to permit State-level schemes such as that providing employment guarantees in Maharashtra, or noon meals for all school children in Tamilnadu – unless the food to back up such schemes had first been created in the form of large food reserves, themselves due mainly to MVs.

A second 'power-related' element in these models relates to the effect of changing food prices upon wage-rates in the formal sector. Such changes directly affect welfare among some poor people, and indirectly affect the government budget. Cheaper food due to MVs means less pressure on governments for wage increases in the public sector. This, in turn, permits more government activity – on health, roads, dams, or 'defence'. The balance of impact of MVs on the poor through such routes is unexplored, but Taylor-type models permit a start to be made [Taylor, 1983].

Taylor argues that, especially in LDCs, major markets – e.g. for labour and for food – often do not respond to big external shocks by either (i) large price changes or (ii) smooth and painless adjustments of quantities supplied or demanded to such changes. Sometimes, huge and socio-politically disruptive 'quantity adjustments' happen. More usually, pressures are applied socio-politically: to raise permitted food imports, or to expand the budget deficit; or the reverse, if the 'external shock' was favourable, e.g. a sharp rise in food output due to MVs. Losers from changes in major micro-markets such as those for food or labour are, in effect, often able to use their power to obtain compensatory changes in government-controlled variables – tax rates, food, subsidies, etc. This leads, for example, to changes in the rate of inflation, which in turn feeds back to the food and labour markets. Structuralist models seek to incorporate behavioural relationships, and to explore the links between MV-induced food output change, food price change, and wage and employment changes by examining how these interact, especially via the Government budget.

(i) Micro-systems

All the GE analyses so far seek to explain the impact of MVs in the context of fairly big systems, with many interacting persons, firms, and groups. A quite different approach is to ask what MVs do to economic and social relationships, and hence to the distribution of incomes and power, in rural microcosms. The two main methods are *village studies* and *farming systems analysis*. These attempts to study relationships within a village or farm household (and, perhaps, how MVs interact with such relationships) are different from enquiries (e.g. into the impact of MVs) located for convenience in villages or farms – respectively 'village surveys' and 'farm management studies'.

For example, a *village survey*, if it enquired into MVs' impact upon savings, would ask such questions as: did the farm households who gained from MVs feature different propensities to save, or different uses of savings, before the MVs arrived, from non-gainers; and was savings behaviour, in each of the two groups, altered after – or because of – MVs? A *village study*, if it homed in on savings behaviour in the wake of MVs, would also look at relationships among village persons, households, or groups, and it would therefore also ask: do MVs change the volume, distribution and use of savings by altering poor villagers' demand for, and richer villagers' supply of, consumption credit? Poor farmers, if they adopt MVs, can meet a larger part of their consumption needs without borrowing; richer farmers will find that MVs render on-farm investment a more profitable use of their savings than previously, and may thus be less disposed to lend them to poor consumers [Bhaduri, 1973].

Several surveys of villages – as opposed to studies of relationships in villages – seek to establish, among other things, the impact of MVs on individual farmers, workers, consumers, lenders, etc., within a single village. The six semi-arid Indian villages, studied for seven years by resident ICRISAT investigators, provide distinguished examples, much cited in this book. However, they do not explore such relationships as that just considered. For example, 'none of the six people who rely on moneylending as their primary source of income [was] in the ICRISAT sample' in one of the six villages, Aurepalle

[T. Walker, pers. comm.]. The model is of a village where individual economic actions can be impersonally treated and added up; it is assumed to matter little, in interpreting a loan, whether it involves a particular pair of households who are, say, landlord and tenant. Where village studies do examine such relationships – e.g. as they affect caste in the context of household behaviour towards MVs [Epstein, 1962, 1973; Rao, 1975; and Bliss and Stern, 1982] – they often prove very important.

The ICRISAT *surveys*, excellent as they are, may therefore be missing important issues that village *studies* would have uncovered. For example, the presence of a local toolbar monopoly proved crucial to the non-adoption of a proposed new dryland farming system [Ghodake, 1983, pp. 23–5]. This suggests that the monopolist's relation with other villagers in many other economic and social transactions was probably important too.

A few distinguished sets of village-study or similar materials [Bliss and Stern, 1982; Hart, n.d.; van Schendel, 1981; Franke, 1972; Richards, 1985] consider adoption, use, and gains from MVs, by various poor and non-poor groups, as in part the effects of a system of village relationships (and of emerging norms and priorities), or as, in part, changing those relationships. One important conclusion is that farmers' own experiments [*ibid.*] – and the presence or absence of rural structures that encourage experiment (especially among women) and relate the outcomes to formal research-station procedures [Jiggins, 1986, pp. 16–17] – greatly affect the pace of varietal change.

As a rule, however, social scientists working on MVs have avoided 'total' village-study approaches. They are sometimes mistakenly dismissed as, of necessity, static and/or non-quantitative. Also, many researchers feel that – because the evidence shows that farmers' adoption (and perhaps gains) from MVs are in the long run little influenced by farm size, status, ethnic group, etc. – intra-village relationships among tribes, castes, or big and small farmers can safely be neglected. But this does not follow: these relationships can affect the speed and smoothness of MV adoption as well as its results, *for everyone*, even if MV-related differences *among groups* are eventually small. Moreover, even if all types of farmers (eventually) adopt and gain, the spin-offs for landless workers and artisans depend on their relationships to the farmer-gainers, and to those who transact with them in land and credit.

One of the difficulties in trying to learn from village studies how MVs affect the poor is that each village is apparently special; we are not dealing with random samples, but sometimes with villages selected because of their particular features (e.g. high emigration), or because of easy access. Bardhan and Rudra [1980; 1981] and Bardhan [1984] overcome this problem by juxtaposing many village studies from one area, West Bengal. A cross-section analysis of these villages showed how 'pre-capitalist', internal, closed labour markets tend to typify only those groups of villages that face unusually high risk. B. Singh [1985] juxtaposes ten village studies from Central Gujarat, and shows that the effects of MVs (and of expanded milk production) on very poor farmers were much better in the villages which had long featured direct and not-too-unequal peasant farming, as compared with the villages where the lands, during the British Raj, had been farmed mainly by tenants of tax-collecting, landlord intermediaries. Both these studies, and others, suggest that each of a manageably small set of distinct types of intra-village power-structure – themselves much influenced by the patterns of village-specific scarcities and risks [Binswanger and Rosenzweig, 1986] – may (as it were) 'splay out' changes in income-distributions, in the wake of a major potential new income source like MVs, that could not be fully predicted *either* by adding up equations describing firms' (and farms') production responses and households' consumption responses, *or* by the various GE approaches of sections c–f above.

A wide-ranging attempt to test the relevance of 'inter-village differences', across some 3,000 Third World village surveys and studies from 1950 to 1975, was made in the 'Village Studies Programme'.[42] Most studies reviewed antedate MVs. However, the Programme's conclusions strongly suggest that particular *village* circumstances (especially inter-household and inter-group relationships) affect the likely impact of MVs, on poor people within a village exposed to them, in ways that cannot be inferred from studying only *individual* households and farms.

For example, Dasgupta [1977a] identified two types of village labour situations. In the more advanced and commercialized villages, greater inequalities in income and land were reflected in higher incidences of landless and near-landless labour, but also in lower participation rates in the workforce, especially by women. Remoter, less commercialized villages showed lower average incomes, less inequality, higher participation rates and more seasonal unemployment. Yet such *village* workforce characteristics – crucial for the likely impact of MVs on the poor – could not be inferred from the *household-level* characteristics: poorer households suffer consistently higher unemployment rates than richer ones, and the poor (though not the very poorest) respond to poverty through higher age-specific participation rates [Lipton, 1984a].

A similar 'village-specific' conclusion exists for rural-to-urban migration. It is quite distinct from the findings at *household* level. There, 'pull' migrants usually come from better-off rural households, and more towards anticipated urban education or formal-sector work; whereas 'push' migrants move from – or with – poorer rural households in a wandering search for work, often informal, taking them in steps towards ever-bigger urban places. However, at *village* level, Connell *et al.* [1976] found that the highest incidences of both types of emigration – 'push' migration by poorer (seldom poorest) groups, and 'pull' migration by better-off (seldom richest) groups – originated from the same villages: the unequal ones. Probably, high intra-village inequality provided better-off groups with (i) an increased economic surplus that permitted them to acquire both the information and the capacity to bear initial costs for 'pull' migration, and (ii) the rural capital and inputs, such as tractors and weedicides, that extruded worse-off labourers as 'push' migrants. So within-village inequality apparently increases emigration from the village. Moreover, the village studies also suggest that such emigration increases within-village inequality [Lipton, 1982]. Although short-run evidence suggests that inter-village migration spreads some of the gains of varietal innovation among *villages*, from those that adopt readily to those that cannot

[Affan, 1982; Kikuchi and Hayami, 1983], this effect of migration in reducing inter-village inequality is offset (if the cross-section village materials are trustworthy) by its effect in raising intra-village inequality.

The 'village structures' into which MVs are introduced, apart from possibly explaining different rates of adoption and benefit among villages, also explain differences in spread to poor people within a village. This is also felt in nutritional outcomes. Malnutrition tends to be somewhat higher in *villages* heavily dependent on selling one or two cash-crops, rather than on growing their own food [Schofield, 1979] – although, given the income-level, *households* with holdings used mainly for cash crops do not feature worse nutrition than households that mainly feed themselves off the family farm [FAO, 1984; Lipton and de Kadt, 1987; Longhurst, 1988].

$$*\qquad\qquad *\qquad\qquad *$$

The studies of inter-village relationships, as they affect response to MVs, leave us in the odd position of having an apparently important conclusion, but too little evidence to know which of two possible, diametrically opposed, inferences to draw from it. The conclusion is that a village's structures – its relationships (of power, employment, income-flows, and rights to assets) among villagers, groups, castes, classes, families, etc. – affect the poverty impact of MVs on the poor in that village; therefore, this impact cannot be wholly inferred from partial-equilibrium adding-up approaches (chapters 3–5) and/or from any or all of the four GE approaches of sections c–f of this chapter.[43] The same MV innovation set, made available to 'equal' and 'unequal' villages, may well have quite distinct results.

But where do policymakers go from here? One possible upshot is that, in villages with very unequal structures (or even histories), assets and power, MV or other innovations will concentrate resources even more in the hands of the better-off, and do less for the poor, than in more equal villages, although such villages have less income-per-person and (for this reason and because of the smaller concentration of income) have less capacity to save and invest. This is what Bhanwar Singh found in Central Gujarat (page 297 above). It would imply efforts to steer MVs towards 'less unequal' villages.

An almost opposite interpretation is placed on the same finding – that village structure affects poverty impact of MVs – by Lynam [1986]. He argues that the evidence of ultimate gains to poor people from introducing MVs to areas like the Punjab – where intra-village assets are unequal, but village infrastructures substantial – may well not apply to initially 'more equal' areas of the Third World (and to MVs of crops such as cassava and sorghum usually grown there). This is because the gain from MVs there will, Lynam fears, accrue largely to farmers enriched by them who acquire control of infrastructure (e.g. transport for inputs,. or water control), private or public, and who thereby greatly widen the economic gap between themselves and the village poor. This would imply steering MVs towards 'more unequal' villages, lest they increase inequality within currently 'less unequal' ones. Major advances with sorghum and perhaps cassava – and ICRISAT's finding that it is probably not feasible to develop semi-arid MV technologies specific to smaller farmers (because they must use similar methods to other farmers) – lend urgency to the question of whether the Lynam view or the Bhanwar Singh view correctly interprets the moving 'equilibrium' of MV innovations in less-advanced villages and crops.

Our grasp of how village microstructures of power (involving tenure and credit as well as land) affects MVs' impact on the poor is at present crude and undifferentiated. We know enough, however, to avoid repeating such disasters as the projects in Zaire, Madagascar, and Ethiopia in the 1970s that sought to introduce MVs, in the context of new farming systems, into rural communities whose pre-existing systems, structures and preferences were unknown to the project managers.

Three misinterpretations should be cleared away about the village study, or more generally the study of a rural community, with regard to the impact of MVs. First, such studies need not be *static*. Village resurveys [such as the ongoing follow-up of Bliss and Stern, 1982]), or even continuous surveys over several years as MVs spread [such as ICRISAT's

in South India], teach us a lot about the mutual accommoda-
tion of new farming methods, village power-structures, and
outcomes for poor people. Second, to study a village is not to
assume that it is an *isolate*. The relationships of a village to the
outside world crucially affect the impact of MVs on particular
poverty groups. For example, the type and organization of
post-harvest processing and disposals in a village crucially
affects the impact of MVs on its female population [Jiggins,
1986, pp. 8, 45–6, 83]. Third, in a village study no less than in
other approaches to measuring the differential impacts of
MVs, groups studied need not be *over-aggregated*. For example, village studies are ideal places to explore how female-
headed households differ from, and relate to, others – and
how that relationship might influence the chances to adopt, for
example, MVs of traditionally 'male-controlled' crops, or MVs
that require early ploughing in circumstances where women
are denied equal access to draught animals [*ibid.*, pp. 23–30;
Bond, 1974].

*   *   *

These considerations apply also to efforts to place the MVs in
the context of a 'farm system' [Collinson, 1982; Byerlee *et al.*,
1982a; Maxwell, 1985; Simmonds, 1985a]. Farming systems
analysis enquires: given the goals and activities of various types
of farmers, what 'recommendation domains' are appropriate
for each type? If this question is carefully approached, some
generally neglected effects of MV options on poor smallholders – via labour available at peaks for other crops; via straw
for cattle; etc. – get examined. Many errors of past MV
programmes could have been avoided, and gains to the poor
increased, with prior understanding of year-round farming
systems for various types of farmers – upland and lowland, big
and small, female and male, etc.

However, nobody lives only in a 'farm system'. Not just non-
MV crops and animals, but also off-farm production, post-
harvest processing, child care and housework, and leisure
activity have relations of competition and complementarity
with MV inputs and outputs in various seasons. Farm systems
need placing in the context of the household's overall activities,
e.g. by time–use surveys.

Also, no real family's production–consumption system is closed, except in one case [Crusoe, 1719, before Friday]; yet it is Crusoe – at best assisted by wholly impersonal market outcomes – who provides the standard economist's model, of 'utility maximizing' consumer behaviour. In reality, the poverty impact of MVs on a farm-based family depends on what happens to other families and institutions with which it transacts. Does the moneylender find his income from interest (on consumer loans) reduced as the borrower is enriched by MVs; and if so does he try to impede them [Bhaduri, 1973], or to switch from lending to investing in production with them? Do MVs accelerate or retard a switch from sharecropping to owner-occupancy? Where do they bring in absentee urban landlordism? Where do they accelerate local self-assertion?

Classic work in development anthropology (e.g. Epstein [1973], Hill [1982], Goody [1980]) does sometimes seek to answer such questions. Many economists, even those seeking to marry farming-systems and village-survey approaches, end up looking only at how MVs affect sampled *individuals* in villages, not at intra-village *relations* as affected by MVs. Yet it is through these shifting relationships that MVs, in the long run, affect land ownership, savings, employment, extra-economic power, and most other components of the societal systems that make people poorer or richer. However, ICRISAT has begun some household modelling to mitigate the Crusoe complex. World Bank analyses of how the behaviour of small farms is affected by the fact that they are also households have also contributed surprising results.[44]

(j) Political economists, disequilibria, and discontinuity

The GE models 'are not capable of handling. . . disequilibrium phenomena, such as the differential adoption' [Quizon and Binswanger, 1983, p. 526] of MVs by farmers facing similar prices and production conditions, but with different assets and access to credit. What fascinates is the concept of disequilibrium here. Imperfect markets in information, credit, or insurance create 'disequilibrium' in a Walrasian GE model, in the technical sense: they prevent it from reaching a stable outcome where (given their tastes and resources) families and firms believe they are performing as well as costs and prices permit. But most non- economists would look for something a bit more dramatic in a 'disequilibrium'. Do MVs change the structure of wealth and power in a society, leading to a qualitative change in it? If so, the impacts on poverty will dwarf even the 'very long-run' Walrasian GE effects.

We shall argue (Chapter 6, k) that historical experience of other major agricultural changes, not GE analysis of MV impacts by economists, provides the best guidance to such long-run effects. Yet most economists have given very little consideration to that experience. *Natura non facit saltum* (Nature does not jump) was the Newtonian motto of the greatest work of standard economics [Marshall, 1890]. But *for a village*, though rarely for a nation, some big changes in agricultural technology either cause, or else require as preconditions, 'jumps': huge changes in the power-structure. Probably MVs seldom entail such a jump (Chapter 6, o). Certainly their massive spread appears to have been consistent with quite different macro-policies, from China to Mexico. But the question needs to be asked. Political economists, including Marxists, try to ask it. There has been a general failure to distinguish between *Marxist analysts* of MVs [such as Cleaver, 1972; Byres, 1972, 1981; Gough, 1977; Rudra, 1978 and Patnaik, 1971] and *general whingers* – some Marxist, many not – who insist that MVs cannot do any significant good.

The Marxist analysts are seldom merely plaintive in this dreary fashion. Instead, they see MVs as part of a powerful modernizing process that destroys 'reactionary' pre-capitalist formations: by formalizing wage contracts, by reducing the role of sharecropping, by strengthening owners of capital against landowners, by polarizing peasants and turning them from 'a sack of potatoes' into clearly differentiated large capitalist farmers and landless rural proletarians, and by commercializing and monetizing transactions in grain, leaving the new capitalist farmers with reinvestible surpluses. There is a lively controversy about the 'mode of production' in Indian agriculture [summarized by Thorner 1982]. Some of the more eclectic Marxists such as Rudra have changed their positions drastically, no longer asserting that MVs in the Punjab were a major component of a transition to capitalist agriculture.

Certainly, MVs have polarized income and power. Within and between villages, districts, developing countries, even continents, the rural people better off initially have gained most and first. This is not at issue between Marxists and other analysts of the 'green revolution'.

What *is* at issue is apparently as follows. (i) Have the poor been made poorer in MV lead areas absolutely, not just relatively to the rich? This was argued for the Indian Punjab in the mid-1970s but seems to have been refuted since. Similar changes of view have been induced by evidence in Mexico [Rajaraman, 1975; Chadha, 1983; Bhalla and Chadha, 1983; Hewitt de Alcantàra, 1972; Byerlee and Harrington, 1978]. (ii) Are poor people's insufficient gains – even losses – traceable to MVs themselves, and to their role in the spread of rural capitalism; or to rising person/land ratios and hence weakening in the bargaining power of workers, small tenants and borrowers? The former position is taken by most (not all) Marxists; the latter by neo-Malthusians or population-orientated neo-classicals, such as Hayami and Kikuchi [1981]. The jury is still out. (iii) Are the slow advances – or retreats – of poor people necessarily associated with the 'package' of capitalist development (including technologies for displacing labour, not only the early MVs); or will the spread to safer MVs, semi-arid areas, and 'poor people's crops' improve matters? (iv) Are 'the poor' in the post-MV rural Third World increasingly labourers (including, perhaps, steadily dis-possessed smallholders) – a proletariat exploited by rural capital? Or is rural poverty a composite creation of technical backwardness, ill-health, population factors, land and water quantity and quality, and pre-capitalist as well as monopoly-capitalist and public-sector institutions – urban as much as rural – for extracting the product of rural labour?

These are unduly global questions. Different histories (and geographies) produce different power-structures and tech-nologies, and hence rather local answers. However, we suspect that, like their predecessors [Engels, 1894, esp. pp. 394–5; Lenin, 1899; Kautsky, 1899], most Marxist commentators on MVs may err by implicitly assuming that there are great and cumulative advantages – especially as new techniques increasingly reward (i) the financial capacity to innovate, (ii) the organizational capacity to engage many workers – for larger farmers over smaller ones, and for owners over tenants. If such advantages also operated in adopting, intensifying, and getting high incomes from MVs, they could indeed play a key role in relative, perhaps absolute, immiserization – especially if pre-MV agriculture really had been fairly equal. Marxists have been greatly influenced by the experiences of the Indian Punjab, but have not always appreciated that repossession of tenancies, polarization of size of holdings, and tractorization all largely *preceded* MVs there [Randhawa, 1974; Chadha, 1983].

Marxist scholars are contributing seriously to the debate about how MVs interact with capitalist development and poverty. For example, Marxists and standard non-Marxist economists have been in creative dialogue in the Indian mode-of-production debates. However – just like marketism – Marxism has a sort of bastard *Doppelgänger* or *alter ego* at the level of cheap rhetorical tricks. On the basis of one or two endlessly repeated anecdotes, it is claimed that the 'green revolution', (whatever that is) will turn red (whatever that means). Careful scholars [e.g. J. Harris, 1977a, esp. p. 35] have found no relation between the incidence of violence and the spread of MVs. The Naxalite rebellions in India were in very backward agricultural areas. The Punjab's recent religious troubles have probably been unrelated to agriculturally based class conflict.

Careful Marxists stress that it is specific to a region and its history whether MV-induced polarization hastens 'class action'; in India this happened in Thanjavur, but not in the Punjab [Byres, 1981]. Such caution leaves Marxist analysts unable – or rather, given their view of history, reluctant – to make general predictions of how MVs will affect power-structures (or social formations, or relations of production). However, neoclassical approaches, including GE, are just as vulnerable to that criticism, if it is a criticism. Magnificent abstractions about history, technology, power-structure and poverty are likely to be less fruitful than the scrutiny of particular historical experiences. What do these suggest about past long-run effects of major farm innovations on the social relations that, together with material options and resources, determine whether poor people stay poor?

(k) Lessons from history and from historians: methods

Historians, looking at the effects on poor people of big agro-technical changes like the Neolithic Revolution or US farm mechanization in the 1850s, seldom use the adding-up approach. Such big changes combine with other technical, socio-economic and political changes to affect jointly the poor, the rich, and social relations among groups. MVs in a few isolated villages, or an innovation raising farm output by only 1 per cent, might be 'near-decomposable' from all these other causal links, so that the effects on poor groups could be analysed in isolation [Ando *et al.*, 1973]. MVs in major regions, raising farm output by perhaps 40 per cent and drastically changing techniques and input-mixes, cannot be 'decomposed" from other big changes affecting rural and urban people of all classes.

However, this also makes it difficult, maybe impossible, to undertake GE analysis of major historical changes, such as the abolition of slavery in the USA, by enquiring how output and incomes would have been different (assuming, for example, Walrasian pricing and employment) if that change *only* had not occurred [see the superb attempt by Fogel and Engerman, 1974]. Most historians object, to the hypothetical method and to econometric history, that one *cannot* take one big event, such as the abolition of slavery or the spread of railways in the USA, and ask what its non-occurrence would have meant (e.g. what it would have done to 'the poor') if all other events had proceeded unchanged. The US railroad map would have looked quite different without the events of 1861–5. Similarly, the rapid spread of agricultural machinery in 1830–60 [Edwards, 1941, pp. 221–9] cannot usefully be treated in isolation from accelerated farm-to-factory migration in the USA. In exactly the same way, if one 'took out' MVs, it is invalid to assume that all other major sequences since 1965 in India – mechanization, population growth, land expansion and distribution, even political structure – would, with their effects on the poor, have been unchanged, or changed only in ways predicted by the maximizing assumptions and equations of short-run micro-economics.

This is the complaint of the historians. Economists try, all the same, to understand what MVs do to the poor via:

(i) *The adding-up approach.* We resort to this in the hope that, even if the impact on poor groups in India of MVs cannot be decomposed causally from that of other great events, the impacts of rice MVs on poor consumers in Delhi, of wheat MVs on poor small farmers in Ludhiana, etc., can be, and that we can then add up all the separate effects, still ignoring other big events such as population growth. This works if MVs have a small, local impact; or if their impact overwhelms all other factors affecting the poor; but not in the majority of cases in between. If one man pushes a great rock down a shallow slope, the interaction with the prevailing wind can be ignored; also, if a million push; but not if a hundred or a thousand do so.

(ii) *GE.* Our analysis of four GE approaches shows that no one equilibrium can be all that general. But suppose that economists one day manage to combine neo-Walrasian, multiplier 'Keynesian', true Keynesian, and Leontief GE analyses, and to tell an agreed story, at least about short-run 'directional effects' of MVs on the poor. Even then, larger 'historical' interactions of MVs with the State, class structures, population change, and land distribution would be left out. And such interactions may be the main way that, in the long run, MVs affect the poor.

This matters to a plant breeder or other agricultural scientist. It shows the inadequacy of efforts to develop MVs that *would* help the poor as consumers, farmers or workers, *if only* such MVs could be isolated from other great currents of history such as population growth. MVs, and associated methods and 'farm systems', have to be poor-friendly in the actual, evolving historical contexts of particular adopting countries. To help us in judging such matters, we now look at what historians have said about other 'agricultural revolutions' that could help us to assess MVs and research priorities.

(l) What is an agricultural revolution?

Historians identify four sets of changes in agricultural technology which, most of them agree, are in some sense 'agricultural revolutions' (ARs).[45] By looking at them, juxtaposed with apparently sensible criteria for deciding whether something is an AR or not, we may cast light on the circumstances under which sequences of events such as the spread of MVs help various groups of poor people.

(1) The Neolithic Settlement, when hunter-gatherers became agriculturists, reached China before 5000 BC, and spread slowly across Europe, from South-East to North-West, in 3500–700 BC [Bray, 1986, pp. 9, 86; Piggott, 1981, p. 30]. This spread sometimes accompanied · invading cultural groups, which brought their newer, settled farming methods to new lands; also, it tended to be a spread from less to more difficult lands for settled cultivation. While an event that took twenty-eight centuries – and involved new farming methods that spread across Europe at below 1.1 kilometres per year [Ammerman and Cavalli-Sforza, 1971] – looks rather unrevolutionary at European level, it had to happen fairly suddenly for any one settling group. Concepts of class organization and property had to change quite sharply – not only where settled farming was spread by invaders as in Wessex [*ibid.*, p. 33].

(2) The medieval AR took about six centuries, from AD 600–1200, to cover Europe. Once again, however, for any one village or manorial farming system there probably had to be several sudden, linked changes [White, 1962].

(3) Faster, though less fast than was once believed, was the 'eighteenth-century AR' (1650–1850!) in North-West Europe [Mingay, 1968, p. 11; Jones, 1974, pp. 78–9].

(4) Finally comes the 'green revolution', surely much the fastest to spread new methods and raise agricultural productivities for wide areas and large populations (1963–?).

There are other candidates, mainly for temperate zones. Big irrigation works, while greatly raising farm output, obviously have potential to transform, or to stabilize, power-structures [Wittfogel, 1957], but have been an available technique for thousands of years, adapted and abandoned in response to socio-economic pressure, rather than an independent AR.

The New World gene pool brought new crops to much of Africa and Europe in 1500–1750, but gradually, and probably without drastically changing power-structures in recipient areas, except for some plantations in West Africa. In temperate agricultures, the mechanical innovations of 1830–60 [e.g. Edwards, 1941], and the biochemical innovations from Peruvian guano through chemical fertilizers in 1870–1915, were clearly evolutionary. Bio-engineering and N-fixing cereals *may* yet be tomorrow's AR. But in what follows it will suffice to see what (1)–(4) have to teach us about MVs and the poor in 'total' historical contexts.

*   *   *

Four criteria are usually suggested for ARs: that the new technologies in agriculture bring (i) accelerated and sustained growth of farm output or 'productivity', (ii) sharp technical discontinuity, (iii) technical change requiring or required by (or, in a weaker version, easing or eased by) social or political transformation, and/or (iv) major change in the incidence or severity of mass poverty. We return to the first three below. Historians usually say little about the impact of agricultural innovations on (iv) directly, not because they ignore the poor, but because sharp discontinuities – or, with the Neolithic and Mediaeval ARs, very long time-periods at national level even if each community changed suddenly – render poor people's conditions 'before' and 'after' non-comparable; even if data were available, poverty is not the same concept for Mesolithic hunters as for Neolithic farmers [Piggott, 1981, p. 31].

Economists dealing with the impact of MVs, unlike historians of earlier ARs, do often assess (iv) directly. Our data and statistical tools are better; the victims of poverty – and agencies funding research on them – live around us. Also nation-states now both proclaim 'poverty focus' and pay for MV development (though the latter was partly true of varietal change in eleventh-century Sung China, and around 1600 in Tokugawa Japan: Bray, 1986, pp. 141, 152). Above all MVs, unlike earlier ARs, have advanced quickly enough, yet with little enough sharp change in life-styles, for direct comparison of household poverty 'before' and 'after' to make sense, at least if sufficiently localised [cf. Aggarwal, 1973, chs. 6–7, on

Ludhiana]. But economists still need to learn from historians that, for the poor, MVs' *systemic* effects can dwarf *direct* economic effects, whether 'added up' or in various (and therefore not truly general) forms of GE. These systemic effects can operate as MVs permit or induce policy changes, which may induce faster GNP growth and lower food imports; or via transformed technologies and associated changes in skills; or via socio-political change induced by, or inducing, MVs and associated innovations. Systemic effects of MVs will normally interact with other great changes such as population growth or, as in much of Africa, privatization of land rights.

(m) How much is an AR? Geography and inequality

An AR attracts attention only by affecting a biggish area. Radical innovation by localized farmers – such as the Asian-style, irrigated, intensive systems that until recently prevailed on Ikara Island, Tanzania; or the experiments with selected rice varieties at different altitudes among the Mende of Sierra Leone [Richards, 1985] – remains of purely scholarly interest, neglected by policymakers, unless such innovation has a clear potential to spread its impact beyond the locality, either through adoption by many other farmers or through substantially easing the access to food of many other consumers. This 'recognition problem' facing localized ARs implies two things.

First, a series of sudden sharp ARs, each in a village or a clan, may take a long time to change a nation's agriculture much; by the time the technology is recognized for its impact nationally, its slow spread leads observers to dismiss it as non-revolutionary even locally. Yet a village may need to adopt swiftly, as a package, or not at all. This need for rapid, total change seldom applies to MVs [Lipton, 1979], but probably did apply to the medieval AR. This involved a shift to (or a spread of) horse-ploughing; improved harnesses, especially horse-collars; a larger ploughed area, made possible by stronger animals and better harness; horseshoes, to permit horses to work well in heavy soils; oats, for more animal energy; and rotations, involving oats, fallows and wheat. For any one rural community – a manorial system or a village – all this implied a revolution: of accelerated growth; of discontinuity in most techniques and in their packaging; and probably of power-structure (pp. 321–2), whether as cause or as effect of the technical change. Yet, to spread across a huge set of rural communities in a nation such as England, the changes took centuries [White, 1962].[46]

Second, as a result, ARs will leave most of a *nation* behind, while transforming *areas* or villages or clans. Those that change first may have agroclimatic advantages that simplify the AR; but, once a few villages have demonstrated the gain from transformation, it is their near neighbours that are likeliest to learn, follow, accumulate, and move further forward. The Punjab led South Asia not only in MVs, but in agricultural innovation at least since the risk-reducing canal irrigation in 1859–1900 [Spate and Learmonth, 1967, p. 522; Singh and Day, 1977; Randhawa, 1974; cf. Lowdermilk, 1972, pp. 15–16]. Similar prolonged innovation leadership had preceded MVs in the lead areas of North-East Mexico – and also in Central Java [Franke, 1972, pp. 63, 189]. In a European context, Norfolk was in the van of technical progress even in the thirteenth century [Campbell, 1983; Parain, 1966, p. 179] and later led the adoption of most new practices in England's eighteenth-century AR [Riches, 1937, pp. 8–17, 34].

However, leaders imply laggards. For the rural poor in areas left behind by the Norfolks, such as Northern England in 1750–1850, absorption by labour-intensive industry was an option. That is far less plausible today, in rural areas without MVs, because their rapid population growth now faces urban-orientated, and not very labour-absorbing, modern capital-intensive industry. Thus Mindanao, Madhya Pradesh and Pacific South-West Mexico feature rapid population growth and no obvious absorbent for extra workers, yet their food production lags far behind MV development in Luzon, Punjab and Sonora respectively. Since the message of historical work, though not of adding-up or even GE economics, is that regional advantage (e.g. from ARs) cumulates, this is very serious [Myrdal, 1958].

(n) ARs, poverty impact, and the nature of science

It is not clear, however, what is to be done about it. On one view, an AR is a response to the increasing scarcity of a particular agricultural resource – land or labour – in which farmers themselves, or scientists, evolve a new farming system that is much more efficient in its use of the newly scarce resource, leading to a prolonged period of renewed agricultural progress [Hayami and Ruttan, 1971; Boserup, 1960]. This neo-liberal view that, in determining the path of science and hence innovation, 'the consumer is king' – that her or his demand for more food, if frustrated by a scarcity of farmland or labour, impels farmers to press innovators (via markets or via politics) for new science, to permit production of more food with less of the scarce factor – is often contrasted with the neo-Marxist view: that innovations, and to a great extent inventions, largely subserve the interests of the ruling class (landlords or capitalists or proletariat) and thus are devised to increase its access to economic surplus.

Despite this apparent contrast, however, both neo-liberals and neo-Marxists agree that demand by users of science – e.g. farmer-producers (whether as individuals through the market, or as one or more classes), stimulated by scarcity, cost, and consumers' offtake – determines the pattern of scientific 'supply' of new methods. Indeed, in normal times of steady change in techniques, we may expect such consequences of 'normal science' [Kuhn, 1973]. However, repeated historical experience suggests that ARs break this mould. Contrary to both neo-Marxist and neo-liberal models, the supply side of science during ARs is mainly determining, not mainly determined.

For example, the view that demand for science – by consumers, by individual producers who satisfy them, or by an emerging 'ruling class' – largely determines the supply of innovation, even in an AR, would lead us to expect science to have generated, for the UK in the eighteenth century, technical change in agriculture that released labour for industry. For a long time, historians believed this, but the facts do not support it. In 1700–50 UK farm output rose by an unprecedented 26 per cent, at least two-fifths of it in the 1740s, probably a record decade up to then. Yet, since the workforce grew as fast, the output growth was almost wholly land-saving [Deane and Cole, 1967, p. 52]; labour-productivity did not grow; labour was not 'released'. Recent research suggests that growth was even faster and more concentrated into 1730–60, and even more based on rising yields, i.e. land-saving [Overton, 1979, p. 375; Turner, 1984, p. 225]. Over the whole period from 1690 to 1831 the application of science to agriculture, far from being induced by economic incentives to save labour and release it for industry, did not in most of England induce significant falls in labour-intensity. For example, the ratio of landless farm labour to farmers – almost always a sign of increasing scale, and thus of labour-saving farm methods – increased only from 2:1 to 2.5:1 [Mingay, 1968, p. 26, citing Clapham].[47]

Land in England was saved by eighteenth-century innovations, not because it was getting sharply scarcer relative to labour (population growth *was* accelerating, but gently), but because the spread of discoveries old and new, such as marling, horse-hoeing, four-course rotations, etc., made it profitable to apply more labour to land. The eighteenth-century and subsequent enclosures in England – while indeed a device, approved by Acts of Parliament, to transfer common land to private uses – were not a way to extrude labour for industry; gentry, not industrial capitalists, dominated Parliament; in enclosing common land, they (like their economic successors today in Rajasthan, India [Jodha, 1983] and Botswana) were responding to the supply of new science, which brought new and profitable chances to *increase* labour per acre [Mingay, 1968, p. 25]. Of course, later mechanical innovations did release large numbers of workers. However, England's eighteenth-century AR (like most accelerations of agricultural growth) is firmly rooted in the spread of discoveries, some made considerably earlier, which were not obviously responsive to factor scarcity, but rather to what the existing conditions of science permitted to be most readily discovered, tried, and spread.

In Japan in 1877–1919, too, new science brought agricultural changes that responded to inventions; but again those inventions responded less to farmers' economic scarcities, than to scientists' intellectual opportunities. Labour supply was growing quite fast, at about 1.3 per cent yearly. Yet the acceleration of agricultural growth, a precursor of industrialization like Britain's land-saving 'AR' of 1730–60, was in Japan labour-saving. Labour productivity rose at 2.6 per cent yearly and land productivity at 1.9 per cent [Ohkawa *et al.*, 1970, pp. 6, 13, 180–1]. There is much debate about what happened in Japanese agriculture in 1877–1919, but the figures do not suggest that inventions are mainly responses to factor scarcities.[48]

Growth *per unit of a factor* does accelerate sharply in most ARs, but the factor saved need not be the obviously scarce one. Instead, it may be the one that science has made it more profitable to save. Mendel, in his Austrian monastery garden, did not study sweet peas in the late nineteenth century so as to respond to land/labour ratios in Asia now; but Mendelian genetics have made the 'green revolution'. If biotechnology produces another, it will owe more to the supply of basic science by Crick, Watson, and their successors, than to any alleged response of applied science to the demand for land-saving innovations. Policy and intellectual history – not just elasticities – affect inventions, and hence not only the scale of technical change in agriculture, but also its path, and in particular the extent to which it reduces poverty or 'saves labour'.

However, deliberate correction of scientific priorities in the interests of the Third World's poor is needed. Such priorities respond, not only to the paths of basic science that occasionally make ARs, but also (in times and places of 'normal science' and normal farm growth) to the demands, political and economic, of powerful and wealthy producers and consumers, who are also best able to finance applied research. Much research money is spent to increase the efficiency of, and hence labour-displacement by, tractors and weedicides; too little is spent to increase efficiency of animal ploughing or hand-weeding. Research, enormously expanded and formalized and internationalized, is the new 'joker in the pack', making the trajectory of the new AR potentially very different from its three predecessors. This can help or hurt the poor. But, as always, the findings of *basic science* – from Darwin and Mendel to Crick–Watson and beyond – have the greatest, most lasting impact. They cannot be reduced to consequences of demand, from individuals or from classes.

(o) Discontinuous technical change

The quality of inputs, or of the skills or methods with which they are combined, is often measured as total factor productivity (TFP). Its rate of increase can be measured as the growth in 'residual' productivity – i.e. the growth in output that would occur with no change in the amount of any input, and that is therefore due to better *quality* of inputs. Such measurements, while difficult, do give a handle on ARs. Did TFP or residual productivity show sharp, sudden acceleration? TFP accelerated sharply in Punjab and Haryana, but not elsewhere in India, between 1958/61–1963/5 (0.5 per cent yearly) and 1963/5–1969/71 (13.4 per cent yearly) [Mohan, 1974, p. A-98]. Similar explosions of agricultural TFP happened in Japan in 1880–1910, and in the USA in 1885–1900 and 1938–60 [Hayami and Ruttan, 1971, p. 116].

However, there are problems with TFP and residual productivity measures. A jump in them need not indicate technical progress of an AR variety. It may signal arrival of new inputs, mistakenly omitted from the measurement [Schultz, 1964]; or of economies of scale. Or it may be merely a 'coefficient of ignorance', an indication that something unknown – improvements in education, or in climate? – has happened to improve the rate of increase in the efficiency with which inputs are turned into farm outputs. Anyway, nobody would call all the four TFP accelerations cited 'revolutions'.

In asking how ARs affect the poor in total systems, however, we are surely looking for, among other things, the effects of discontinuous, or at least sharply accelerated, technical change. It is a necessary (but not sufficient) condition for an AR. Can we find an objective, testable indicator of it that does not – as TFP and residual productivity do – involve dubious economics and statistics?

In the ideal type of an AR (or any revolution), we would expect exceptions to *natura non facit saltum*. Techniques, at least, jump. In such cases, innovations are *not seriable* [Shackle,

1952]; in other words, they must be tried out on a farm system (and change it), because piecemeal experiment is infeasible. They are *not separable*; the package cannot be unpacked. They are *not single-unit*; adoption involves each unit in relationships with neighbours and/or authority structures.

All four so-called ARs – Neolithic, medieval, eighteenth-century, MV – meet the criterion for an AR of dramatic acceleration in rural growth. Indeed, each was quicker than the last. But the key fact about long-run 'effects on the poor' is that only the Neolithic and medieval revolutions involved non-seriable and non-separable innovation; these features, as we shall see, were strongly associated with a transformed structure of power. Unlike these ARs, the eighteenth-century and MV experiences, although more rapid in their impact on GNP, could be taken gradually, piecemeal, and individually: could be tried out by one or two farmers in a village, and by any farmer on a small part of the farm. The poor must anticipate that this smooth, gradual process will 'feed' MV benefits into *existing* power-structures.[49] To understand what that means, we should look at the opposite: the truly non-seriable, multiple-unit, discontinuous ARs, Neolithic and medieval.

Neolithic settlers required security against animals and their hunters; sufficient settled workers to achieve scale-economies in land clearance, and to discharge simultaneous tasks in a highly seasonal activity, cultivation; and, fairly soon, households (not necessarily specialized ones) that made farming implements, storage devices, etc. while their neighbours cultivated – and also households to exercise authority, to protect the goods of persons developing 'delayed-return systems, (p. 320). Therefore, Neolithic settlement involved several families: *non-single-unit* decisions. It was *non-seriable* too, in view of the labour and time costs of clearance, the food foregone, and the scanty stocks in pre-settlement hunter-gatherer societies. Also it was *non-separable*, involving a package of practices: all European establishments carbon-dated as Neolithic were 'stone-using and all showing the essential features of cereal cultivation (wheat or barley) and animal husbandry based on cattle [plus?] sheep and/or goats, pigs and dogs in variable proportions' [Piggott, 1981, p. 31].[50] Animals and stone tools would have been needed to clear, and probably to plough, heavy hardpan soils.

Europe's medieval AR was also *non-seriable*; a whole community needed to adhere to a rotation, balancing fodder crops (so animals could over-winter and plough) and food crops. Since the seigneur's 'demesne. . . was made up as a rule of various fragments. . . mixed up with' peasant lands, he too had to observe the three-field rotation. In many areas 'collective grazing rights over the stubble, and the compulsory rotation. . ., were binding on all, often even the seigneur' [Bloch, 1966, pp. 242, 276]. Unless farmers synchronized their rotations, animals would more readily eat standing crops, and farmers' access to their fields at harvest-time would be impeded by the immature crops of others. Farmers in a village (or manorial) system could not try the new methods at different times and speeds. Also, the methods were *non-separable* [White, 1962]. The package of practices, permitting major rises in TFP, centred upon horses. They replaced oxen and permitted more land to be ploughed, but needed more fodder cropland. From this package, the three-field rotation of fodder, food and fallow, which spread cultivation across seasons and area, was inextricable. For horses to plough the extra land, this AR also required – in heavier soils, anyway – blacksmiths, to work iron into horseshoes (and increasingly into improved ploughs with mouldboards or wheels); and leatherworkers, to make improved harnesses with breast-strap and stiff collar [Parain, 1966, p. 144]. Plainly this medieval AR, which required social control of rotations, of grazing, and (where markets were primitive) of availability of work in leather and iron, must have been *non-single-unit*. What a contrast with the separable, seriable, evolutionary non-package of inputs [Lipton, 1979] and practices around MVs!

Similarly gradualist were the technical changes of England's 'eighteenth-century AR'. Hundreds of years of gradual farm enlargement, probably with only a minor contribution from enclosures [Mingay, 1968, pp. 15–17], had created some large capitalist farms,[51] often well before the acceleration of technical change around the 1740s. These farms could take decisions as *single units*, and thus could often act as lead innovators. Moreover, the alleged 'package' comprised practices suited for different environments, or alternatives in the same one; for example, horse-hoeing and marling were alternative ways to reduce seed rates [Riches, 1937, pp. 5, 16, 77–81]. Also four-course rotations, turnips as a clearing crop, etc. were *separable* and each could be tried, *serially*, on a tiny part of one of those big farms. Historians increasingly see this 'AR', even for one farm, as continuous, technique-by-technique, in essence evolutionary [Jones, 1974, p. 88; Mingay, 1968, p. 11].

The MVs may transform GNP, imports, and hence economies and the position of the poor in them, but the technical changes involved are in essence evolutionary. They rest on long histories of 'proto-MV' seed releases, and successive waves of seed innovation [Hayami and Ruttan, 1971, pp. 158–9, on Japan; Dalrymple, 1985, on Japan and Taiwan; Bray, 1986, on Mainland China; FAO, 1971, p. 6, on the Philippines; Kaneda, 1973, p. 169, on Pakistan; Saxena and Jadawa, 1973, p. 65, on India]. Importantly for the poor, many 'proto-MV' rices rested for their main appeal not on dramatic yields but on robustness: against wind damage for ponlai for Taiwan in 1911–24 [Carr and Myers, 1973, p. 32], or overall in the early 1960s for H-4 in Sri Lanka [Peiris, 1973, pp. 2–3] and ADT-27 in Tamilnadu, India [Frankel, 1971, pp. 90–1]. Farmers' own selections, and researchers' efforts to improve on these, had long prepared farmers for the MVs; manure and compost, for chemical fertilization; dug wells and canals, for 'fine tuning' of water from tubewells.

MVs themselves are *seriable*. For instance, they make feasible the approach, ever since the late 1960s, of the extension services in Sri Lanka. 'Progressive' smallholders there are invited to receive a free 'mini-kit' and to try out, typically, three proposed MVs, at each of three levels of urea use, on nine tiny micro-plots. Next year, many farmers in the village purchase, from among the nine combinations, the full-scale 'production kit' that seems to fit their circumstances best.

In this procedure – and similar ones used later in India – MV-linked inputs also have to be *separable*. A precisely mixed and timed package of non-separable practices and inputs, almost useless unless adopted as laid down, would make life very hard for poor farmers and unskilled labourers, but is in most cases fortunately mythical. Appropriate input-mixes vary with soils and terrain, and each farmer knows his or her micro-environments best – although even the *myth* of a non-separable package may, if believed, harmfully delay adoption among the poor [Lipton, 1979].

MVs and linked inputs are also *single-unit*. Unless a farmer depends on others for timed water, his or her net gains from MVs are seldom much affected by neighbouring farmers' decisions. Thus MVs neither represent technically, nor (by increasing the discipline that each adopter expects from his neighbours) require from social systems, a sharp discontinuity. This eases adoption for poor people and places. But it also increases the chances that the power-structure within existing social systems – without sharp challenges emerging from any requirements, in the nature of the technical change itself, for new forms of social organization – will steer the fruits of MVs largely to the entrenched better-off. So we should expect, in MV areas, that inequality increases but absolute poverty declines.[52]

(p) ARs, power-structures, and the poor

In our increasing unease with the usual bitty approach to the question of how vast technical changes, such as MVs, affect the poor, we have been pushed from the 'adding-up approach', via GEs that turn out to be partial, to less rigorous but more realistic historical accounts of 'general disequilibrium': of how major changes in total agro-rural technosystems, as related to political and social structures, affect the poor. The paradox is that ARs that have been progressively 'faster' in the sense of their scale and speed of impact on agricultural output and techniques at national level – Neolithic, medieval, eighteenth-century, MV – have nevertheless involved progressively 'smoother' (more seriable, separable, single-unit) technical progress. Over the four successive ARs, the increasing speed – and effects on output – of technical progress seem to raise, but its growing smoothness to lower, the prospect of bringing radical changes in power relations that affect poor rural groups. What in fact happened?

If a technical AR were either one-way cause or one-way effect of a new power-structure, there would be no room for incremental policy, let alone for fine-tuning the MVs. The view of such an AR as one-way *cause* of a new power-structure is criticized by Anderson [1974, p. 183] as a 'fetishism of artefacts'. Institutional gaps meant that 200 to 300 years elapsed 'between [the] initial sporadic appearance' of improved ploughs and harnesses (to simplify horse-ploughed three-field rotations) and 'their constitution into a distinct and permanent system'. Indeed [Dodgshon, 1980, pp. 2–3] 'manors and villeinage [were] present during the earlier. . . Saxon settlements'; the need to organize 'plough technology' cannot, therefore, have been *necessary* to cause that system.

Just as unacceptable is the view that a transformed public policy, or a new ruling class, must consciously form, in order to cause an AR. The Neolithic settlement was 'in no sense a *conscious* exploitation of resources by means more effective than those of the hunter-gatherers' [Piggott, 1981, p. 35]. Nor, later, were the field systems of the medieval AR 'consciously contrived institutions of field layout and husbandry. . . . It was not a case of early communities [deciding] how they might best farm their lands, and then devising [field systems] as the answer' [Dodgshon, 1980, p. viii].

However, though technical change is seldom simply either cause or effect of institutional change, the two are closely linked. All acephalic societies and most very equal societies, some (like the few remaining hunter-gatherer clans among the !Kung San of Botswana) observable still, seem to be pre-Neolithic: non-settled hunter-gatherers. Agriculture and authority emerge and grow together, for four reasons. (1) Settled societies are *'delayed-return systems'* with investments made before harvests – even, if land must be cleared, before sowing. These societies therefore need 'ordered, differentiated, jurally defined relationships [to secure] binding commitments' [Woodburn, 1982, pp. 431–3]. (2) The shift from nobody's to communal *property rights* requires increased authority [North and Thomas, 1977, pp. 229–31], even if Kennedy's [1982] caveats are correct. Indeed, (3) the growing need for *group security* for standing crops and settled investments, and hence for property rights 'to be sustained by. . . public and collective goods. . . defence. . . dispute regulation, law enforcement' [*ibid.*, p. 384], also advances authority, agriculture and 'States' (even if village-States) together. (4) The complex of *non-separable, non-seriable, multiple-household* processes of land settlement and clearing pushes a clan towards formal structure of authority, valued as a 'public good' and thus able to secure widespread consent when it extracts surpluses to reward well those high up in the structure, thus encouraging them to maintain both it and the new techno-system.

Settlements, and (much later) complex and integrated rotations, vastly increase and reward such legal and governmental hierarchies. However, no 'law' tells us whether the technical transformation precedes or follows the new social structure. Interactions matter. The matrix is not near-decomposable [Ando *et al.*, 1963], with some variables entering almost entirely into 'social', others almost entirely into 'technical', equations. (One ideologue's 'simultaneous determination' is another's 'dialectic'.)

Whatever the causality, agricultural settlement and authority-structure arrived more or less together. About 2200 years after it was being practised in China, settlement, with ploughing, reached England (in Wessex) somewhat before 2810 BC. A complex hierarchy, from clan chiefs to Wessex provincial chiefdoms, has been inferred from carbon-dated implements, burial grounds, and ceremonial places [Renfrew, 1973, pp. 597–8]. At this time elsewhere in England, where hunter-gatherers had not yet turned into farmers, only 'presumptive small kinship groups' are traceable [Piggott, 1981, pp. 55, 58; Case, 1969].

We see this process again in the second AR, bringing fully settled, crop-rotating, horse-centred medieval agriculture. To reap the benefits from the new rotations, each peasant required manorial courts and enforcement officers to compel his neighbours to observe those rotations, e.g. to separate grazing animals from standing crops. It thus paid *each* peasant to give up some surplus in order that legal officers should be paid enough to administer and enforce the system on *all* peasants.

The medieval AR provided farmers with further rewards for accepting locally authoritarian power, though this implied a larger share of a growing economic surplus for power-holders. However, seigneurial and not slave systems were needed for efficiency with the new farming, 'regulatory authority' in policing rotations had to be combined with non-slave incentives to efficient work [Bloch, 1966, p. 276]. 'Disciplinary assemblies, notably manorial courts' were required to regulate not only three-field rotations, but also use of the shrinking common claims upon grazing and stubble, as well as timings, partitions and disputes resulting from inter-mingled lands [*ibid.*, p. 242; Dodgshon, 1980, pp. 17–18, discussing Thirsk's work].

*   *   *

Can anything be learned from the Neolithic or medieval ARs about MVs' likely effects on power, and hence on the poor? These two early ARs required much more strengthening of authority at local level than do MVs, because the early ARs were much less separable, seriable, or single-unit than MVs. The Neolithic Settlement may have involved local transitions from acephalic clans to slave or serf systems; the medieval AR, from slavery to seigneury. MVs require no such *large* change in local power-structure.

The reason why all three ARs – Neolithic, medieval, MVs – might alter power-structures locally (though MVs far less so) is not scale-economies, but the benefits of new public goods. Neither the medieval nor the MV innovation sets involved economies of scale in farming. The medieval AR 'increased productivity on the small units [giving] them an advantage over the larger estates' [Parain, 1966, p. 125], just as today's MVs should, via labour-intensity, favour small family farmers, with their lower search and supervision costs. But just as Mexican small farmers do better with MVs when they have co-operative *ejido* institutions to finance and manage common irrigation investments [Burke, 1979], so small medieval farmers required shared institutions of security, law and settlement. Now as then, requirements for packages with precise timing or (fortunately rarely) fixed proportions, for water from central suppliers, or for non-competitive deliveries of credit or fertilizers, could increase the vulnerability of poor farmers to rural and urban extractors, seigneurs, and other recipients of economic rents, tributes, or bribes. This is not about big vs. small, or public vs. private input supplies. Rather it is about competitive or farmer-controlled supplies vs. external non-competitive ones.

Unlike Europe's Neolithic and medieval ARs, but like the MVs, the technology of the AR in eighteenth-century England, even during its most dramatic advance in the 1740s (Chapter 6, 1), posed few requirements for a change in authority structures at local level. Both the eighteenth-century AR and the MVs bring rapid growth in agricultural output, but are seriable, separable, and single-unit. At local level, therefore, these two later ARs seem ideal for 'standard' economic analysis of the effects on the poor, because the institutional impact or requirements seem relatively small. In both these later ARs, the local structures of power do not need to change much. Indeed, they are reinforced as richer farmers gain from adopting the AR innovations first (Chapter 6, o).

*   *   *

Yet this view is *too* local. The eighteenth-century AR in England (later in most of Europe) generated four major pressures towards industrialization, and hence changes in *national* structures of power. First, better-off rural people acquired or increased cash surpluses over current consumption needs and – after a quite long period, in most countries, of reinvestment in agriculture and forestry – began to find that it paid to place part of their savings in support of the new techniques in textiles, ironmaking, railways, and other branches of industry. Second, this AR, by increasing food surpluses sold by rural areas to the towns, increased the number of industrial workers who could be fed, and therefore the pace of attainable industrialization. Third, industrial development was favoured in areas unable to benefit greatly from the new techniques, and hence with increasing comparative disadvantage in agriculture. Fourth – because the core innovations of the eighteenth-century AR (four-course rotations, marling, turnips, etc.) did not save labour – agriculture faced pressures to do so by mechanizing *later*, thus raising demands for industrial outputs (and supplying labour to the cities).

Thus North-West Europe and North America now – with their massive food surpluses; with well below one in ten workers in agriculture; with capital, labour and politics largely urban; yet with rural bias, even rural veto – embody political transformations initiated, in part, by the take-off in agriculture in England around 1730–60. There are several countries in Latin America and South-East Asia where MVs appear to be the mainspring of similar ongoing political changes at national level, although there are complicating factors: power is much more urban, and population growth much faster, than in eighteenth-century Europe. We should be rather sceptical of claims that MVs transform local power-structures, and hence relations of production; MVs are too seriable, separable and single-unit for that. For example, the dramatic changes to capitalist wages systems in Javanese rice farming probably owe much more to the post-Sukarno changes in the central polity of Indonesia than to MVs [Bray, 1986, pp. 187–9]. However, this does not render MVs apolitical.

*   *   *

Today's AR may be affecting the poor most, not via adding-up effects (except in non-MV areas!) or GE economics, nor by inducing new local structures of power, but by feeding new resources into old local structures – while changing the *national* structures, of work as well as power. The greatly increasing speed of these four successive ARs – Neolithic, medieval, eighteenth-century, MVs – also means that each AR is likelier than its predecessor to create growing urbanized surpluses of rural food and agricultural raw materials (and to turn the internal terms of trade against these products). Thus each AR is likelier than the last to strengthen national, urban-based power than its predecessor was, although also to create increasing pressure on that urban power to 'do a deal' with the rural élites who, increasingly, provide rapidly growing surpluses of food, etc., for urban-industrial expansion. Since later ARs do progressively less to strengthen local power and organization, their relatively enhancing impact upon *central*, urban power also increases. However, big surplus farmers eventually become fewer, more coherent, and more readily organized [Olson, 1982], and transform Third World urban bias into the strong rural bias typical of Western politics.

(q) International agricultural research: a new fact in history

We have tried to show that, despite some major reservations (which suggest new directions for IARCs), without MVs the Third World's poor would in the short run have fared worse. However, in the long run it will be difficult for the poor to maintain, let alone to increase, their gains. Both GE effects (like the passing on of benefits from food price restraint from employees to employers: Chapter 5, b), and the historical evidence that seriable and separable innovations strengthen rather than challenge local power-structures, suggest this. To help poor people get and stay significantly less poor, research must seek sets of innovations which help them to gain options, assets, or power in their changing and differing political contexts. In Bangladesh, most poor people are employees; in West Africa and some parts of semi-arid India, they are small farmers; in most of Latin America, they are townspeople. Efficient pro-poor innovations will need to be more sharply pointed towards these groups. Thus in Bangladesh a suitable innovation in many areas is the hand-pump, which is likely to be substantially used and owned by employees [Howes, 1982]. In much of semi-arid Karnataka, India, MV finger-millet is well designed to help poor subsistence farmers [Rajpurohit, 1983]. In Colombia, MVs of corn and cassava home in on the consumption requirements of the urban poor [Pinstrup-Andersen, 1977].

Deliberate, centralized seed research and innovation are at least nine centuries old in China, and four in Japan [Bray, 1986, pp. 141, 152]. However, international research institutions such as the IARCs are a new fact in history, unknown at the time of the earlier ARs. Their greatest comparative advantage is relative immunity from the pressures upon researchers in national public research systems – and in the private sector – to respond to the factor scarcities [Grabowski, 1981] and crop priorities, not of the poor, but of the powerful.[53] But IARCs, though new as makers of AR history, need to learn its lessons if they are to serve the poor. The innovations of North-West Europe in 1730–1850, like the MVs, raised labour requirements per acre, and did not possess economies of scale. Yet their main benefits, for many years, accrued mainly to those who held political power, rather than to labourers and small farmers. Now as then, mere passive reliance on the pro-poor micro-economics of particular innovations, such as MVs or four-course rotations, is unlikely to overcome the political reality. This is that the channels of benefit from such innovations are controlled by the non-poor, and that the innovations (at least in the context of rapid growth of labour force, and – in our time – of other, simultaneous labour-saving innovations transmitted from North to South) do not suffice to restructure the channels or the control in ways that help the rural poor.

What IARCs do (and what, of course, was not on the agenda in 1730–1850) is to make it at least conceivable that an agency can have the power and the incentive to select and design sets of MVs, other inputs, methods, and outputs that will redirect research benefits – despite national pressures – towards assets, control, or employment incomes for the poor. In the closing chapter, we ask how this might be done in the context of some emerging issues for the MV-based agenda of international agricultural research: the lean to Africa; the impact of population change; the scope of biotechnology, and for improved biological nitrogen fixation; the role of the burgeoning food surpluses of Europe and the USA. In each case, we shall need to ask whether the lessons, learned from MVs' impact so far and reported in this book, apply to the options of policymakers and research managers for handling the new issues. Chapter 7 therefore begins by briefly rehearsing and developing these lessons.

Notes and references

1 While not yet available in full, the large 1983 Indian National Sample Survey shows that the proportion of persons below a fixed 'poverty line' – and the gap between their real average monthly outlay and that 'line' – did show some decline in 1977–83, but showed no trend from 1963–5 to 1983 [Ahluwalia, 1985; Rao, 1985; Minhas *et al.*, 1987].

2 Although MVs raised labour *demand* per acre – as did the shift from pasture and fodder crops to wheat, due to MVs – the *supply* of labour per acre rose proportionately more, due to growth of the domestic labour force, immigration, and decisions by small-holders to rent out land to middle farmers and seek work. Given

the elasticities of labour supply and demand (especially in peak seasons), this is not, as it might seem, inconsistent with the observed rise in real wage-rates – an unusual feature in MV areas, and not large even in the Punjab.

3 In other words, a small rise – say 1 per cent – in the real wage (due, for instance, to cheaper food as MVs came onstream) would, within a year or so, permit or induce a rise of well over 1 per cent in the supply of labour on any particular day, at least outside the brief seasonal peaks. Workers would be able and willing to work longer; also, new workers, perhaps from further away, would find the costs of seeking employment were justified by the higher real value of the wage on offer.

4 For example, if MVs mean that food price inflation is cut from 8 per cent a year to 3 per cent, and if 70 per cent of unskilled wages are spent on food, the annual restraint on price rises, and hence the rise in real wages, to begin with is about 3.5 per cent for unskilled workers; however, once labour supply has increased, the real wage rise could be trimmed to 0.5 per cent or so.

5 Some of the simplifying assumptions behind Walrasian GE can be relaxed. It is compatible with uncertainty – e.g. due to climate plus a very imperfect credit market – if there are markets in 'contingent commodities', e.g. if I can now buy 'one ton of wheat to be delivered and paid for next 1 January if and only if the rains fail'. A considerable degree of monopoly (and of economies of scale) is also consistent with the GE, provided that each firm with monopoly power (i) correctly perceives the level of demand that will prevail for its product at the equilibrium price (not necessarily at any other price), and (ii) does not change its behaviour in response to the past or anticipated behaviour of any single rival (collusion or strategic action) [Arrow, 1983, pp. 172–85, 220–2].

6 Binswanger [1980, pp. 203–4] confirms this: MVs, as a 'labour-[using] technical change, will always [improve] the growth rate of labour incomes and [worsen growth in] the rewards of capitalists and landlords, compared to neutral technical change'. In this respect 'there is not much difference between partial and GE analysis and between the open and closed economy'.

7 Compare 'regression' in the sense of Hicks [1946, pp. 93–6]. This contradiction – labour clearly gaining from labour-using technical change in a two-factor model, but not if a third factor is introduced – is not invoked to attack Walrasian GE models. These are so complex that they have to be drastically simplified if we are to understand the results (i.e. it is not just a difficulty due to small computer capacity, bad data, etc.). The trouble is that, as yet, we

are not clear about what sort of simplifications are innocent; 'sensitivity analysis' in its usual meaning is not much help.

8 And hence the GE effects of successive 'rounds' of (i) MV-induced output changes, (ii) food price falls, (iii) responsive changes in food and other output levels, (iv) responsive price and wage changes, etc.

9 The authors assume a 10 per cent 'exogenous farm output increase' – e.g. as a result of a costless shift from TVs to MVs, so that then 110 units of farm output can be produced with the same amount of inputs of land, labour, fertilizers, etc., as previously produced 100 units. If prices are flexible (i.e. if net imports and stocks do not change, so that the extra supply cuts consumer prices), this raises real income for landless rural labourers by an amazing 40 per cent (as against only 7 per cent with fixed food prices) in the Indian model. That may be because the model appears not to allow for the fact that the big fall in food-crop prices will later cause farmers to cut output, and hence employment of landless labourers. In South Korea there is an 11 per cent short-run fall, but a 3 per cent long-run rise, in landless labourers' real income – presumably because lower farm-gate food prices initially depress farm employment, but labour-productivity increases later raise it. In Sri Lanka, there is a 2.5 per cent long-run rise. Urban workers always gain. MVs 'with downward flexible prices have highly progressive effects. . . if the poor are principally landless rural workers, small net buying farmers, and urban workers' – though money wage-rate response to the lower food prices, (Chapter 5, b) and their later effect on food output and hence farm employment, need review. With prices unresponsive to MVs, i.e. in an open economy with low food-transport costs, 'wages need to respond to value-productivity gains' – as in South Korea, where rural labour is becoming scarce, but not India – 'for farmworkers to share in the economic gains of farmers' from MVs [de Janury and Sadoulet, 1987].

10 The assumptions are closely linked. A big rise in farmers' income, and hence in their demand for some consumer goods, would on Walrasian assumptions quickly shift price-incentives to encourage the production of such goods, and, in response, fully employed labour would be quickly reassigned towards those product lines. A rise in real wage-rates might cause a rise in the supply of labour seeking (and, by assumption, finding) work – a higher *level* of 'full employment' – but only thus, not through 'fuller employment', could farmers' extra demand in a Walrasian model call forth more (as opposed to *different*) national product from the non-farm sector.

11 A full economy-wide multiplier analysis would also consider the effects on successive rounds of spending of (i) higher real income for consumers (as MV prices fell), and (ii) lower income for those farmers whose yields rose more slowly, in the wake of MVs, than their farm-gate sale prices fell. No such analysis yet exists.

12 Keynes [1936] fully acknowledged that it was first systematically expounded by Kahn [1931]. In fact its origins are in Kautsky [1899]; see Lipton [1977], pp. 117–18.

13 It is usually not feasible, in these studies, to separate MV-induced from other changes in *initial incomes*. However, in all three studies of 'consumption linkages' so far, the large majority of such changes can reasonably be attributed to MVs. Without MV rice, most of the increase in fertilizer and irrigation use would not have paid in Malaysia and Nigeria. In North Arcot, though rice yields from MVs may not have gone up by quite enough to make up for the cost-price squeeze, the main extra farm income sources – groundnuts and sugar – were typical 'Type IV' regional crop-shift effects, due to MVs elsewhere (Chapter 3, i).

14 Because local products must compete with 'imports' from other parts of the country – which also rules out substantial rises in local costs of production.

15 This assumption of the work on consumption linkages – that extra output, *ceteris paribus*, is 'better' if incomes received by its producers are spent on local products – also has efficiency implications. In presenting the work-in-progress on North Arcot, Hazell has stressed that consumption linkages were weaker than in Muda largely because transport into and out of North Arcot was better, so that extra income could more readily be spent on non-local commodities – presumably on a wider range of items, or on lower-cost items. 'Weaker' local linkages thus meant stronger outside linkages, i.e. wider, and less expensive, choices for consumers; and 'preference' for local spending would impede efficiency. Also, the definition of 'local' is inevitably rather arbitrary – often it is determined by the administrative units for which data are available.

16 Once again, the ongoing IFPRI-TNAU study of North Arcot will remedy this, showing the different impacts of an extra Rs. 1000 of MV income in the hands of (i) big or small farmers, and also, separately, (ii) well-off and poor farm households.

17 Strictly, they do this only in Kalecki's version [Robinson, 1965].

18 At least in the short run, a 5 per cent wage fall is very unlikely to raise the amount of labour bought by as much as 5 per cent, as would be required to maintain aggregate demand out of wages; figures of 1 to 3 per cent are usual.

19 'Modest' in the sense that all products are 'gross substitutes' – a lower price for one means less total demand for all the others. Food cannot be assumed to be a gross substitute for other consumer goods, nor labour for other producer goods – after effects of price changes via income are allowed for. These effects are not 'modest' enough, i.e. cannot be relied on to be 'not too large in the aggregate'. Hence a Walrasian competitive equilibrium cannot be shown to be unique, nor probably stable [Arrow, 1983, pp. 123–6].

20 These look only at the effect of food price changes on the supply and demand for food, as if all other prices were held constant. Walrasian analysis looks at the effects via markets for all outputs and for producer inputs, allowing prices to vary to a new full-employment GE. Keynesian analysis looks at income adjustment, normally between underemployment GEs.

21 The sketch is terribly spotty and incomplete. Walrasian and 'Keynesian' attempts to estimate MV impacts in GE exist, i.e. real-life numbers have been used to test models. This is not true of the Keynesian model itself. Also, it remains an open question whether such a model needs to be much more specific about expectations (and money) than we are here.

22 Many commentators deny that Pigou held this position, and/or that his views were typical of neo-classical analyses of unemployment.

23 There are of course much more sophisticated neo-Walrasian views, allowing for prices, including wage-rates and interest-rates, to equilibrate economies at full-employment GE despite the major income effects on demand – and despite the price rigidities – stressed by Keynes. For an excellent account of these neo-Walrasian attempts (from 'real balance effect' to 'rational expectations') – and of their ultimate inadequacy – see [Dernburg and McDougall, 1963].

24 It is not clear whether both labour and capital markets must fail, in one or more of Keynes's three senses, to perpetuate an under-employment GE. The issue is very important in assessing the likely impact of MVs on employment, and hence on poor people's income, in developing rural areas. Standard Walrasian models in this context, if they are to be consistent with lasting unemployment, appear to require serious failure in at least two factor markets, e.g. labour and land, or labour and capital [Binswanger and Rosenzweig, 1981].

25 Keynes assumed that each level of national income entailed a fixed level of employment. The jobs/income ratio was constant. He

measured national income in 'wage-units'. This assumption greatly simplifies his exposition, but can be dispensed with.

26 His controversial remedy – State responsibility for maintaining full-employment levels of investment ('the socialisation of investment') via public works – does not concern us here.

27 Some 'saving' is put into non-interest-bearing accounts, or under the bed. This does not affect the national-accounting identity: investment equals saving (plus, in an open economy, the current balance-of-payments deficit).

28 Some 'investment' is financed from retained profits – in-firm saving. That does not affect the argument; see fns. 27 and 29.

29 If firms finance investment from retained profits, the rate of interest still measures the costs to them, since they could *lend* at this rate if they did not invest in equipment.

30 Recall that Keynes assumes that employment is in a fixed ratio to national product.

31 Normally defined as 'money interest rate deflated by overall rate of price-inflation', but to the borrower it is his or her input costs, and product prices realized by sales – not overall price levels – that matter.

32 Lower interest-rates are supposed to reduce savings by making the rewards, per unit saved, *less*; but they also mean that, to obtain a given income-stream (or a given asset value at the end of a period), *more* must be saved. Precautionary and transactions-orientated savings, therefore, could well rise when interest-rates fell. So could speculative savings, if lower interest-rates (which raise the value of assets bearing a fixed, unalterable interest) create the expectation that further falls in rates will bring further rises in the value of the assets, inducing households to forego consumption so as to save more, via speculative purchase of such assets.

33 Not only the end of the Bretton Woods arrangements in 1971, but the real-income effects of the 1973 oil price shock, and the inflationary attempt by each group *within* Western countries to avoid those effects.

34 Keynesian approaches to a poor, mainly agricultural, economy have been suspect mainly because many people use them to advocate Government deficit spending to reduce unemployment. Whether or not justified in developed monetary economies, such spending in the Third World is widely believed to induce more inflation (or bigger balance-of-payments deficits) rather than growth of real incomes and employment, because much of the extra income, generated at successive rounds of the multiplier (Chapter 6, d), is spent on extra food – which is often in short-run

price-inelastic supply. However, MVs have in many countries made food supply much more price-elastic. So a more discriminating approach to Keynesian expansion is needed, distinguishing circumstances in which it is likely to increase mainly real income and employment, from circumstances where it increases mainly inflation and imports. In any event, we use Keynesian approaches mostly to see where they lead in *explaining* the aftermath of MVs, rather than to *advise* on expansionist or other policies.

35 This is a deliberately vague formulation. A worker has more subsistence costs if (s)he has more dependants, yet does not command a higher wage-rate. However, rates in many circumstances do seem to stick at levels just covering, for an average landless family with average worker/dependent ratios, the basic costs of subsistence and of doing the work [Lipton, 1984a].

36 Because farmers with small deficits (i) being better off due to MVs, substitute leisure for income from work, and (ii) divert some of their work to their own farms, away from the hired-labour market.

37 The distributions of extra incomes from producing rice, wheat, clothes, etc. between labour and non-labour incomes – and of each type of extra income among outlays on extra rice, wheat, clothes, etc., and extra imports and savings – are estimated empirically, and are assumed to be roughly constant during the process.

38 Alternatively, input costs might stay constant in the process because competing imported inputs prevent prices of domestic inputs (such as fertilizers) from rising when MVs increase demands. This assumption requires negligible transport costs, low and steady levels of protection, and *either* squeezable domestic wages and profits *or* no problem in affording extra imports!

39 In other words, increased supply of these inputs (i) is price-elastic, (ii) does not unduly draw resources away from product-lines in price-inelastic *demand*.

40 In principle, MV expansion in each of several developing countries could help poor people in the others, by expanded trade in inputs among them.

41 In several developing countries, 'social accounting matrices' or SAMs have been constructed, to help combine the sets of information consistently. This important tool was developed mainly by British economists: Richard Stone, and subsequently Tibor Barna and Graham Pyatt.

42 In 1960–75, a team at IDS, Sussex University, (i) collected all available intensive single-village studies, many unpublished, done

since 1950 in developing countries; (ii) visited persons and institutions responsible, to discover methods used; (iii) published several bibliographies, methodology guides, and substantive cross-section comparisons of villages in respect of labour use, migration, and nutrition.

43 Such analyses are almost invariably at the level of the nation, or at least the region, not just of the village; villagers have, almost always, too large a part of their transactions with non-resident households and firms for a village-level GE (whether Walrasian, matrix multiplier, Keynesian, or input–output) to be a sensible object of enquiry.

44 Thus Singh, Squire and Strauss [1986] show that family farms' price-responsiveness, in several countries, is well below that of commercial farms (though the latter are seldom more, often less, 'efficient': Berry and Cline, 1979). This is mainly because family farms are enriched as households when crop prices rise, and thus substitute leisure for income, partly offsetting the effect of higher prices in encouraging more effort and output; whereas commercial farmers rely much more on hired labour, and are more prepared – once MVs increase the rewards – to accept the resulting search and screening costs. Conversely, price *falls* in the wake of MVs could well leave poor family farmers much less 'discouraged' (and output-reducing) than big farmers – especially if the latter sell their MV outputs, while the former eat them.

45 Not all involved *agrarian* revolutions, which are changes in the structure of rural power – normally, in the class that controls most farmland.

46 The standard-shaped 'logistic curve', showing the proportion of adopters on the y-axis and time on the x-axis, should probably interpret 'adopters' as rural communities, not as individual farms, in the case of ARs like the medieval, where all (or most) of each community must adopt together. Especially with low literacy and poor communications, the spread of adoption is almost certain to take longer among communities than among farmers in a single community.

47 By avoiding search costs, etc., family farmers almost always find it pays to farm more labour-intensively than commercial farmers working their land mainly with hired labour; see Chapter 4, b, above.

48 (i) This is despite the statistical analyses of Hayami and Ruttan [1971], which – while ingenious, clear and honest – rest on particular assumptions about 'lags' between changing factor scarcities, changing innovation responses, and adoption

sequences. The crude data (pp. 313–14) show that invention did not respond to a demand for land-saving innovation. (ii) Nakamura [1966] greatly reduces Ohkawa's estimate of Japanese farm output growth in this period; but this does not affect the view that such growth was largely caused by scientific innovations that ('perversely', if one believes that economic demand rules scientific supply) raised productivity of labour rather than of land. It does suggest that scientific output did not respond as effectively, even in raising agricultural growth, to economic need as had been claimed before Nakamura wrote.

49 Especially so to the extent that – if the whole community need not innovate at once – the early innovators are likely to be those with sufficient information, access to inputs, and financial cushions to accept (and to be able to reduce) the risks of the unknown.

50 It has recently been argued by Shrire that current pre-Neolithic societies are revivals – to permit existence on the periphery of capitalist agriculture – not survivals; and correspondingly, by D. R. Harris, that even Neolithic settlement was seriable and separable, being based in techniques (of settled agriculture) known to hunter-gatherers and gradually intensified by them [Richards, pers. comm.]. Yet the decisions to wait in one place for the crops to grow, to defer gratification and invest until they do, to live on stores of purchases meanwhile, and to trust some common authority to guard the crops, can hardly be taken in isolation from one another; or gradually; or by one or two households alone, in isolation from the community.

51 These were exceptional. The low overall ratio of labourers to farmers (p.313) shows that even as late as 1830 there must have been many small farms, often with little or no net hiring-in of labour, for each large capitalist farm.

52 This contrast between the 'first two' and 'last two' ARs is a little too polarized [Clay, pers. comm.; Richards, pers. comm.]. Neither the Neolithic nor the medieval AR was quite as much of a non-separable package, as non-seriable, or a demanding of adoption by the entire community, as we suggest; and there were genuine complementarities, among components and nearby participants, in the eighteenth-century AR, and also (especially for users of gravity-flow irrigation from the same source) with MVs. Nevertheless, the first two ARs were much more 'packaged' – in components, among members of a local community, and as regards the need for whole-farm adoption – than the last two.

53 The IARCs' only absolute disadvantage over national researchers,

perhaps, is excessive closeness to the international scientific establishment, rather than to the *local* needs of poor peasants, workers or consumers.

7. The Future of Modern Varieties

(a) From past to future: four challenges

Agricultural researchers have produced brilliant successes with tropical food staples. Largely thanks to MVs, average yields of Asian and Latin American wheat and rice have probably increased by more in the past 25 years than in the previous 250 years. Hybrids are in some developing areas verging on comparable achievements: maize in Africa, sorghum in South Asia, perhaps wheat and rice in China.

All this has saved the lives of millions of people. Almost all were saved in infancy. However, the levels of living in the very poor households into which these children survived are in most poor countries little or no higher than before the MVs arrived.[1]

Agricultural researchers, on the whole, misinterpret the reasons for this latter, and disappointing, outcome – if they even recognize it. This is, in part, because they do not survey the lessons from past agricultural revolutions. Hence, in planning future research, they lack the appropriate weaponry to deal with the four main challenges of the 1990s. These challenges present puzzling, because conflicting, problems to research planners.

Persistent rapid *rural population growth* (section 7, c), because associated with increasing dependence among the poor on hired labour as a source of income to buy food, requires a shift in the emphasis of pro-poor research from small farmers towards employment generation. The growing *lean* to *Sub-Saharan* Africa (SSA) (section 7, d), assumed to be an area of past research neglect, seems to shift the emphasis the other way – towards smallholders, who in SSA (only, and perhaps not for long) remain the majority among the hungry poor. *Biotechnology* (sections 7, e–i), and many older initiatives from weedicides to slow-release fertilizers, provide 'cloth' that could help the poorest most, but is normally 'tailored' by and for the conditions of agriculture and research prevailing in temperate rich countries; such initiatives, therefore, even in tropical poor countries, tend to be steered by applied researchers so as to lower unit costs in ways relatively more helpful to big farmers, and to employment-reducing processes. Finally, there are *international issues* (sections 7, j–k). Rich countries' agricultural surpluses, although artificial and unreliable, appear to threaten the case for all agricultural outlay (including research) in, by, or for poor tropical countries. Should not these industrialize, and buy cheap imported food, especially as Western surpluses are increased by rich-country agricultural research that is largely inappropriate for LDCs?

How should planners guide agricultural research – international and national; formal and farm-level; MV, hydraulic, and chemical – to produce pro-poor outcomes in face of these challenges? For it to do so, should its institutions be reformed? These questions, and others about its future impact of agricultural research on poor people, are reviewed in Chapter 8. We cannot answer them unless we better understand its past impact. This cannot usefully be 'reduced' to shifts in production functions and consumption functions – nor to seeing MVs as helping the poor largely via more food availability from small farms. Economic benefits to the poor arise mainly via food price restraints and higher levels of hired employment, rather than via 'small farm' income; and the benefits are embodied mainly in extra entitlements to food, rather than in extra availability of food. The extent of such extra benefits and entitlements for the poor is, in turn, largely determined by the balance of politico-economic power, as it is either reinforced or weakened by the results of research, and as in turn it affects the composition of research activities and clientèles. New basic science has considerable autonomy; its application and adoption, however, are determined by 'what pays' potential adopters. But what does pay, and who are potential adopters, depends not only on capital/labour ratios and scarcities, nor only on consumption functions, but mainly on who has the power to seize the gains of new techniques. To borrow a celebrated pun [Schaffer *et al.*, 1981]: in assessing the impact of agricultural research on the poor, appropriate technology matters, but the power to appropriate technology matters more.

Does this leave agricultural researchers as powerless pawns? Not at all. They can do much to plan their work so that it helps the poorest. But, to do so, they need to recognize political forces, options and constraints in planning their research.

(b) Lessons of the past

To say that, with today's population but 1960's seeds, millions of poor people now alive would now be dead, is not to take a crude neo-Malthusian view that poverty is caused by an imbalance between food availability and population. Indeed, it has not been mainly by their direct effects in making more food available that MVs have saved lives. Such effects have been important where MVs came to dominate small family farms of poor households. However, the main life-saving effects have come as MVs have provided poor workers with more hired employment, and have restrained the price of food to poor consumers. These effects have enabled poor households to feed infants better, thus reducing infant mortality rates. However, while that is a tribute to the MVs' capacity to create productive work and to restrain food prices, the stagnation in most LDCs of poor people's levels of private consumption – both of food and of other purchased consumer goals – is disappointing, for it suggests that MVs have seldom overcome the institutional obstacles directing their benefits to the better-off. We have seen that the physical features of MVs could reasonably be expected to provide a bonanza for the poor, not merely a lifebelt. Yet, even in MV lead areas such as the Punjab, levels of living among the poor have increased surprisingly little.

MVs have reached *the poor as farmers* in lead areas; but poor farmers adopted late, when benefits to adoption had been reduced by the price-restraining effects of the marketings by the earlier, better-off adopters; and farmers in non-MV areas sometimes got poorer (Chapter3, i). MVs have raised employment per acre, with given methods of farming. However, MVs have also provided surplus farmers with the cash to change those methods through labour-replacing inputs (weedicides, mechanical threshers) – and with the political clout to persuade the urban users of the food surpluses to subsidize such inputs. Hence the response of farmers' demand for hired labour (i.e. for *the poor as workers*) as crops are switched from TVs into MVs, while still positive, has declined over time (Chapter 4, b). Meanwhile, burgeoning populations (and hence labour supply) have prevented both the extra labour demand, and the restrained food prices, due to MVs from raising real wage-rates much; thus *the poor as consumers* have gained less than expected (Chapter 5, b).

All these effects, however, do not suffice to explain the near-stagnation of poor people's living standards in the wake of MVs in South Asia, nor the failure of MVs to spread among poor family farmers (despite their deepening poverty) in most of SSA. The GE effects outlined in Chapter 6, b, help to explain at least the first of these happenings (or non-happenings). However, 'the lessons of history' provide a more rounded and convincing explanation. MVs are not a 'transforming AR' like the Neolithic or medieval ARs. MVs in Asia and Latin America – like the European ARs of 1730–1850 – have not transformed the rural power-structures, but rather were used by the rural élites (the early adopters) to strengthen their economic positions. As for the urban élites, ARs in recent centuries have strengthened them by increasing their access to a price-restrained 'wage-good', through higher urbanized food surpluses. Urban élites since 1965 have often found it a politically paying proposition to maintain the price of a particular MV-affected crop mainly grown by large farmers as an urban wage-good (e.g. wheat in South Asia, maize in parts of Africa), and this has had spin-offs to the poor, some positive and some negative. However, such political alliances between urban élites and some surplus farmers have seldom placed either group under greater pressure than previously to transfer resources towards the poor, least of all the rural poor, since urban groups now enjoy far greater concentration of power than did similar Western urban groups during early development [Lipton, 1977].

The interplay between rulers and ruled, not just between prices and incomes or factors and outputs, will affect the handling of the four challenges to MV research (pp. 336–7). All, too, need review in the light of the regionally different, and changing, nature of poverty. Today in South and East Asia, a growing majority of the poor lives mainly off agricultural labour, not off farming. In Latin America, perhaps a majority of the poor is urban, and a much smaller proportion than in Asia is so poor as to be at nutritional risk. Only in sub-Saharan Africa are most of those so poor as to be at nutritional risk still small farmers. Increasingly, even in many parts of Africa – Kenya, the Sudan, Rwanda, even some of the Sahel – many of the poor are agricultural labourers. These distinct regional perspectives are one reason why the issues for future MV research may produce conflicting priorities. A lean to Africa, for example, could push researchers more towards MVs aimed mainly at enriching the poor as smallholders, who are a much larger part of SSA's poor than of Asia's or Latin America's. Yet growing concern for the effects of population change on the structure of poverty would push research much more towards seeking MVs that increased income from hired employment.

The goals and methods of MV research were designed to remove poverty (i) among small family farmers, implicitly for those in MV-prone areas; (ii) among consumers, implicitly constrained from an adequate diet by 'food shortages'. However, the main problem of poverty is that population growth in most poor areas is inducing, not so much a vulgar-Malthusian 'crunch' in which poor farmers need to grow more food, but growing labour supply, increasing risks of unemployment [Lipton, 1984a] and static real wages, especially in areas where MVs have not reached. The impact of MVs in restraining the price of food (Chapter 5, b) may be of limited help in the medium term; if rapid growth of the workforce renders the supply of poor labourers very real-wage-elastic, food price restraint merely enables employers to restrain money wage-rates. And without extra wage demand from the poor, extra output from MVs (unless it induces a price collapse that discourages future production) must mostly go into reduced net imports or increased stocks – not improved nutrition. Despite the major breakthrough in wheat, rice and sorghum yields in India in 1964–87, some 20–30 million tons of grain are held in Government stores for want of *effective* demand[2] from the hungry poor.

In large parts of Africa, a research policy that benefits 'small farmers' will still help to alleviate poverty. Elsewhere, however – and even in a growing proportion of African countries – *all the other poverty impacts of MVs* – via biology of plants, on farmers, on consumers – *will increasingly be subsidiary to their impact on incomes of hired labour*; and probably, in most cases, via employment rather than via wage-rates. Research that concentrates on helping small farmers will increasingly be useful to the rural poor *either* only to the extent that unskilled workers own land (or other assets that rise in value with agricultural growth), *or* by incidentally raising hired employment.[3]

Yet, as we have seen, the employment impact of MVs, initially substantial, is declining; IARCs (by research into labour-displacing weedicides, direct rice seeding, and mechanization) sometimes make matters worse. Also MVs' linkages to off-farm employment, inherently promising [Bell *et al.*, 1982; Hazell and Roell, 1983], are seldom helped, sometimes impeded, by research in the IARCs; apart from research into mechanical threshers, the shatter-prone nature of many thin-husked MVs accelerates the shift to mechanized milling, directly displacing poor landless people (especially widowed women) from custom-hulling with traditional techniques [Greeley, 1986]. Of course, these nagging worries should not obscure the major contribution, to extra employment, from MVs via bigger harvests, more use of fertilizers and water-control, and double-cropping.

An important implication of this book – especially of the possible conflicts between farmers, employees and consumers; and of some of the paradoxes of GE – is that MV policy for the poor is easiest where they are mostly small, not-too-unequal family farmers. Rice probably benefits from, and helps to preserve, small *operational* farms, even if land *ownership* is communal or commercial [Bray, 1986, pp. 177–96] – but high rents and credit payments (increasing with MVs) mean that farm operators often face surplus extraction that causes deep poverty; and there are many desperately poor, near-landless, rice labourers in Java, Bangladesh, etc. Rice apart, it is exactly where not-too-unequal family farms still predominate – in much of SSA and in some of semi-arid South Asia – that much less MV success has been achieved so far. Where it has, as with hybrid sorghum and finger millet in SW India (despite low levels of water control and soil enrichment), the favourable impact of success with MVs on the poor is at least as vulnerable as in rice and wheat areas to the growing labour supply, and perhaps – it is too early to tell – to concentration on an increasingly polarized group of bigger farmers with access to scarce inputs and infrastructure [Lynam, 1986].

Probably the top priority for anti-poverty work in the IARCs and the CG system, therefore, is to raise yields in ways that substantially raise the demand for labour. Attempts to save labour with research into direct seeding, mechanical rice transplanters, weedicide screening, and mechanical threshing are natural for European agribusiness, but are conducive to despair as a use of aid funds in Asian research centre, and may be shortsighted even for SSA. It is quite true that much *past* research has been rejected by African farmers as too labour-intensive [Pingali *et al.*, 1987]. But populations (and rural workforces) are growing at 3 per cent per annum or more; prospects for urban labour absorption are dwindling; and these effects, during the 10–15-year time-lag between initiation of research and its adoption, will typically add a further 30 per cent to person/land ratios. This casts doubt, for example, on research to produce quicker-maturing varieties to provide more food per unit of labour-time – not of land – in drought-prone areas of Africa. More directly labour-saving objectives in MV research (especially if the result is readily transferred from the target area to a more densely populated one) are especially inappropriate for an aid-financed institution – *even if they enrich farmers* as they impoverish employees.

However, in drawing such 'lessons from the past', social scientists should recall another: that recommendations to

breed types of plant varieties to accentuate pro-poor characteristics can quickly degenerate into a free-for-all. After an
MV has been introduced, its faults as well as its advantages to
the poor become clear. The benefit from avoidance or
correction of such faults is clearer to social scientists than its
scientific feasibility, time-requirement, or cost (in terms of
cash, land, and reduced research skills to achieve other
research goals). For example, a robust, high-yielding, but full-
duration rice for the second season may (i) induce tractorization, because the turnaround time between harvesting the
first-season crop and planting the MV second crop is shorter
than was the case for the quicker-maturing second-season TV;
(ii) suffer heavier losses in storage. To advise breeders 'to
breed MVs that resist storage losses', or 'to select quicker-
maturing second-season MVs', is a natural reaction by social
scientists who observe the impact of the selected MV, without
the desired characteristics, on the poor. But social scientists'
natural reaction is misplaced, if they have not set the expected
benefits against the costs (and the chances of success) of the
proposed research. A request for pre-release *screening* of MVs
for the desired, but missing, advantages is usually more
responsible.

To help natural scientists to select goals and priorities for
MV breeding, social scientists must be involved in planning
and implementing even the apparently 'agronomic' parts of
the IARC research process. Yet IRRI had one social scientist
out of seven senior professionals at its inception in 1960 – and
this was regarded as radical. IRRI and several other CG
institutions have now enormously improved in this regard, but
there remains a tendency for social scientists to do economic
(or occasionally anthropological) research in isolation from
plant breeders, plant pathologists, and other key persons in
the research process. ICRISAT's experience, in which social
scientists prevented a drastic and costly mis-specification of the
nutritional problems of the poorest (Chapter 5, h) by means of
joint work with natural scientists, remains exceptional. Even at
ICRISAT, the strength and integration of the social-science
programme have much declined of late.

At several centres the 'farming systems research' (FSR)
programmes, while inducing socio-economic and biological

scientists to work jointly on poor people's problems (and to learn at first hand from poor people about them), have also brought fresh problems, and have themselves been, to some extent, marginalized in research planning. FSR diverts economists from both macro-economics and research planning. FSR also, obviously, concentrates researchers on households which have farming systems – i.e. on the poor as 'small farmers', rather than as landless labourers or as consumers.

*   *   *

How have MV breeders' objectives, more or less influenced by social-science research, changed in response to 'lessons of the past' about poverty impact? Unfortunately, only IRRI has published the information to permit an assessment and that only for rice breeding by national researchers in Asia. A comparison of rice breeding objectives at some twenty-five experiment stations in ten Asian nations [Hargrove *et al.*, 1985, pp. 6–7] showed that, between 1974–5 and 1983–4, breeders came to selected parent plants less for yield potential, and more for growth duration, grain quality, and insect and disease resistance. These four goals dominated breeders' selection priorities in both years, however. Tolerance of drought, cold or adverse soils, *or* genetic diversity was sought from at most 6 per cent of semi-dwarf parents in 1974–5 (and at most 18 per cent in 1983–4); the comparative percentages for yield potential were 85 (51), and for grain quality 40 (54). Taller parents were somewhat more sought for drought tolerance than semi-dwarfs, and 19 per cent of *all* parents were selected for this key 'poor farmers' need' in 1983–4, as against only 6 per cent in 1974–5.

Yield performance locally is not the same as 'yield potential' (to do best in competitive trials); yet the latter was sought in 93 per cent of all rice parents in 1983–4 [*ibid.*, p. 6]. Robustness against insects, diseases and droughts is clearly vital to poor farmers, workers and consumers. Genetic diversity may be of crucial long-run importance, and surely merits more emphasis than these data suggest. On the other hand, 'grain quality' in many cases can harm the poorest (Chapter 5, h); the very heavy and growing emphasis on it may suggest undue attention to the demands of higher-income consumers (or of the farmers, often not the poorest who supply them). Also, we have pointed to the relative neglect of some research issues of uspecial concern to poor farmers: birds, rats, weeds (Chapter 2, i).

However, if there are n goals in a breeding programme, each with probability 1 in m of attainment, the chance of releasing a successful MV is 1 in mn. For example, if each variety has 1 chance in 10 of meeting any goal, each new goal multiplies the cost of a successful release by 10, even if, as is unlikely, the new goal is no more costly to achieve than were the old goals (and makes it no harder to achieve them). Therefore – though this book has been critical of gaps in MV breeding (and, even more, screening) programmes from poor people's point of view – caution is needed in proposing new goals; and research to meet any such goals needs more resources. It is not sufficient, though it is necessary, to seek more questioning of scientific fashions; more integration of social and biological science into research *planning*; more estimation of gains and losses to poor groups from MV options; and, above all, more awareness among research planners of the power-structure into which control over such options is, in fact, 'released'. Criticisms of MVs' past performance for the poor – and expectations for the future – need to bear in mind the costs and complexity of the alternatives, and the delays involved in breeding extra characteristics into MVs.

Of course, population trends (and foreign trade possibilities) may change the required characteristics. Biotechnology may well shorten the breeding period. And it should be feasible, in breeding MVs for Africa, to learn from the successes and failures of MV research elsewhere. To these issues we now turn.

(c) MV research, population, and entitlements of the poor

The absence of linkage between MV research and demographic policy is strange. At least for Asia, the major initial thrust in the early 1960s towards MV research came from a perception – right or wrong – that both the survival of the poor and the stability of the polity were threatened, as population

growth pressed against the extensive frontier of usable farmland, then being swiftly approached. That is, it was believed –
especially by the Ford and Rockefeller Foundations, who
largely financed the forerunners of the IARCs in the 1960s,
but also by most Third World experts – that the main threat
was *neo*-Malthusian (Malthus himself would have warned that
extra food might well set population rising faster).

Promoters of MV research in the 1960s saw the problem as
follows: that growing rural populations could no longer
feasibly expand onto new farm areas of a quality close to those
already farmed;[4] that the rates of increase of population (rural
plus urban) would therefore outpace those of national food
supplies,[5] given the technology; that this would greatly
increase poverty and hunger; and that the chain could be
broken if, and only if, some sort of 'green revolution' were
brought about, to gain time for family planning to spread.

The problems of national hunger and rural poverty, in this
approach, were perceived entirely in 'Malthus versus Boserup'
terms. Malthus had seen population growth – absent natural
checks, virtuous or vicious – as ultimately destructive of any
initial improvement in the level of living standards, as 'hungry
generations tread them down': people breed up, and breed
levels of living down, to subsistence level. Boserup [1965] saw
population growth as the necessary stimulant to technical
progress that intensified land use and actually *raised* levels of
living. The whole MV strategy was conceived as an effort to
push the outcome from a Malthusian to a Boserupian 'victory'.
This was to take place via exogenous intervention in agricultural research. New technology in agriculture was to create
a breathing-space for education, rising welfare, and con-
traception to bring down family size.

Given that approach, it is quite amazing that there should be
almost no demographic content in MV research. But it is not
obvious that the Malthus–Boserup debate suggests the right
way to fill that research void, especially if we seek to help the
poor. There have been major developments in socio-economic
research and theory about populations–land relations since
Boserup [1965]. Hence the appropriate question is not, as it
might have seemed twenty years ago, how research and
interventions involving MVs can best increase total food

supply per person. This is no longer credible as the key to the poverty impact of such interventions. Instead, the appropriate question is: what pattern of MV research and interventions will best help poorer families (and the society that surrounds them) to improve their per-person entitlements to food? The new research in no way means (as is sometimes suggested) that more food doesn't matter; it does prove that the types and locations of that extra food may matter as much as its amount.

These research findings – which should help to determine what sort of crops, MVs, and farming systems are researched, in order to help the poor – are of four types.[6] (i) It has been shown that famine, and probably hunger, is more often due to inadequate 'entitlements', *relative to need*, than to a small or dwindling supply of food per person – or, therefore, to the sources of pressures upon adequacy of supply, e.g. from growth of population [Sen; 1981; 1986]. (ii) The complex demographic characteristics of poor households – very different now from what they were in previous ARs – have been clarified; large households with many children are very heavily over-represented among the Third World's poor. So are the landless and near-landless. But these two groups tend to be separate (i.e. landless households do not tend to be relatively big – rather the reverse [Lipton, 1983a]). (iii) Population behaviour has been shown to be usually a rational response, by each married couple, to opportunities and risks confronting them (e.g. to the need to safeguard security in old age by having several surviving children). But this response, though individually rational, soon 'adds up' to huge numbers of underemployed or ill-paid adolescents – an outcome that all parents (and children), could they have acted together, would have wished to avoid [Becker, 1981; Cassen, 1978]. (iv) For all these reasons (and for others), growth in the level of living – and in particular the more equal distribution of several sorts of benefits, including those from agricultural growth – appears likely to slow down population increase [World Bank, 1984; Repetto, 1979], not to accelerate it as Malthus believed.

How, if at all, should this new demographic knowledge affect MV researchers? Population growth, at rates that double rural person/land ratios every 25–35 years, still seems likely in most poor countries;[7] especially in Africa, the rural dependency ratio will continue to rise,[8] and agricultural workforces will grow at over 2 per cent yearly. Do the new findings destroy the neo-Malthusian concerns that rendered MV research a plausible response to such population growth; or do they change the appropriate responses of such research to a persistent, but redefined, 'population problem'?

All the above research findings point in the same direction. MV research needs to place reliable extra resources, especially the command over food, directly with households threatened with loss of nutritional entitlements by population change: big families and/or landless and near-landless families. In appropriate policy contexts, and provided that the resources are not offered in ways actually rewarding big families as such, this will retard, not advance, population growth.[9] But what can MV researchers do about any of this, given that seeds do not grow differently for big families – and do not grow at all for the landless?

*   *   *

It is perhaps easiest to answer this question in the context of farming systems research (FSR). MVs are increasingly tested for safety and profitability at smallholder level not just in themselves, but in the context of farming systems (Chapter 6, i). The suitability of the MV is thus tested in the context of the farm household's aims and resources – especially timed labour resources – prior to FSR. If the FSR and the subsequent tests are done well, they also take account of 'non-own-farm' incomes, and labour requirements of the household, for its own enterprises other than farming (e.g. carpentry); for hired work on other people's enterprises (including farms); and for domestic work.

Evaluation of MVs in this way – and design of new MVs, to provide incomes in the context of past FSR analyses of poor people's known systems, preferences, and work patterns – can help people, under population threat to food entitlements, in several ways. *First*, before an MV is chosen for release, its timed labour requirements can be compared with those for other varieties, in order to discover which MV (or farming system) is most suitable for large families; these are the likeliest to be poor, and their small children can suffer if MVs sharpen the peaks in demand for mothers' work away from home. *Second*, it could make sense to choose MVs (or crops for MV research) with a view to providing adequate, inexpensive food for lactating women and weanlings; families at special risk (i.e. at a specially poor time) within their life-cycles may well contain both. *Third* and perhaps most important – since population growth on a limited land area is likely, even for poor people who do some farming on their own account, to increase the proportion of time and income *not* associated with own-account farming – the MVs on offer to the poor should increasingly be 'adoptable' with labour peaks that do not coincide with those of hired labour opportunities.[10]

Of course, FSR has its limits (Chapter 6, i). The sort of features, in each of the above three categories, that should be sought in the screening of alternative MVs, before one is selected for release on an area where rapid population growth threatens poor people's food entitlements, can also be detected by other sorts of surveys, e.g. village studies (Chapter 6, i). And the agricultural scientist has to tell the socio-economists whether their desired features of 'MVs for population-pressured families' are likely to be found through screening (or even breeding) – or whether socio-economists should specify second-best desiderata. Dialogues along these lines are urgent.

One consequence of population growth as a threat to poor people's food entitlements (not mainly to food availability, as in crude interpretations based on Malthus's early work), however, cannot be handled by FSR, nor indeed by research approaches emphasizing farmers. This consequence is the increasing proportion of poor people dependent mainly or wholly on hired farm labour, or farm-related labour, for their income-based entitlements; and, alongside this, the weakening of rural labourers' position in three senses.

- First, via standard economics, the growing numbers of person-hours on offer by rural labourers overall both depresses the wage-rate and makes it less likely to rise – in any sustained fashion – if MVs bring about increased, but localized, demand for labour (e.g. for double-cropping).

- Second, via routine politics, the bargaining power of labour is weakened in its conflict with larger farmers, who are sources of jobs – and of tenancies and loans – as a growing population of labourers brings ever-increasing competitive demand for work; labour becomes especially weak if decent arable land is in more or less fixed supply, if big-farm employers control their population (and hence reduce their competition) while employees do not, and if there are few off-farm employment prospects.

- Third, via the political economy of groups, the growing population of rural labourers in an area can, paradoxically, make it harder for them to finance effective organizations for economic and political bargaining; small groups can often agree on, supervise and enforce common actions where the costs are shared by each member, but a member of a large group is prone to leave the payment of such 'subscriptions' to the others [Olson, 1982].

What can IARCs or national researchers do to help the 'new poor', who are not as such created by population growth, but who are made by it to be increasingly dependent on labour incomes, and decreasingly able to increase them? We see three options, but first make two preliminary remarks.

- The MVs, crops, and farming systems, likeliest to help the rural poorest if they obtain even 15–20 per cent of income from own-account farming, may be different from those best for wholly landless labourers.

- Researchers involved mainly with Africa rightly note that many past research outcomes were not accepted by African farmers owing to labour *shortage* [Pingali *et al.*, 1987]. It may sometimes be inferred that agricultural research there should not worry that population growth now means an impending labour *glut*. However, person/land ratios are doubling every 20–30 years in most of Africa. MV research takes 5–15 years from design to widespread adoption. It is increasingly unlikely that 'labour-saving' MVs or techniques are sensible research goals. Labour-using goals will be best – by the time the new MVs are in use – for poor people, even in most African areas, as food entitlements are threatened by population increase.

There seem to be three options for MV research to help farm labourers.

- First, researchers might aim at farm systems (and crops and MVs) that, while safe and profitable to farmers, greatly raise the demand for labour. For this to be a reliable source of income to the labouring poor, it is important that farmers not be encouraged to avoid employing them, e.g. via subsidies (or free research) for tractors, threshers, weedicides, mechanized milling systems.

- Second, where Governments are able and willing to undertake large-scale food-for-work schemes,[11] researchers can screen MVs, crop-mixes, and farm systems for their capacity to support such approaches. This requires crops and varieties that, in good seasons, provide cheap, readily stored foods, well adapted to such schemes (in particular to the needs of small children, and mothers of big families dependent on income from the schemes).

- Third, researchers could concentrate on finding varieties, crop-mixes, etc. for the areas where, even with substantial population growth, the poorest will for some decades rely on own-account farming, not on labour incomes. However, as a *sole* strategy for IARCs or other researchers, this would be a counsel of despair, since it would leave a large majority of the world's rural poor unassisted, except by accident.

There has been almost no impact, on appropriate poverty-orientated research into MVs, of the new knowledge on population growth. This section has been written to begin a debate that might put that situation right. We do not pretend to present a definitive or researched conclusion to that debate.

(d) Agricultural research, MVs, and sub-Saharan Africa

Most discussions of these issues begin by apologizing for – or denying – any attempt to generalize about an area as vast as sub-Saharan Africa (SSA), and as diverse agronomically, economically, culturally and politically. The discussants then forget such denials or apologies, and treat all the countries of SSA as a homogeneous unit, with a similar history of ill-luck and policy error, and with similar remedies in similarly improved agricultural research. We shall try to avoid this path. SSA is no less, but also no more, diverse than India. The persistence of generalizations about SSA – by Africans as well as foreigners; among the wise as well as the foolish; and on price policy, nutrition, or democracy as well as on agricultural research priorities – strongly suggests that there *is* some set of 'African' experiences, problems or opportunities that is worth generalizing about, however cautiously this is denied by the generalizers.

We suggest *seven generalizations*, relevant to research priorities for MV work seeking to remedy African poverty, listed below as G1 to G7 (with 'exceptions' where relevant). We next draw five *conclusions for MV research strategy* for poverty alleviation in SSA, numbered *R1* to *R5*. Then we compare these conclusions with some 'stylized facts' about agricultural research in SSA, and derive some proposals for changes (*C1* to *C4*) in SSA policy for MV research. As in section c above, we have no space for detailed reasoning and evidence; interested readers are referred to Mellor *et al.* (eds) [1987]; Eicher and Baker [1982]; Eicher and Staatz [1984]; Lipton [1985]; Swaminathan [1986]; Thirtle [1987] and sources cited therein.

G1: The key food crops in SSA are not the classic MV breakthrough crops. In 1982 in South and East Asia (excluding China where the proportion is probably even higher), wheat and rice comprised over 80 per cent, by weight, of estimated cereals harvested; in SSA, the proportion (much more unreliable, but probably below estimate) was 13 per cent. For millet, sorghum and maize, the corresponding proportions were 18 per cent and 77 per cent. For every kg. of cereal grains produced, South and East Asia produced only 0.2 kgs. of roots and tubers, but SSA 1.9 kgs.[12] (However, the major breakthroughs achieved for rainfed sorghum in India since 1972 may refute the view that rapid, MV-based growth is infeasible for semi-arid, rainfed coarse grains in Africa.)

G2: SSA lacks the artificial water control typical of most MV breakthrough areas in Asia; yet, in agro-climatic zones otherwise comparable to Asia's, SSA suffers from somewhat greater and less predictable variation in natural water supply from rains and river flows [Delgado and Ranade, 1987, pp. 124–8; Harrison, 1987]. In 1981, Asia (excluding Japan and Israel) irrigated over 30 per cent of arable land, but SSA below 3 per cent. The exclusion of a couple of atypical African countries (Sudan, Madagascar, parts of Ethiopia) would strengthen further this picture of SSA as dependent on unirrigated coarse grains, roots and tubers.

G3: Deep ignorance surrounds smallholder food farming in SSA. (i) The above African statistics are far less reliable than for Asia, and have been deteriorating; only for three or four countries of SSA, not including any of the four largest, can we be 95 per cent certain that food output in any given year lies within 40 per cent of the official figure.[13] (ii) African small-farming systems, especially in humid and semi-humid areas and among herders, are little understood by researchers; the proportions of output and area accounted for by distinct systems are seldom known even roughly; and most formal researchers are as ignorant or dismissive of on-farm experi-menters [as shown by Richards, 1985] as vice versa. (iii) While colonial research was not quite as neglectful of food crops as is often alleged [Arnold, 1982], only recently have researchers begun to acquire systematic knowledge about smallholder-level micro-environments, varieties, practices and prospects, especially in mixed cropping, which dominates most African small farms and the process has not been helped by financial retrenchment, heavily affecting research, by many African governments in the 1980s. (iv) The resource base is substan-tially unresearched: large statements about the presence, absence or depth of groundwater, for example, or about the quality or regenerability of soils, have in most of SSA no basis in serious research.

G4: Though *output* data (and therefore trends) are of little value for most of SSA, more reliable post-1975 data show that *net food imports per person* have been rising sharply almost everywhere [World Bank, 1986, p. 190]. The reason is not that resources have been successfully switched to the production of agricultural exports to pay for imported food, for export production (except in a few cases, such as the Ivory Coast) has done badly too; and almost nowhere in Africa has industrializ-ation provided much extra net foreign exchange to import food. Growing reliance on food imports has accompanied worsening in chronic undernutrition and in acute famines.

G5: The great majority of SSA governments devote much lower proportions of domestically financed public investment, personnel, and probably current expenditure to rural areas than do governments in comparable[14] Asian or Latin American countries. Partly for this reason, the ratio of income-per-person in agriculture to that in other activities – typically 1 to 1.5 in the West during early modern development, and 1 to 3 in Asia and Latin America today – is currently around 1 to 5 in most of SSA.

G6: Agricultural exchange (as regards both the marketing of outputs and the provision of inputs) in most of SSA faces a severer problem than in most other developing areas. SSA's agriculture is typified by areas of rather low productivity, separated from each other – and from towns and ports – by long distances, bad terrain, and (usually) land not cultivable except at very high cost. Thus agricultural exchange requires transport and other infrastructures that are very costly per unit of farm population or of farmland; yet those units are each producing rather little output by value – and much less that is surplus to immediate needs – to finance these high costs. This probably cannot be much improved without extra direct farm investment; to *start* with big outlays on transport infrastructure, before farms have the resources or technology to greatly expand their surpluses for exchange, would be to overstrain the rural resources available from scarce savings and aid.

G7: Because land shortage is fairly recent, except in a few areas such as Kenya and Rwanda, inequalities in land rights are much less extreme than in Asia or Latin America. Most of the rural poor have significant rights in land, i.e. are not 'landless labourers'. If inequality helps keep them poor, it is much more rural–urban and much less intra-rural than in Asia or Latin America. However, in more and more of SSA, growing rural populations create, or presage, the initial signs of land scarcity and polarization; already in the mid-1970s, 40 per cent of SSA populations lived in countries where land was judged 'scarce' or 'moderately scarce' (given its quality) in respect of capacity to feed their populations, and even this classification [FAO, 1983] questionably classifies land as 'moderately abundant' in

Tanzania, the Sudan (including Nile Valley agriculture) and Chad.

What do these seven 'stylized facts' suggest about priorities for poverty-orientated agricultural research in SSA? How do these suggested implications compare with the realities, trends and main current proposals for such research?

*   *   *

R1: *G3* above implies that, *in order to formulate a research strategy for MVs in SSA, much more needs to be known*: about the farming choices made by African smallholders; about the physical, economic, and community environments that affect those choices; and, perhaps most urgently, about the results at national level, and for main regions within countries, for crop inputs and outputs. No sensible strategy for publicly supported food-crop research in India or Australia could be designed, without some reasonable approximations of the levels and trends in main crop outputs and inputs, by major regions and soil-water conditions. Yet no such numbers exist in SSA save for a handful of countries (including Botswana, Rwanda, and probably the Ivory Coast and Kenya) – though official series often create a dangerous illusion that they do. Major research strategies are accordingly prepared in virtual ignorance of, for example, whether cassava production (or area) in Nigeria, let alone under different agroclimatic circumstances within Nigeria, has in the past year or decade been greatly outpacing population increase or falling far behind it; or doing better or worse than sorghum or maize production; or becoming much more unstable and risky, or much less so.

R2: *G7* implies that, in most of SSA, provided it substantially raises total factor productivity in agriculture – a huge 'if'! – *a small-farm MV-based research strategy still has a better chance of benefiting the great majority of the poor, even the poorest, than in Asia or Latin America*. This is because most of SSA has gone much less far down the path of intra-rural polarization and landlessness – or of commitment of the poor to permanent urban settlement. However, the speed of rural population growth, the spreading of land shortage, and the scarcity and cost of productive infrastructures (*G6* and *G7*) mean that 'any old' small-farm MV strategy might enormously increase rural polarization. Strategic carelessness could well help or stimulate only those 'progressive' farmers who first obtained MVs and linked inputs, inducing them to greatly increase their shares of land, of control over employment (and employees), and of credit infrastructure, inputs and income. For the rural poor, even this process might be better than nothing,[15] i.e. than continuation of the absolute immiserisation that they have suffered in most of SSA during the last 10–15 years; but it could easily be even worse. In any event, a sensible MV-linked strategy – along lines we shall suggest – can reasonably expect to do far better than this, for Africa's poor. Also, we need to recall that Africa's rural poor (being mostly small farmers) tend to eat most extra food they grow; that urban élites are disproportionately strong in SSA (see *G5*); and that such élites will often seek strategies permitting gains, especially extra food, from MVs to be extracted for urban use. Such a strategy will *either* concentrate MVs and linked inputs on big and capital-intensive farms, which will sell the extra food to the towns (as with hybrid maize in Kenya and 'Southern Rhodesia' in 1960–75), *or* adapt the sort of 'small farm strategy' that tends to steer MVs to leading smallholders, polarizing them rather than reaching the poorest, but again generating food surpluses for the towns.

R3: There is unlikely to be an MV-based strategy 'on tap' in most parts of SSA that can rapidly offer prospects of advance. (i) Colonial research seldom pointed to *major* advances, profitable to *smallholders*, in *food* crops (*G3* iii). (ii) Neither crops nor water conditions in SSA (*G1, G2*) are often those for which MVs are likely to be directly transferable from elsewhere. Even where this seems feasible, yield gains for sorghum varieties 'improved for Indian conditions' proved low in Zambia and Zimbabwe – although at first observations in Southern and Eastern Africa suggest that Indian-style millet MVs may 'be more directly useful. . . than in the Sahelian region' [ICRISAT, Annual Report 1985, pp. 41–2, 83]. Yield gains in the field in India have been much less with pearl millet, leaving only sorghum and the relatively minor crop finger-millet as areas where major yield increases seem directly feasible from screening, testing for farm-level profitability, and extending non-African MVs to apparently comparable SSA environments. Greater water risks (*G2*), and perhaps greater physical fragility of soils [ter Kuile, 1987, p. 103], will in most SSA environments require locally specific, intensive breeding efforts, incorporating both robust local and high-yielding exotic germ plasm, and testing alternative crosses for safety and profitability within smallholders' complete systems, usually with several crops mixed in one field. This is more difficult than breeding an improved wheat or rice variety for big areas of irrigated land – though the experience of ICRISAT (and much else) refutes the statement that 'all they had to do in Asia was to come up with new varieties of the main cereal' [Hartmans, quoted in Harrison, 1987].

R4: To meet the three requirements that new MVs should not exhaust fragile soils, should be readily usable where there are seasonal labour shortages (much of SSA now), yet should enhance opportunities for employment income where landless and near-landless labourers form significant proportions of the poor (significant parts of SSA now; most of SSA by the year 2000), *MV releases should improve conversion efficiency of land, water and chemical inputs into outputs with the help of higher amounts of off-peak labour*. Both for this reason and because – despite at least twenty years of effort – really high-yielding varieties have seldom been bred for marginal *and* risky water environments in any continent (not least because in such environments fertilizers, needed for really big yield increases, seem too risky for poor farmers), a successful MV strategy for the very poor in SSA almost certainly requires *much more attention to prospects for improving water control (G2) in ways complementary to the development of appropriate MVs*. The African record with (i) major irrigation is, with a few exceptions, discouraging, but better prospects are (ii) farmer-controlled micro-irrigation, both traditional (Nigeria's *fadamas*, Botswana's *molapos*, and much else) and modern (Africa's 'boreholes', used to water cattle, are similar to South Asia's tubewells, used for higher-value multicropping), and (iii) re-timing of planting (and here ploughing) to reduce rainfall risks – a hardy perennial of paternalist extension with TVs (where farmers reject advice that seems to offer only marginal extra profits for much more work), but much more appealing with appropriate MVs.

R5: Research seeking rapid, major MV-based gains, supported by much higher inputs of water control and fertilizers, must in most of SSA form a large part of any strategy for avoiding further, and disastrous, worsening of poor people's conditions. Wider access to farmland renders the prospect for spreading the benefits of MV progress – if attainable – to the poor as producers better than it was in Asia or Latin America; though, also, that careful socio-economic screening of releases will be needed if this prospect is to be realized (*R2*). However (*R3*), the physical environments – the quality and reliability of water, of soils, and (*G6*) of built infrastructures – is in most of SSA much less favourable to rapid local or transferred MV progress than was the case in most of Asia or Latin America; and the knowledge base – information on output data; prior agricultural research; understanding of smallholders' systems – is much weaker (*G3*). Yet an MV-water-fertilizer research strategy, seeking rapid advance, remains the only imaginable way to create enough rural livelihoods to prevent continued reductions, and deepening fluctuations, in poor people's food *entitlements* (section c) – not simply availability – as population grows, and as the prospects of farming new land become worse.

*   *   *

A contrary view has been forcibly expressed [Hartmans, 1985,p. 12]:

None should advance the hope that Africa will experience a 'green revolution' [like Asia's]. Africa's traditional agriculture is much more complex. The dramatic breakthroughs in Asia with rice and wheat were based on a backlog of knowledge in the plant sciences that is not so readily applied to roots, tubers and grain legumes. . . spread across a wide variety of ecosystems.

We accept all these statements except the first.

To insist on the need for rapid MV-based rural transformation as an African research priority is not naive optimism. The present signs are that few African governments are ready to allow rural areas to receive the major domestic resources needed for anything like a 'green revolution'. But what is the alternative? (i) Better incentives for farmers in SSA, while highly desirable on equity grounds, are not an alternative. They will not enable farmers to produce much extra *total* farm output – as seasonal labour constraints, and increasingly land shortages, bite harder – unless dramatic technical progress creates new possibilities, within or near the margin of profitability [Lipton, 1987]. (ii) More export-crop outputs, while usually easier to achieve in SSA than more food output (owing to a better base in research and infrastructure), are seldom an alternative. In most major export crops, for SSA as a whole, such strategies risk glutting markets, so that world prices decline and, producers (taken together) do not acquire much – if any – extra revenue with which to buy food [Godfrey, 1985]. (iii) African industrialization appears a long-run prospect, anyway unlikely to create much income for the growing population of poor labourers. (iv) Finally, gradual improvements in traditional farming systems [Richards, 1985] – while unduly neglected by formal researchers, and while capable of responding to quite rapid past rises in urban food demand [Hill, 1977] – cannot suffice, as land reaches the extensive margin, to raise food-crop output at 3–4 per cent annually for 10–20 years. Yet such progress is mandated by rising, ageing, and urbanizing populations – especially if income-per-person grows, and as grains shift from food to feed.

Only a combination of MVs, improved water control, and fertilizers can approach that sort of growth rate in food output. Admittedly, the term 'green revolution' is perhaps inappropriate in Africa (as elsewhere) for the impact of MVs. Admittedly, too, MV progress for African smallholders will not everywhere be as dramatic as (in India) for wheat in North Bengal, rice in parts of Tamil Nadu, sorghum in Maharashtra, and finger-millet in Karnataka; or, in Africa, for hybrid maize in substantial parts of Kenya, Malawi, Zambia and Zimbabwe. A third necessary admission is that, even in SSA, extra food production by 'small farmers' – even if MVs can help make this substantial and sustainable – overlaps more and more imperfectly (though, still, much better than in other continents) with extra food entitlements for the undernourished poor. Nevertheless, MV-based farm research is, for SSA's poor, the only real hope. It is the only game in the countryside.

How can its potential be realized? Two illusions must be removed [Lipton, 1985]. First, for at least ten years, it has been an illusion that African agriculture suffers from research neglect overall. Per agriculturist, per farmed acre, or per unit of farm output, more is spent on agricultural research for SSA than for other poor continents. But research into smallholders' food crops – despite large outlays and large claims in Africa, and continuing large successes elsewhere – has in SSA been rather unproductive. This is due partly to concentration on crops covering small proportions of farmed areas; partly to rapid turnover of costly researchers; partly to dispersion of effort among too many research stations, crops, and problems; but mainly to absence of local, especially governmental, support.

This brings us to the second illusion. Despite popular belief, there are, in agricultural research as in agricultural policy generally, few signs of commitment – especially to small-scale, poverty-oriented agriculture – by most African governments. The money and experts, for research and for development projects, come mostly from abroad. Pressed by debts (and sometimes by drought and by worsening terms of trade), few African governments devote much of 'their own' money (from taxes, domestic borrowing, etc.) to the support of agricultural research. Indeed, well over half the value of public and private gross investment of African low-income countries in agriculture in the broadest sense, including irrigation and rural development, is accounted for by net inflows of foreign aid to the sector; for capital *and current* support of food-crop research, the proportion is probably higher still. Nor is there a clear personnel commitment; few of the increasing numbers of good African agricultural scientists stay long in post with MV research systems in SSA (despite pay levels far better than in Asia), probably in large part for want of the sense of achievement that can come only if governments – not foreigners – implement and finance the necessary current research support. The IARCs spend much more – per acre, per farmer, and per unit of agricultural GDP – on and in Africa than for other developing areas; much capital aid, too, has gone to build up national food-crop research systems. However, with a few partial and threatened exceptions (such as Botswana, the Cameroons, the Ivory Coast, Rwanda, Kenya and Zimbabwe), those systems are constrained by lack of current government support: constrained, quite frequently, to the point of non-performance because salaries are not paid.

*   *   *

C1: Our first conclusion on agricultural research in SSA is that, to help the poor (or to achieve anything at all), most national MV research systems require much more commitment by national governments. Lacking this, foreign aid to such systems can achieve little – indeed, may merely help to tempt the best local scientists away from such systems (e.g. out of the Ivory Coast, into the excellent international network of 'francophone' agricultural research). It sounds harsh, but foreign support of MV-based research in SSA should probably be confined mainly to:

• national systems in those few countries with Governments able and willing to guarantee, and demonstrate, corresponding sustained commitment of necessary cash, personnel, and logistic support, and

• international activities likely to produce results (such as MVs) directly helpful to farmers in ecosystems found in several SSA countries, of which at least one or two possess public-sector, NGO, or other, applied field research agencies committed to the ancillary national efforts required to test those results for farm-level acceptability, and if appropriate to diffuse them, and to deliver the requisite inputs on time (or to enable the private or co-operative sectors to do so).

Often, the two overlap, as with the promising ICRISAT/ SADCC/SACCAR programmes based on Zimbabwe and Botswana, for millet and sorghum research for the semi-arid lands of Eastern and Southern Africa. However, if neither a government nor a national private agency – profitable or charitable – can provide steady current funding and staff to the national agricultural research system, aid donors cannot make that system work. Nor can they 'push on a piece of string' by repeated, unsupported donations.

C2: Where the condition of adequate local support is met, extra aid to African national MV research needs to concentrate on greatly increasing the research effort in relatively neglected 'poor people's crops' that support, both as producers and as consumers, large numbers of nutritionally vulnerable farming households: cassava, yams, cocoyams, sweet potatoes, bananas and plantains, sorghum and millets. Past research efforts have greatly emphasized high-cost production of rice on large irrigation schemes, of tropical wheats, of fashionable crops such as soybeans, and of hybrid maize, a crop much eaten by the poor (though more in urban than in rural places), but vulnerable in years of low rainfall, and competitive for land with safer crops that have suffered from relative neglect by researchers.

C3: In view of the ignorance of smallholders' practices in SSA, especially in humid and semi-humid areas – and the consequent disastrous failures of numerous MV-based agricultural projects – research into 'farming systems' is a necessary *part* of the socio-economic knowledge required by agricultural scientists in developing MVs, and by governments designing rural policies. However, such research must be based on reasonably trustworthy *macro*-data, at national and regional level, on outputs of major food crops; else the contributions, relative to costs, of the various farming systems so meticulously analysed will (as in Liberia) be quite unknown. Furthermore, the *micro*-research into farming must rest on in-depth year-round analysis of the total earning and consuming systems of a few communities, preferably containing several typical households of poor farmers, set into the constraints and opportunities offered by distinct social systems and ecologies. Quick visits to larger numbers of farmers, while an understandable approach (in view of the wish for sampling typicality, and for swift understanding of several distinct types of farming in a country), can achieve little, except in the hands of very experienced and exceptionally skilful analysts, of whom only a tiny handful exists. Normally, short visits will not unveil complex patterns of seasonality – affecting nutritional risk, family behaviour, and often even tenure and household size – that affect a family's farm and non-farm options, needs, and incentives. 'Rapid rural appraisal' [Chambers, 1981] is a splendid approach to FSR where (as in much of Kenya) the basic facts are known. Elsewhere, the establishment of 'recommendation domains' [Collinson, 1981] of farmers – or rather farming households – in various circumstances requires slow and clean, not quick and dirty, research, if socio-economists are usefully to help in the design of an MV work programme and above all if they are to trace, understand, and follow through the farm-household systems, and the community options and constraints, of the very poor.

C4: MV research planning for any particular country, to succeed in most of Africa, requires more from socio-economists than that they help research planners to a better micro-level understanding of smallholders' circumstances. (Indeed, agricultural researchers have been rather slow to respond to such understanding in the past, e.g. by greatly expanding work on mixed cropping; on *quelea* and other bird attack; or, until very recently, on *striga*.) Socio-economists also need to provide – and agricultural research planners, to respond to – better macro-level links between agricultural research and public policy priorities. Four areas stand out.

- The policy judgment on the desired *balance among major food and other crops* 5–15 years hence, when the MVs to be researched should be widely adopted, needs to iterate with agricultural researchers' best judgement on what is achievable in these alternative crops, and with socio-economists' analyses of likely border prices, rates of return, and poverty and nutrition impact of these alternatives.[16]

- A policy view on the role of *agricultural exchange* – i.e. of the affordability of big public subsidies to long-distance transport – as against production for local use, needs to be iterated with research judgements on the prospects for the (often quite different) crops and MVs involved.

- The view on *population policy*, much less clearly specified (let alone implemented) in most LDCs of SSA than elsewhere, needs to be worked out, and conveyed to MV researchers. This is especially important where – as in Zambia, Zaire and parts of West Africa – aggregate land shortage seems a long way off; just what are researchers doing when they give priority in such areas to raising per-acre yields? There *are* sensible answers, but the choice of appropriate yield-raising MV research depends on the likely course of population change in the various rural areas.

- Perhaps most pressingly, policymakers need to take a view on *the role of irrigation* as a potential source of food and income for the poor, and to iterate that view with MV researchers' judgement of comparative agro-economic prospects for rainfed and irrigated crops. Plainly, MV researchers need to press policy planners for reliable basic data to analyse these issues.

(e) Biotechnology and other new research directions[17]

Many good descriptions of new research prospects for tropical food farming exist [Norman, *et. al.*, 1982]. However, in applying the lessons from our book to incipient new techniques (in farming or in research), we are concerned to see these techniques as means to an end: sustainable reduction of poverty and hunger. Therefore, the last two subsections have tried to focus on the main *issues* that face planners who want to steer the benefits of agricultural research to the poor.

Rising populations, and the limited capacity of urban activities to generate employment-based livelihoods, mean that such research should seek MVs, inputs and methods that not only reduce smallholders' average cost of food production, but do so in ways that increase average employment, especially per hired worker.[18] This is desirable even, perhaps especially, if research planners accept the new view of population and nutrition: that the former threatens the latter, not as in a vulgarly interpreted early-Malthusian way (by reducing food per person), but – as Malthus himself [1826/30] later recognized – by reducing entitlements, as more workers compete for fewer jobs or dearer food: Chapter 5, b.

Increasing emphasis on sub-Saharan Africa in agricultural policy and research (section 7, d) – and, more generally, on tropical and sub-tropical areas left out of progress so far with MVs – requires 'pro-poor planners' to shift emphasis, in MV development and selection and in overall policy, towards new issues:

'subsistence' or deficit smallholders, with major off-farm flows of labour and income; mixed cropping; options in water control; inputs and procedures robust against (or helping to minimize) delivery failures, whether due to the 'transport infrastructure dilemma' (p. 354) or to scarcities in public-sector capacity or will to assist rural development; and sufficiently regular and accurate information on output and rural systems to permit sensible choices on all these matters, and also on the mix of research – among crops, topics and regions – that is likeliest (given the prospects of scientific success) to offer substantial benefit to large numbers of poor farmers, workers and consumers.

Given these two priorities, what have the main 'new research prospects' to offer? In discussing biotechnology (BT), we present the 'likeliest-case' issues, prospects and futures in a more clear-cut, less open way than most scientists would regard as permissible. This is done so as to concentrate the mind on the issues of poverty impact – not feasible if a large, almost indefinite range of outcomes has to be reviewed.

Of the many proposed definitions of BT, the most relevant may be 'industrial or agricultural use of biological agents – specifically microbial, plant or animal cells or enzymes – [in, respectively,] fermentation, cell culture and biocatalysis' [Faulkner, 1986]. The main applications of BT, and the great majority of funds spent on it, have been by the private sector and on human health. In agriculture, applications have sought to improve industrial processing of (or, in the case of sugar, substitution for) farm products first; animal production (via improved bovine reproduction, vaccines, and growth hormones) second [Joffe, 1986, pp. 53–62]; non-food crops, notably tobacco, oilpalm and cotton, third; and food crops fourth and last, though with help from BT on soil bacteria, and a recent boost from Rockefeller's rice BT programme.

These priorities within commercial BT are not good for the poor. (i) If BT increases the attractiveness to farmers of animal production relative to crop production, employment per acre will fall, costs per dietary calorie will rise [Buttel *et al.*, 1985, pp. 46–7], and coarse grains will shift further from food to feed. (ii) Industrial substitutes via BT cheapen or displace labour-intensive Third World exports, notably sugar. (iii) Many non-food crops from LDCs, if productivity rises somewhat (due to BT), will glut the market, and face a much lower price; yet, as with tea, they often occupy land with few attractive alternative uses.

We revert to possible responses, by public-sector and other researchers who can afford to be poverty-orientated, in Chapter 7, i. Here, we concentrate on the substantial, and in tropical and sub-tropical areas potentially pro-poor, contribution that BT might make via its 'fourth priority', food crops. Can that potential be realized? Might it be subverted?

(f) Today's BT and the poor: tissue culture

It is useful to look at two phases of BT in respect of its contribution to breeding improved food crops:[19] tissue culture and gene transfer. Tissue culture (TC) involves (i) removing a surface-sterilized part of a plant, (ii) placing it on or into a semi-solid medium, where it grows rapidly as a disorganized 'tissue' or callus, (iii) transferring this to a liquid culture (which is then agitated, both to move the tissue around and to let air get at it), with a view to (iv) regenerating a new plant with desired characteristics [Locy *et al.*, 1984; Joffe, 1986, p. 17]. The *intermediate aims* of TC – as compared with traditional plant breeding (TPB) as a means of producing plants with a particular set of *final aims*, or desired features – are to reduce the time; to increase the precision and control; and to extend the range.[20]

TC is already widely practised, but faces problems. First, as in other forms of plant breeding, regeneration of plants, especially some key cereals, is a difficult, expensive, and often lengthy process. Second, desired plant characteristics – above all yield – are usually polygenic, so that the effect of changing particular genes in culture is seldom precisely known. Third, plant performance depends on environment as well as geno-type – for maize yield, only about a quarter of yield variations among plants are heritable [Janick *et al.*, 1981], i.e. environment appears to be about three times more important.[21] Hence, and for other reasons too, even when plants are regenerable from TC and look promising in the dish, they must next be tried out in a range of field environments in the context of TPB programmes (and perhaps further crossed within them).

This is not only in order to multiply sufficient plants for on-farm testing. Before this happens, TPB programmes have to screen whether virtues apparent *in vitro*, such as salt tolerance or rice blast resistance, (i) carry over to assorted farm-like situations, and/or (ii) are revealed in such situations to be associated with other virtues, or vices, of the genome produced by the TC scientists. Such screening may lead to rejection of the plant produced from TC, to its multiplication, or to its use in further TPB and/or TC before a variety can be tried out by farmers.

The effects of TC outputs on the poor, like those of TPB, depend quite heavily on the characteristic products, which in TC depend on the types of tissue used. The six main types are the *meristem* (apex bud); the *anther* (pollen sac), cultured mainly to obtain rapid homozygosity; *individual cells*, sometimes 'peeled' to expose the protoplast; [22] *protoplasts* themselves, cultured to obtain more rapid mutations; *ovaries*, fertilized in the dish; and *embryos*, rescued so as to be cultured and regenerated as 'wide crosses' [Sondahl *et al.*, 1984; BOSTID, 1982; Joffe, 1986, pp. 38–40]. All six methods are practised now; new knowledge in BT is enormously expanding their use and scope.

          *   *   *

Meristem culture, combined with 'heat therapy', rapidly produces disease-free clones of roots and tubers. Such clones are increasingly exempted from national regulations designed to exclude import of TPB clones, which may carry undetectably low levels of infectious 'foreign pathogens'. By early 1985 IITA had distributed such disease-free planting materials to thirty African countries for cassava, and forty world-wide for yams and sweet potatoes [Joffe, 1986, pp. 47–8, citing Ng and Hahn, 1985, and Withers and Williams, 1985].

Apart from the great gain from such readier exchange of plant materials (especially for African countries, some heavily dependent on clonally reproduced roots and tubers, but almost all short of TPB or BT capacity), reduction of 'sub-clinical' infection in a clone can greatly raise yield, even without genetic change. Thus 'plants of traditional cassava cultivars grown from meristem cultures yield over 50 per cent more roots and planting material than plants grown from stakes, [probably] due to eradication of diseases and viruses [that] affect plant vigour. . . but are not expressed physically' [CIAT, 1985, p. 27], nor otherwise detectable.

Reduced disease risk and easier access to better planting material have to help the poor people, if available at low cost and directed, as in this case using meristem culture, towards their subsistence roots and tubers. Such approaches turn the normal trade-off – 'as the number of resistances increases. . . the more likely [is it that] yield potential is being sacrificed' [Lynam, 1986, p. 9] – into a complementarity.

Poor people face only two dangers from meristem culture. First, privatization might restrict new, cost-reducing materials to better-off, large, or otherwise privileged users – unlikely, since improved clones, once released, can be simply propagated. Second, as the costs of root-crop production fell, it (and land for it) might shift towards larger farmers, producing capital-intensively (as root sources such as cassava chips outcompeted other sources of animal-feed) increasingly for animals, rather than for cheap root calories eaten by the poor directly. This risk is greater because BT does so much more, in general, for animal than for plant productivity. However, meristem culture as an aid to disease-free root-crop cloning appears the most strongly pro-poor of the BTs, and – perhaps for that reason – currently the least privatized. Experience with MVs from TPB, however, shows that in unequal societies even an inherently pro-poor technique need not, in fact, benefit mainly the poor. Further, if meristem culture were to spread a single clone – because disease free – over large areas, and the clone proved either highly soil-extractive or vulnerable to a new pathogen, the effects would be worst for the poorest; but at present the risk seems fairly remote.

*   *   *

Somewhat more recently than meristem TC, *anther culture* has become almost routine, producing homozygous parents swiftly for over twenty wheat and eighty rice hybrids in China and others at IRRI [Han, 1985, and Abrigo *et al.*, 1985, cited in Joffe, 1986]. At CIAT, although in 1984 only 40–5 per cent of rice varieties responded to the procedure (i.e. plants could be regenerated) – and for these only 46 per cent of plants were fertile (diploid) leading to 'homozygous or purebred lines' – such lines were produced in nine months, 'as compared with 3–6 years using' TPB [CIAT, 1984, p. 32]. This must give a big push towards hybrid rice and wheat MVs, and also away from population approaches towards hybrids with natural out-breeders such as maize.

Breeders' and TC scientists' burgeoning options from hybrids, due to anther culture, will raise yields – but will also compel farmers increasingly to rely on seed suppliers (Chapter 2, d), just as BT may be increasing private control over patented MVs, a process for which hybridization from pure lines is a prerequisite. There must be a risk here that new techniques are leading professionals into innovations that may exclude poor farmers – especially in some African countries with limited or uncompetitive capacity to produce and dis-tribute high-quality, timely hybrid seeds each season. This could concentrate progress upon bigger, less employment-intensive farmers who can more readily buy the rapidly changing hybrids – and perhaps herbicides that complement these hybrids (which may be bred to tolerate them) – to cut costs, and to out-compete poor farmers seeking to sell crops grown from today's MVs.

Even if we diagnose this risk correctly – and as non-biologists we could be quite wrong – it is not the fault of anther culture. (This also has promise in breeding cereals for tolerance to climatic and soil stresses [Joffe, 1986, pp. 45–7], crucial to spreading MVs into neglected and impoverished areas.) The fault, if any, would lie in a use of BT – as of TPB – to overemphasize hybrids, perhaps as part of an approach to science (i) mainly technique-led, (ii) to the extent that it is objective-led, attaching too much weight to 'privatizable' research products, i.e. to profitability, *before* poor people have acquired adequate effective economic or political demand for such products.

*   *   *

Embryo rescue and culture, and *ovary culture* for fertilization in the dish, are TC methods to achieve 'wide crosses' between species. Wheat-barley and wheat-rye crosses have been achieved at CIMMYT, and bean-pea-peanut crosses elsewhere, via embryo rescue [*ibid.*, pp. 47–8, citing Mujeeb-Kuzi and Jewell, 1985, and Raghavan, 1985]. However, TPB is also good at wide crosses, and has produced the only significant economic success with food crops to date, triticale (wheat × rye). Although such research may well be a justified use of general tax resources for wealthy countries, it is not clear that the same applies to scarce aid-financed research in IARCs.

This doubt applies more strongly to *protoplast culture*, directed at faster mutations through 'peeled' cells more readily prone to somaclonal variation. Few plants have been regenerated; by 1985 none had led to significant economic success [Torrey, 1985, p. 362]. The earlier history of mutations in TPB is similarly discouraging. Scarce scientific resources for BT, like scarce land and expertise in TPB, are perhaps better used in planned assaults on specific problems, rather than in inducing and scanning mutations in the hope of finding one that looks promising. Non-biologists like us see this, perhaps unfairly, as science led by techniques rather than by objectives, for the poor or otherwise.

(g) The next phase of BT: gene transfer

So far, we have raised issues about BT viable now, and certain to be of increasing commercial importance in 1987–94 or so. The next phase of BT aims at *in vitro* genetic manipulation, based on operating via rDNA transfer and other techniques to transfer foreign genes.[23] In plant breeding proper, these advanced techniques are generally agreed by natural scientists to have great potential importance, but little prospect of producing big changes in food crops for ten years or more. In breeding of soil bacteria, however, such changes are possible now; present dangers (as well as opportunities) to poor people; and interact with MV issues discussed in earlier chapters, for example with the improvement of poor people's capacities to farm, work, and eat cereals grown with low-cost nitrogen.

However, in this context also, we must be impertinent enough, as non-scientists concerned with the economic analysis of pro-poor innovation, to warn against technique-led approaches to science. Although the latest fashion is in this case clearly of great potential importance, poor people's access to cheap nitrogen is not sensibly reducible to issues of BT. Also BT may not be the most 'cost-effective' use of scarce research resources – especially by IARCs, or national crop research institutes in poor countries – to help poor people via extra nitrogen delivered to crops. This caution applies, more generally, to BT (as opposed to TPB) for all purposes. For less expensive (and less speculative) forms of TC research, the case for work in IARC, is stronger than for gene transfer. In both cases, the choice is very difficult: Total neglect by the public-sector research agencies could render poor farmers reliant on private monopoly alone; high spending on high fashions could use up resources better devoted to TPB for neglected 'poor people's crops' and areas.

*   *   *

Gene transfer via rDNA is technically possible now for maize (and almost so for rice). However, commercial application is generally agreed to be 10–20 years off for monocots (including all cereals). Even if specific genes can be isolated, inserted into vectors, and expressed in a newly regenerated monocot plant, 'most plant breeding requires manipulation of polygenes, which seems beyond the skill of even genetic engineers' [Simmonds, 1983, pp. 21–2]. Moreover, 'desired characters expressed at the cellular level' are neither necessary nor sufficient for their appearance in the whole plant [Hansen *et al.*, 1986, p. 33]. Gene transformation for major food crops, then, seems some way off.

However, there is a major exception. Massive corporate money has been invested in gene transfer to create resistance to specific herbicides in corn, soybeans and cotton. Calgene has cloned a gene for resistance to Roundup, the most commercially valuable herbicide produced, and a resistant cotton variety is due for marketing in 1989–90; work is well advanced to achieve similar results in maize [Joffe, pp. 44, 84; Kenney and Buttel, 1985; Doyle, 1985, pp. 119–20; Sin, 1986, p. 1361]. This means that only a farmer who can obtain such varieties –

each year, since they are of course hybrid, and often male-sterile, so as to prevent on-farm seed retention – can grow the crop and apply the stated herbicide. Could this harm the poor?

- If the herbicide greatly cuts a user farmers' unit production costs (or raises crop supply), farmers unable to use (or afford) new seeds *and* herbicides lose as their output prices are competed down.

- If the herbicide is used or transferred to fields without the herbicide-resistant crop variety, crops will die.

- If fields without herbicides suffer increased weed growth (e.g. because pollinating agents move there, from the herbicide-affected fields), yield there falls.

- Weeding labourers may be displaced.

In such cases, poor rural people could well lose absolutely from the transfer of a herbicide-resistant gene. They are safe only if they are (i) not greatly dependent, for income, on working as hand-weeders, or (ii) small farmers, able to buy and use both the hybrid and herbicide as cheaply as other farmers, and to obtain similar reductions in unit production costs. Even then, poor people's interests would probably have been better served if the costs of overcoming the problems of gene transfer had been devoted to ends other, and less labour-replacing, than higher sales for particular commercial herbicides.

What of *gene transfer via cell fusion*? Given the failure to regenerate plants leading to commercial MVs from 'ordinary' protoplast culture, prospects with fusion might seem remote. As a route to wide crosses, it is less promising than TPB or embryo rescue [Torrey, 1985]. Success in rapidly breeding blight resistance into fused potato cells remains to be followed up for whole plants (let alone commercial varieties), and has not been paralleled for cereals; fusing the genomes of two complete cells, in one or successive cultures, to regenerate (one hopes) plants lacking exposure to the evolutionary challenges of the field, may well be a costly way to create complex, delicate, but unviable outcomes [*ibid.*; Yoxen, 1983, pp. 149–50; Burgess, 1984]. Yet there is a striking exception.

Monoclonal antibodies (MABs) are proteins that recognize, and fuse with, specific substances foreign to an animal or plant.

The fused cell (called a hybridoma) multiplies rapidly in culture, and each resulting cell secretes antibodies of determined specificity. Assays using MABs have been devised to provide precise assessment of the host's resistance to given levels of disease attack, *or* of the levels of attack given the known resistance. This was not possible with standard, less-specific antibodies, and is a major step forward, for resistance breeders, from the imprecisions of 'moderately resistant', 'moderately susceptible' and so forth.[24] More excitingly, subsequent breeding (TPB or BT) may perhaps permit this resistance to particular diseases to be combined with other desired plant characteristics [Joffe, 1986, pp. 22–3]. Rapid testing for rice plant viruses via MABs is already feasible [Hibino, 1985].

*   *   *

BT, especially the new phase of gene transfer, is generally more difficult for monocots (including all cereals) than for other plants, and for plants than for bacteria. Therefore, commercial applications of gene transfer are likely to affect cereals indirectly, via 'BT bacteria', first. Monsanto is developing a soil bacterium that expresses an insecticidal toxin, which was due for release in 1987 but has been delayed by the US Government's Environmental Protection Agency [Kenney and Buttel, 1985]. Such bacteria could well be developed or adapted for various LDC agroclimates. They would certainly spread to poor people's soils, and reduce disease there. However, direct resistance to the toxin could develop among the target insects, or among other insects.[25] It is not hard to imagine uncontrollable ecological side-effects – especially in LDCs where environmentalist lobbies are weak compared to those who stress the need for quick gains for the poor (or quick profits for the less-poor). The poorest would be least able to escape the effects of any possible 'genetic Bhopal'. Yet the poorest also are in most need of extra food at low cost; bacterial BT that killed harmful insects (often carrying harmful viruses) could provide it.

(h) BT and plant nitrogen

Plant BT to make 'cereals fix their own nitrogen. . . would rest on the transfer. . . of a multiplicity of genes' [Simmonds, 1983, p. 22]; this is several decades beyond present levels of genetic knowledge and technology.[26] However, *bacterial* BT presents one promising route to cheap nitrogen for poor farmers. Salt-tolerant genes have been transferred into nitrogen-fixing bacteria from others [*Diversity*, Winter 1985, p. 12]. Work is afoot to overcome problems with transfer of BT-improved bacteria to field conditions, notably competition from other soil bacteria.

It is important to recognize five limits to bacterial BT as a source of plant nitrogen. (i) BT is only one (fashionable) route even to improved bacterial nitrogen fixation. (ii) There are non-bacterial routes to fixing more soil nitrogen for plants. (iii) Such fixing is not always a clear gain. (iv) There are other ways to provide extra nitrogen to plants than by fixing it from soils. (v) Extra nitrogen from any source may not be the most cost-effective way of improving a crop's contribution to poor people's incomes.

(i)   There is great variation in the nitrogen-fixing efficiency of any given strain of bacterium, among varieties of a crop. Of four millet cultivars, two showed 30 per cent yield rises (unfortunately from a very low base), and two no significant rise, in response to identical amounts of the same bacteria; rises with three sorghum cultivars ranged from 2 to 10 per cent [ICRISAT, *Ann. Rep. 1984*, pp. 38, 102]. Thus TPB, or simple varietal screening, may improve bacterial nitrogen fixation more than a similar research outlay on BT. So could research into adjustments of the soil-water environment. So could conventional screening and selection of appropriate strains of *azolla* fern,[27] for combination with appropriate *anabaena* cyanobacteria, to fix nitrogen for rice. IRRI is carrying out research on the recovery of symbiosis from endophyte-free azolla [IRRI *Annual Report* 1987, p. 24–5]. Separation of the complex has been reported, but not recovery of symbiosis. IRRI has reconstructed symbiosis in anabaena-free azolla with the endophytic anabaena developed to fix nitrogen.

(ii) Blue-green algae can fix nitrogen without bacteria; blue-green algae, organisms competing with them, and plants (mainly rice) benefiting from them, can all be selected or altered by breeders.

(iii) If more nitrogen is removed from the soil by successive crops through enhanced fixation, soil fertility can decline, or may require expensive restoration – infeasible for poor farmers – later.

(iv) The IARCs and national researchers are engaged in research, to improve MV-nitrogen relationships, which may well offer routes both more labour-intensive and less soil-extractive than bacterial nitrogen fixation. (1) IRRI is seeking to improve the efficiency of nitrogen release, to the plant, by chemical fertilizers by such methods as mudball technique, sulphur-coated pellets, and root-zone placement. (2) IITA has shown that 2–4 tons/ha. of cuttings of *Sesbania rostrata*, an African shrub, provide nitrogen to raise yields of unfertilized flooded rice by 42–60 per cent [IITA, 1986, p. 98]; transport and application costs with organic manure are heavy, but involve mostly family labour, so that MVs[28] with good base yields and nitrogen responses could justify them for farmers lacking cash or access for fertilizers. (3) Agro-forestry shifts the emphasis, from improving the interactions between bacteria and plants, to improving the interactions among the farmer's various nitrogen-related activities.

(v) While nitrogen shortage constrains crop yields in many circumstances where other constraints (e.g. water shortage: Chapter 2, g) had earlier been blamed, it does not follow that providing extra nitrogen is the best way to relieve such constraints. It could be more cost-effective to improve – or to research – the plant's absorption or conversion of available nitrogen to it, and/or the rivalry for nitrogen between healthy crop plants and competing unhealthy crop plants, or all-too-healthy weeds. And there *are* many conditions where Nitrogen is *not* constraining yields.

*   *   *

These remarks are not meant as a discursive attack on BT, in general or for the improvement of bacterial nitrogen fixation. The latter issue, however, illustrates how questions, about *alternative uses of scarce research resources* to benefit poor people, are raised even more sharply by BT than by conventional MV research. That is because BT absorbs such large amounts of cash resources and top-grade scientists; tends to transfer them to the private sector (pp. 379–82); and thus leaves scanty resources for BT in the public sector, which alone can – though may well not – pursue the 'public good' of poverty alleviation as such. (Public authorities can, of course, contract out BT work to private firms; but if this becomes the main public approach to BT, severe conflicts of interest and access are likely.)

Plant-related, food-orientated, Third World, public-sector, poverty-conscious BT is of the fifth order of smalls. Even this tiny effort is divided between IARCs (under peer-group scientific pressure to show they can perform in this fashionable field) and national research systems (in many LDCs unable to sustain the inputs, of cash or of 'frontier' scientific expertise across several disciplines, needed for effective BT applications). In what areas is BT research the most cost-effective route to MV-linked goals? Does BT's choice among such goals correctly match poor people's concerns for cheap food, rather than scientists' concerns for technical advance or intellectual beauty?

(i) BT and the poor: four issues

In respect of poor people's concerns, four issues arise. First, does BT continue the evolutionary (though recently rapid) paths of TPB, or does it seek higher yields in ways more likely to bring fundamental changes in farm techniques and hence perhaps in power-structures? Second, how does BT affect diversity – among MVs available for a given environment, or among environments where MVs as a whole can help? Third, what are poor people's actual and potential gains and losses from the privatization of research, associated with BT and perhaps with its political context? Fourth, as BT improves yields, will it greatly raise input requirements – or will it improve food crops' (sustainable) conversion efficiency, (possibly sustainable) partitioning efficiency, or (probably unsustainable) extractive efficiency; and will the upshot be atypical for *some* needed plant inputs, especially sunlight? This issue leads us from BT into international influences on future linkages between MVs and the poor.

*　　　　*　　　　*

(i) If somatic hybridization and gene transfer became common in major food crops, one might speak of a revolution: in the speed (and range) of technical progress and yield increase; in their cheapness and accessibility to commercial farmers; yet also, perhaps, in their remoteness from subsistence farmers – let alone landless labourers – and their concentration on expensive, largely private, research centres in (mostly) rich countries. Cost-cutting innovation in food production would, on balance, surely be transferred to richer countries, and to the end of the farming spectrum with low labour-inputs, both per acre and per ton of grain. Subsistence farmers who could keep their land would increase their self-consumed food output, as the new 'bio-engineered' MVs trickled down to them; but both they and landless workers in LDCs would lose chances to convert effort into saleable grain, in competition with ever more dramatic, BT-induced cost-cutting at the capital-intensive end of the farming spectrum. But all this is rather a remote prospect.

However, major and commercially 'farmable' progress via gene transfers in tropical food crops (via 'advanced' BT) are probably 10–20 years away. BT's main impact on farming, especially in LDCs, is for many years likely to come from TC. This earlier phase of BT is really an adjunct to TPB. It can be argued (we believe rather tortuously) that by about 1985 TPB had perhaps reached a plateau: rice MVs are stuck around the yield potential of IR-8, released in 1964; wheat MVs cover most of the appropriate areas; and recent TPB in sorghum, millets, and cassava has left few important crops to attack. Also, the argument continues, much TPB keeps food output in LDCs as a whole barely abreast of pest evolution plus population growth.[29] To the extent that all this is valid, TC was needed, not as a new 'green revolution', but to avert a brown counter-revolution by restoring some momentum to TPB. Plant breeding will be accelerated [Kenney and Buttel, 1985]; more precise methods of introducing disease resistance and environmental tolerance will sharpen the need for careful planning of research priorities; but BT will not 'revolutionize' TPB [Simmonds, 1983].

However, though 'rDNA [and somatic hybridization will have] little impact on agriculture in the short or mid-term, [at least] in the poorer areas of the Third World', some argue that this 'misses the point [that] gene transfer techniques [for] agro-industrial [processes] will completely change linkages within and across sectors' [Joffe, 1986, pp. 2, 26]. Labour-intensive LDC farm products may be undercut by BT-based industrial processes in rich countries – following such precedents as maize-based syrups, to replace sugar – long before LDC cereals benefit from gene transfer. This would certainly induce a (counter-)revolutionary change in labour's already limited power in some LDCs, though the process will also raise 'world GNP'.

(ii) It seems obvious that both TC and gene transfer, by easing scientists' task in planning and changing a plant's genome, must help to diversify plant populations. This reduces the threat of resistance breakdowns, a threat from which poor people have least access to agrochemical protection (Chapter 2, j). It also helps MVs to spread to less-affected, and generally poorer and more equal (Chapter 3, i), areas; notably, unlike TPB, BT can rapidly and precisely assess a cultivar's tolerance of various soil chemistries, and looks likely to be better at improving MVs in this respect, e.g. inducing salt tolerance in millet [Joffe, *ibid.*, pp. 18, 45–7, 111].

Both advantages of diversity are genuinely promised by BT for tropical food crops, and both should help the poorest most. However, three warnings are needed. First, as gene transfer becomes more precise – and as the chemical impacts of TC are more exactly related to changes in the genetic structure of the plant cells – so the apparent prospects for altering single-gene characteristics, such as most vertical resistances, increase. Both forms of BT appear likely [Hansen *et al.*, 1986; Doyle, 1985] to do little for – and may thus produce MVs that occlude (Chapter 2, j) – the more diffuse, less well-understood, but 'safer' gains from horizontal or polygenic resistances. Then a crop's sources of disease resistance might become (even) less diverse, and more exposed to new pathotypes (Chapter 2, k), with the poorest at most risk.

Second, although it seems obvious that more, genetically more diverse, MVs 'must' spread more readily to poor people in remote areas with 'difficult' soils, the search of TPB scientists for wide adaptability in a MV was not absurd. It used scarce research resources to seek MVs useful, directly or via local screening or breeding programmes, to many poor people. Moreover, diverse MVs, each tolerant of a distinct soil deficiency, are readily seen, by eager local extension workers in hitherto unpromising environments, as a chance for quick results at the cost of extracting and dispersing, in one or two seasons, the limited amounts of various scarce micro-nutrients – here zinc, there chlorine – from particular problem soils.

(iii) The other limit upon BT, as a source of diversity for poor farmers, is that such diversity may not be accessible if the new MV is privatized. They then face some danger of being confined to 'traditional' MVs, while only better-off farmers get the 'BT-MV' (often tied to agrochemical supplies), and enjoy both safer and more diverse pest resistance and lower average costs. We do not believe that this gloomy scenario is likely, because seeds are usually a tiny part of production costs, and because the 'BT-MV' or a close relative will in practice often be available from alternative sources. However, the effect on the poor of BT's thrust towards privatized seed research do need watching.

The huge potential market for crop BT has induced large investments in such research by several major corporations. In the USA, by 1980, five companies owned 30 per cent of all US plant variety patents, and in 1980–2 alone some 13 per cent of scientists in US State agricultural experiment stations moved into the private sector [Hansen *et al.*, 1986]. Private funding also covers a rapidly growing proportion of university BT research [Joffe, 1986, p. 80]. Of course, private companies investing in BT do so in search of profit, and therefore seek to protect their products – the new MVs – by prohibiting access and genetic information, and by breeding to reduce risks of replication by other breeders or by farmers (e.g. via male-sterile seeds). As the story of hybrid maize in the USA shows, *competitive* private enterprise in seed-breeding, especially alongside an innovative public research sector, can both make profits and help farmers and consumers to gain from a stream of alternative, inexpensive, and steadily improving MVs.

However, farmers – especially small farmers in LDCs – may experience less happy results from BT privatization. First, the high fixed costs of such research render private competition in particular lines less likely. Second, legislation in the USA, permitting varietal patents and safeguarding plant-breeders' rights, has increased the private returns to secrecy. Third, public-sector competition has been weakened, both by short-ages of government funding for university and similar MV research, and by 'ideological' thrusts to privatize it and force it to pay its way (as has happened to the Plant Breeding Institute, Cambridge, England), which it usually does extremely well, but at the expense of shifting priorities away from areas other than profit-making. Both in developed countries and in advice and lending to LDCs, the privatizing and monopolizing thrusts of the costlier forms of BT have been supported by legal and political trends.

*   *   *

All this need not harm poor people in LDCs. After all, it is the smaller farmers who now use purchased inputs least, saturate them with most labour, and should thus show the best returns on them; is this not a tailormade market, in which private enterprise – including large corporate sellers of BT – can do well by doing good? In SE Asia, there are signs of this.[30] In most areas, however, high distribution costs to dispersed LDC smallholders – coupled with the greater similarity, in LDCs, of large-farm requirements to the approaches and technologies of large and corporate farmers in the USA, Europe and Australia, who form the bulk of market demand for BT services – render it rather unlikely that private firms will steer such services substantially towards small LDC farmers, low-cost food staples, or employment-intensive farm-household systems. If these are policy goals, appropriate incentives – or substitutes – for private BT corporations will need to be put in place. Otherwise, they will probably focus initially on export crops of processing interest; on BT applications complemen-tary with agrochemicals, especially herbicides, rather than with labour; and (in most places) on more accessible, bigger farm enterprises.

Decision-makers in a developing country, either themselves or via IARCs, will need to understand private BT if they are to alter, control, or complement its activities. IARCs allow open access to both germ plasm and research methods and results; private BT firms cannot do this, yet have access to IARC research, enjoy very much more generous funding, and can pull away top IARC and other public-sector scientists. Hence a growing part of agricultural science, and germ plasm, is becoming inaccessible to LDC customers – an effect that partly offsets their potential for gains from BT.

The USA, a major funder of IARCs, having legislated for both plant breeders' rights and the patentability of new BT organisms, is not likely to share these privatized rights with IARCs. Nor is it at all obvious that IARCs – with their brilliant past record in TPB, its still major potential contribution, the limits of cereals BT, and the concentration of BT science elsewhere – have comparative advantage in developing a major, innovative BT capacity. However, IARCs certainly require (i) to build on their own substantial achievements in TC, so as to provide non-privatized sources of some of its main products to LDCs; and (ii) to understand 'frontier BT', so that opportunities – and threats – posed by it to poor people may be anticipated and publicized in good time.

LDCs themselves also need policy responses to the privatization of BT. Only a handful of larger LDCs, including China and India, can afford the scientists and cash to build up comprehensive BT research programmes. However, 'while genetic engineering [proper is] expensive, plant TC is not. A fully equipped laboratory (excluding land and buildings) might cost $250,000. . . . Also the basic skill requirements involved in the simpler clonal propagation techniques are not great, and could be learnt quite quickly' [*ibid.*, p. 108, citing Wilkes, G., 1984]. This might be consistent with a 'small science' approach, directly involving and responding to small farmers' own preferences – and to their long traditions of on-farm experiment – from the beginning of TC research design and priority setting [Bell and Richards, 1986, p. 20].

Private BT is not to be assumed hostile to LDC smallholders. With public-sector competition and involvement, the interests can probably, in most cases, be reconciled. After all, private BT can make money only by selling products that cut unit costs, or increase safety, for farmers; and these effects should also help many labour-intensive, input-deprived farmers. However, private BT also depends on 'privatizing' information, seeds, and the collection of rewards – and responds mainly to affluent markets. Quite determined actions by IARCs and LDC governments (which both have interests that include, but are not identical with, those of the poorest) are needed if private BT outcomes are to provide net gains for the poor in LDCs.[31]

(iv) BT, as a market-responsive and largely private research area – in which the leading research companies also sell farm inputs – is under strong pressure to develop in ways that raise demand for such inputs. This can best be done by raising the efficiency with which they create outputs. Thus – although rich farmers often obtain inputs more readily and cheaply – there is potential gain to small farmers, and (since they are labour-intensive) also to their employees.

Moreover – although a largely private BT research effort is likely to be more responsive, in crop-mix and technology types, to the demands of larger farmers and industrial processors in rich-country markets – tropical and sub-tropical countries, when farmers apply BT, should find that access to year-round warmth and sunlight provides an enormous natural advantage over temperate countries. This is because BT is unlikely to be able to do much to improve photosynthetic efficiency, though it may greatly increase efficiency of other forms of input-to-crop conversion. To improve photosynthetic efficiency, it would be necessary to increase light absorption or the electron transport rate, or to convert *C3* to *C4* plants (maize is the only really important *C4* food staple). Basic knowledge about which genes perform these complex and almost certainly polygenic roles is at present far too small to achieve such goals via BT [Herdt, pers. comm.]. Hence both BT and TPB have been rather unsuccessful in improving photosynthetic efficiency.[32]

TC (and TPB), and later perhaps gene transfer, will continue to improve staple food plants' capacity to make dietary energy out of soil nutrients, water, and the land, capital, and labour that supply and enhance them. However, TC and other research will probably do very little to improve photosynthetic efficiency; and there are limits (obviously reached for perennials, but also approached in many food-cropping irrigated areas) to attempts to 'capture' more sunlight by growing crops for more of the year. So BT will multiply the natural advantages, for food staple cultivation, of areas that enjoy a lot of sunlight. This will be even more the case – assuming BT does not make vast advances in cold tolerance – if such areas also enjoy high enough soil temperatures to permit year-round growth of staple food crops.

Plainly, then, BT – by doing more for food staple crops' nutrient efficiency, disease resistance, adverse-soil tolerance, and possibly water-use efficiency, than for their light-use efficiency – increases the advantages of tropical and subtropical LDCs in food-crop farming. Provided BT concentrates on raising conversion efficiency of soil nutrients, rather than extraction efficiency,[33] this should be to the long-run advantage of the countries containing the great majority of very poor people: Asian and African tropical LDCs.

Whether this biophysical prospect is realized, however, may depend on the balance of international pressures upon the food output in *developed* countries. Such pressures originate mostly from big, capital-intensive, and heavily subsidized Euro-American farmers. We now turn to the interactions of this, and of related international issues, with the prospects for future benefits to the poor from MVs, whether achieved via TPB, BT, or both.

(j) Food surpluses and MV research: redistributing world supply and demand

The apparent lunacies of world agriculture are familiar. In Europe, North America and Japan, massive food stocks, expensively kept by the taxpayers, swell each year to support rich farmers – and are, in part, disposed of by encouraging people to consume, largely in ignorance, products that damage their health. In Africa and parts of Asia, numerous hungry people cannot afford enough to eat (and partly for this reason must watch their children die), while the State (in search of rapid urban development) neglects, impoverishes, or otherwise discourages producers of food.

Many people blame 'distortion of agricultural incentives' for these absurd and tragic outcomes [Schultz, ed., 1976]. Indeed, that is part of the story. But we shall not understand the whole story, nor learn how to change it, by concentrating on misincentives. The distortions of the rural-urban power balance, and of the determination of priorities for scientific research – within the affluent North, within the poor South, and to a lesser extent between North and South – are responsible for these wrong[34] incentives; more important, for wrong allocations of public resources between city and country, and thus for Northern surpluses, Southern hunger, and the needless deaths associated with both. Excessive and mal-structured food supply in the North, ineffective and mal-distributed food demand in the South, mutually supportive through restricted and distorted agricultural trade [World Bank, 1986a, chapters 6–7], are intimately linked to the pattern of *scientific* supply and demand.

Scientific supply, fortunately, does have some autonomy, because discoveries in pure science are not mainly determined by impure economics or politics. That is why the impacts of MVs, or of BT, on the poor cannot be crudely attributed to the self-interest of powerful economic entities, or to the distorted balance of effective (cash-backed) demand, at national level. However, to complete the picture, we need to look briefly at likely trends in *world* food supply, demand, and research.

(k) The international context

The poorest quarter of people in Africa and Asia *use* at least two-thirds of their subsistence and employee incomes for simple food staples – prepared cereals, roots and tubers, and pulses. These people also *obtain* at least two-thirds of their incomes from producing these basic foods (as owner-farmers, tenants, processors, and/or employees). Thus the conditions of food production and consumption are the most important determinants of whether this 'poorest quarter' can escape from poverty.

These conditions are created by the contexts of ownership and power in LDCs, not only by the technology of production. Yet these contexts usually change only slowly; and new technology in agriculture, although itself changing fast, seldom upsets them much. So it is not ownership or power, but the technology of food production – notably the spread, crop-mix, distribution, and regional focus of MVs – that is likely to prove the main *domestic* source of change, in (say) 1965–2005, in poor people's lives as producers, consumers, and employees in LDCs.[35]

But are not such domestic technical changes much less important, as determinants of events in LDCs, than *foreign* forces or events? Or, if LDCs' domestic agro-technical change is important, does change abroad largerly determine it – or its impact on the poor? Is it not US, European and Japanese agricultures – their supply and cost conditions; the demand for their products; the political, as well as economic, pressures from their farmers (and the associated providers of industrial inputs and processes) – that dominate world food markets, trade, and hence the generation of supplies and demands both for world food and for technology in food production?

The tortuous formulation of these questions indicates their complexity. We are concerned with effects on supply and demand; at home and abroad; for agricultural products, trade in them, and technologies to grow them; as mediated by politics and by economics. Also, we are concerned not just with 'Northern' effects on 'Southern' trade and technology, but with the 'Northern' impact on how any *given* pattern of trade or technology in LDCs affects the poor.

We shall glance at MVs' role in relation to two main international issues: Western food surpluses, and world research biases. First, though, we recall four facts about food production and trade. (i) Foods, especially poor people's cheap foods (and above all roots and tubers), mostly have high weight/value ratios – increasingly important when fuel price rises further deter long-distance trade. (ii) There is increasing national emphasis, especially in LDCs, upon security and self-sufficiency. (iii) Staple foods are dominated in world production – and even more in exports – by large countries, but in imports increasingly by small ones. (iv) Poor people's well-being depends on household food entitlements rather than on national food availability as influenced by food exports and imports.

All four facts somewhat isolate 'poor people's foods' from trade. Such products – cassava, yams, sorghum, millets – are imperfect substitutes for, and less traded than, wheat or even rice (let alone vegetables or animal-based foods). Even when traded, they tend to move from less-poor growers, via capital-intensive processors (e.g. into cassava chips or cattle cake), towards Western cattle – not from poor farmers to direct consumers.[36] Hence, for poor people's food staples, production and demand conditions within LDCs are of dominant importance, as compared to international trade. It is on products that they themselves farm, work on (or exchange locally), and eat that the poorest people, and to some extent countries, depend. These four facts, cutting trade/production ratios for poor people's staples, also reduce the direct impact of 'Western' *trade* on food technology for poor people in LDCs. However, the four facts do not themselves reduce the impact on such people of changing 'Northern' *technology* in food production.[37] The development of MVs (which has spread significantly to hybrid sorghum and maize, finger-millet, and perhaps cassava) was, in some ways, an astonishing exception to the process of international transmission of inappropriate, labour-displacing farm technology from North to South; are the South's poor threatened, as agricultural research is privatized, by the renewal of that process?

In 1995–2005, as the results of MV research begun in 1988–90 spread to farmers, international trade in cereals may offer different opportunities from those of 1985–7, when such research was planned. If internationally traded food staples are likely (for reasons largely unconnected to agricultural research in or for LDCs) to get much more expensive, compared to the other products than a LDC can make, that strengthens the case for such an LDC to undertake its own research – and to support and press for overseas and international research – into the production of such staples. Conversely, if they are likely to get cheaper, the case for an LDC to promote research into them is weakened, since when such research comes onstream the LDC would be able to import the food staples for fewer of its exports, or (if an importer of food staples) sell them in return for fewer imports.

Price expectations for staple foods should from an efficiency viewpoint affect an LDC's research plans – or those of an IARC concerned to help LDCs – in the same direction, therefore, whether it imports or exports such foods. As for the equity viewpoint, the conclusion is the same, though for different reasons in importing and exporting LDCs. In importing LDCs, the poor gain more, especially as consumers, from local research that cheapens domestic production of staple foods, if import prices of such foods are rising. In LDCs exporting food staples, employment gains are likely to count for more, in amplifying the effects of changing world food prices upon gains from domestic agricultural research, if export prices of such foods are rising.

In both sets of LDCs, moreover, poor and hungry people will probably be affected even more in 1995–2005 than today by the price level of internationally traded food staples. In almost all LDCs, for poor and rich alike, a growing proportion of food consumed will be purchased; and the share of food consumption based on growing, or buying, little-traded roots and tubers is likely to fall. This will raise the importance, in poor people's budgets, of *internationally traded* foods and their very close domestic substitutes. On the other hand, poor people in most Asian and Latin American countries, and in a few African countries, will enjoy somewhat higher real income per person around 2000 than they do today. For the poor, though not for the very poorest 10 per cent or so [Lipton, 1983], this will somewhat reduce the ratio of *all* foods to income. On balance, we anticipate that – for most poor people in most LDCs – importable or exportable foods (and very close domestic substitutes) will absorb a larger share of total income and consumption in 2000 than today.

The effect of international price expectations on the proper scale of poverty-orientated MV research plans also depends on the likely *stability* of world food-crop prices. A food-importing LDC, if its planners reasonably expect that prices of imported food staples will fluctuate much less in the future, can rely with less risk on extra imports in bad years – rather than on MV research, either to build food stocks through higher average levels of domestic production, or to stabilize such stocks – and cut their required average level and cost – by reducing its year-to-year fluctuations. A food-exporting LDC, on the other hand, should plan to undertake *more* research into crops whose international prices are expected to fluctuate less, especially if it is poor people whose employment or farming incomes (when expanded by such research) are made more reliable by greater price stability.

We see, then, that international price projections – if even roughly reliable – should affect MV research policy. Such projections are difficult, uncertain, and shifting. However, ever since the early 1980s [World Bank, 1983], the outlook for primary commodities in general, and for food products in particular, to the mid-1990s has been one of gradual price decline, relative to manufactures and even fuels.[38] This would continue a steady trend of over thirty years. In 1986 and 1987, it has become increasingly difficult to obtain approval of agricultural-sector projects in leading aid agencies, despite the recognition of past rural neglect and favourable poverty impact, because – in view of these gloomy price forecasts – the expected economic rate of return on such investments appears low. The slight decline in real terms since the early 1980s of aid-financed IARC research, i.e. of *international* research into LDC food production [Joffe, 1986, p. 89; CGIAR, 1987, esp. pp. 66–7] – alongside the very weak current funding of *national* agricultural research by African Governments, even in support of major aid to capital funding [Lipton, 1985] – may indicate a spreading sense that international factors, principally the effects of Western farm subsidies (and dramatically successful Western research) in raising food stocks to near record levels, may reduce the need for LDC-orientated research into major food staples.

A secure prospect of substantial reduction and stabilization of world prices of food staples – relative to LDCs' other products – would justify some slowdown in the growth rate of food research for LDCs. However, this prospect is much too uncertain to justify any such slowdown, for several reasons. Poor people's access to food – their food entitlement – depends increasingly on *labour-intensive* progress in food production. Except in a few regions and crops, such research has been neglected in and for most LDCs. Expected trends in trade and Western agricultural research are anyway insecure grounds for projecting declines in world prices of food staples, let alone in the instability of those prices.

As regards the impact of trade on prices, much depends on the future of agricultural policy in the EEC and Japan. Both use internal incentives and external protection to stimulate farmers to produce much higher cereals supplies than would be the case if public policy were neutral among lines of production.[39] Therefore, such policies considerably reduce *world* cereals prices. The net impact of EEC's cereals policies on poor countries is hard to assess, but, we believe, on balance slightly damaging.[40]

The main point, however, is that the projections of continued falls in cereal prices – and hence the alleged weakening of any case for expanded MV research in LDCs – depend on assuming the maintenance of Western policies, especially in EEC, that encourage massive overproduction of cereals. This assumption is questionable, because the very large and rising costs, to EEC taxpayers – and to EEC, US and Japanese consumers, who pay food prices above world market levels – from supporting a dwindling number of farmers, have come under increasing attack. The progressive implementation of uniform agricultural regimes for EEC's new members – Greece, Spain and Portugal, all relatively big, poor and 'rural' newcomers – will under the existing Common Agricultural Policy provide steadily increasing support to larger amounts of relatively high-cost agricultural output, and will encourage the newcomers to grow even more; all this will further raise the cost to EEC's taxpayers and consumers of the very policies by which it depresses world cereals prices.

We do not wish to be unduly speculative. Those who hope for rational pricing and investment policies for food, in face of the rich world's rural and the poor world's urban bias, have seen many false dawns. However, it is possible that the costs of excessive Western incentives to expanded farm output, especially the fiscal costs of EEC's agricultural policies, are becoming prohibitive. If so, one key assumption underlying the projected real falls in world prices of traded food staples – continued oversupply due to unchanged Western policies on agricultural protection – is too gloomy.

Other assumptions, underlying standard price projections, may well also overestimate growth in world food supply, or underestimate growth in demand. On the supply side, although much research 'in the pipeline' will lower *Western* farmers' average production costs and stimulate further output increases, projections of medium-term growth in commercial output due to BT and nitrogen fixation may have been oversold, at least for cereals. Few *African* countries have followed up improvements in farm price incentives[41] with increased readiness by the public sector to steer its resources of capital or skilled personnel towards agriculture; there is little sign of sustainable growth of food output per person. In *South Asia*, the newer rice MVs have much the same yield potential as those of the 1960s; the wheat MVs, still not very diverse genetically, are at risk from rusts; and the spread of MV sorghum and finger millets to semi-arid areas is limited.

On the demand side, the shift of grain from food to feed continues. Population projections for Africa appear to increase steadily [UN, 1985]. In Latin America and parts of Asia, urbanization (and the beginnings of a wider spread of benefits from growth) are increasing the demand for food – and, once again, shifting it towards tradeables and animal products.

Combinations of demand and supply trends are shifting several types of country from net exporters to net importers of food staples. In East Asia and Latin America, this happens benignly, as growth and urbanization boost demand and pull workers rapidly out of agriculture. In much of Africa, the shift to net food imports is malign, as rapid population growth outpaces food production systems with deteriorating land and almost stagnant technologies.

A final factor, raising *delivered* prices of internationally traded food staples, is the secular uptrend in transport costs. This was disguised from food-price projectors in the mid-1980s by the weakening of the OPEC cartel. However, the trend weakens the incentives, otherwise provided by food prices on their own, to both producers and consumers of food staples to rely on international trade.

A reasonable case can be made that real prices of traded food staples will not fall in 1987–97, as current projections assume.[42] There is too much uncertainty for LDCs to be tempted, by such price projections, to scale down food-crop research, relying instead on low-cost food imports; price 'blips' upwards after the 1987 Asian and 1988 US droughts under-line these risks. In the six to twelve month run, a case *can* be made that many an LDC could achieve a given degree of food security more economically if it were to replace some of its publicly-held grain stores by various forms of insurance – larger 'stores' of foreign exchange, or operations in forward or even options markets in internationally traded cereals. However, the longer the time-horizon, the greater is the uncertainty, and the costlier, therefore, is such insurance. Greater outlay on MV research to *enable* a country to meet its poor people's needs – not simply for food supply (domestic or foreign), but for income from food-related production activities – is, for most LDCs, sensible even if such countries intend, where comparative advantage so indicates, to rely on short-term increases in the level of food imports.

That is especially so since, apart from long-run uncertainty about trends in real food prices, there is substantial, perhaps increasing, instability around those trends [Hazell, 1984, 1987]. The experience of 1974, when many rich and poor countries suffered bad harvests together, and when LDCs (several already facing foreign-exchange crises) found food aid scarcer just as commercial grain prices rose sharply, will not be quickly forgotten. Since 1981, the IMF's 'food financing facility' has provided some assurance, but nobody is sure how this would work under real stress, such as that of 1965–6 or 1974. In particular, the EEC's agricultural policy powerfully destabilizes world grain prices;[43] reforms of that policy, as since mid-1987 have been recommended by the Commission, would reduce this destabilizing effect, but nobody is sure whether such reforms, or any reforms, will be chosen, or adhered to if chosen. Meanwhile, the steady extension of EEC agricultural policy to producers in the new member States – and the persistent rise in EEC food output and supplies, due to the incentive effects (on researchers as well as farmers) of producer prices well above world market levels – continue to increase the destabilizing impact of EEC farm policy on world food prices.

*　　　　*　　　　*

Thus LDCs, and their donors, would be ill-advised to cut back on research into tropical and sub-tropical food staples in the expectation that average import prices of such products – including transport costs and allowing for possible weariness among food aid donors – would fall substantially, or become more stable, during the 5–15 years between the initiation of such research and the widespread adoption of its results by farmers. Poor people would lose most, if such reliance on good fortune in respect of world food prices proved misplaced. Most LDCs will find the long-run fate of their poor people determined significantly by prospects for farming, working on, or consuming nationally produced food staples – and therefore, not so much by international trade in food, but by relevant national or international research into MVs of such staples.

To some extent a big LDC, and small LDCs acting collectively, can use appropriately poverty-orientated MV research – especially if adequate international support for it is forthcoming – to reduce poverty even if world food prices change drastically and unexpectedly. The crop-mix, regional emphasis, risk impact, and above all the labour-intensity of production, induced by such research, are crucial to poor people's food entitlements. Such goals can, consistently with reasonably efficient economic policies, be significantly promoted by expanding research into major staple MVs – not of course in disregard of a county's comparative advantage, but not, either, in blithe hopes that imports can permit major economies in such research. LDCs' governments are not simple victims of output and price irrationalities created by Western farm policies. There is, however, more reason to worry about LDCs' increasing dependence on the results of *research* systems designed chiefly for the needs of rich countries.

The real costs of hiring extra rural labour in the USA, Europe and Australasia, per unit of extra food output per acre attributable to such labour, are for various reasons – and despite substantial unemployment – much higher, relative to the real costs of producing the same amount of extra food-per-acre from tractors or weedicides, than is the case in India or Nigeria. Research in 'the West' therefore seeks to reduce unit costs of labour, rather than of capital (or even land). Such

Western research – and the prospects for BT are at once increasing it and making it more directly and single-mindedly responsive to private profitability in the West – (i) is much greater than in or for LDCs, yet (ii) has major impact on costs in LDCs too. Tractors and weedicides, per acre ploughed or weeded, become less costly relative to hoe work, not only in Europe but also in LDCs – for research to cut per-acre costs of hoeing is tiny, compared to such research for tractor or weedicide methods. Hence the total pattern of world agricultural research probably tends to produce an innovation-mix that is responsive mainly to rich people's demand in rich countries, not least also because rich farmers, consumers and governments can afford higher research/output ratios in agriculture. World research output is thus unduly labour-displacing for LDCs.

MV research, partly because it was a catch-up operation based on earlier work with temperate-zone crops, has been a splendid exception to this rule. IARCs have helped it to be so. BT, for all its fine prospects, may vigorously revive the problem, just as poor people in LDCs come increasingly to depend for livelihoods on labour-using innovations. In the light of this book's findings, what can LDCs and international research centres do to improve these prospects?

Notes and references

1 MVs, by raising poor infants' chances of survival into poor adulthood, have indirectly helped to maintain the proportions of South Asian *living* populations that are poor.

2 There is some analogy between the serious though controversial possibility that English living standards fell in 1815–47 [Hobsbawm, 1957] despite the resumption of rapid agricultural advance, and the likelihood that Indian poverty incidence did not change much between the early 1960s and the early 1980s despite the MVs [Rao, 1985].

3 Small-farm AR helps poor labourers by raising labour use per acre more than does large-farm AR. This help, however, is somewhat reduced by small farmers' lower ratio of hired to family labour. In the long run, the rising demand for hired labour induces a quest for labour-saving technical change – usually available readily from developed countries.

4 This was apparently supported by the sharp decline, to well below the rate of agricultural or rural (let alone urban) population growth, in the rate of expansion of cropped area in India in the early 1960s. Not only the bringing into cultivation of new land, but also the further expansion of double-cropping (via irrigation), as ways to maintain per-person food supplies in face of population increase, appeared to face very sharply increasing average costs per acre, and even more per unit of food output.

5 It was also believed – seldom explicitly – that there were poor prospects for food imports to 'fill the gap' between growing population and slower-growing food supplies. LDCs' export outlook was viewed with gloom, and food aid was seen as unreliable. We question the validity of the 'filling-the-gap' specification of the hunger problem on p. 347.

6 We cannot, for reasons of space, do more than telegraph the main results, and refer readers to one or two summary sources.

7 The enormous population dynamic can be seen by trying to put numbers to the worst-case medical predictions of the effects of AIDS in the most ravaged of African countries. At the very worst, we are told, this scourge could kill one in five people in such countries in the next twenty years. On plausible assumptions, this would reduce population growth 'only' from about $3^1/_2$ per cent to $2^1/_2$ per cent!

8 Because rural-to-urban emigration – in permanent forms an exaggerated and probably declining phenomenon – is heavily selective of adult (male) workers; because life-saving medical advances affect infants and children much more than adults; because the generation *first* affected by these advances is, for the next 15–20 years, going to swell the number of births; and, on a pessimistic view of AIDS, because many adults (but a much smaller proportion of children) will die of this disease.

9 Several processes operate. As incomes in poor families rise, fears for security in old age – a major motive for producing many children – decline. So does expectation of, and new conceptions to replace, child deaths, as better nutrition reduces them. Also, as family incomes progress, children move from farm labour to school, so that large numbers of children become a burden rather than a source of quick income. Finally, poor parents, as they acquire income, tend to have more education – especially female – which is also associated with reduced birthrates.

10 It is also desirable that inputs purchased for MV cultivation should require payment – if this cannot be deferred – at times

permitting financing from seasonal peak incomes normally accruing to poor people from their hired employment, remittances, or other sources likely to loom larger as person/land ratios increase.

11 That is, *both* to help private agents who undertake (or to undertake competitively themselves) substantial and appropriate rural non-farm activity; *and* to distribute food payments to workers on it, or let 'the market' do so.

12 Millets, sorghum, maize and roots and tubers are everywhere – but especially in SSA – likelier than wheat and rice to be grown in mixed stands.

13 One result is that we cannot be sure to what extent the surging food imports of 1975–85 were due to unfavourable trends in domestic food output; to unusually deep downward fluctuations possibly due to climatic bad luck; or even to *rising* food output outweighed by rapid increases in population and in urban offtake per person.

14 That is, with roughly similar real income-per-person, and proportions of population reliant mainly on agricultural activities in rural areas.

15 The economist Joan Robinson once remarked, 'One thing is worse than being exploited: not being exploited.'

16 This may seem to readers to be formalistic, State-centralist, and unduly demanding skills in policy analysis that are scarce everywhere but especially scarce in most of SSA. A formal procedure for 'connecting' MV researchers, policy analysts, and policy-makers is indeed essential, given the extreme lack of clarity among MV researchers in most of SSA about the activities of the other two groups. However, this neither implies any particular degree or type of public-sector intervention, nor requires any complicated or time-consuming techniques.

17 Sections e–i inclusive owe much to Joffe, 1986.

18 Such extra labour requirements should, to the extent possible, (i) be in slack seasons; (ii) involve types of work that do not use up large amounts of energy; (iii) create independent incomes for women; (iv) involve 'learning-by-doing' of new farming or processing skills. To combine all these goals in any given MV, of course, is a counsel of perfection, especially across many different agro-climatic and socio-economic conditions.

19 (i) Faulkner [1986] refers to a phase, prior to modern BT, in which fermentation was used for many centuries, without theories and on a small scale, to produce drinks, food and fuel. (ii) The two phases of modern BT mentioned here are not specific to work on food crops, or even on plants.

20 Improved precision and speed have been achieved for some forms of *salt tolerance* breeding in millet [Nabors and Dykes, 1985], an area where TPB is weak; and of disease resistance breeding, through the use of monoclonal antibodies (p. 373). Extended range has been achieved via *wide crosses*. In these two areas, TC is a supplement to already strong TPB performance.

21 (i) This constrains progress with farmers' seed selection, TPB, and screening, as well as with BT. However, the field problems are less likely to leap to the eye with BT than with other methods of seed improvement. (ii) Division of responsibility for yield variation between genetic and environmental features understates the importance of seed improvement, by excluding 'GE interactions' [Simmonds, 1981a] – genetically better plants stimulate selection and improvement of environments via fertilizer, irrigation, etc.

22 This last method, so far reveals few, if any, cases of successful regeneration to produce plants from which commercially successful MVs could be produced. However, potato regeneration seems close.

23 Variously known as somatic hybridization, protoplast fusion, and *in vitro* hybridization.

24 MAB techniques also allow the separation of degrees of inherited resistance from variations in the level of attack or exposure due to varying phenotypes, positions, etc. of the plant. This again improves on TPB for resistance; measured damage caused by a pathogen can vary greatly (especially within a small number of experimental plants), partly due to phenotype (non-genetic) variations that cannot be extrapolated to large plant populations in the field.

25 Just as a long series of insecticides, designed to safeguard plants against ever-shifting pathotypes, has caused the evolution of more and more robust sub-species of *Anopheles* [Bull, 1982].

26 However, improved interaction between legumes and *Rhizobium* bacteria is being achieved, by BT and other methods. ICRISAT has developed procedures greatly increasing the rate of such nitrogen fixation [Joffe, 1986, p. 49].

27 In 1982 a Nottingham University/IRRI research project was attempting to improve *azolla* through BT (cell fusion), but selection and TPB of *azolla* remain options here too.

28 The IITA report cited does not specify the rice varieties.

29 Better in China, worse in SSA. Of course, this is a big achievement, and it is neither necessary nor sufficient to prove anti-poverty impacts from TPB (which we think are clear); but it does suggest that TC may merely restore momentum to TPB, not revolutionize it.

30 In 1982, the then biggest US firm devoted solely to plant BT, IPRI, and Sime Darby Berhad, the large agro-industrial conglomerate formed two new companies: ASEAN BT Corporation, to apply BT to several tropical crops, including rice and cassava; and ASEAN Agro-Industrial Corporation, to manage and market the results [Joffe, 1986, p. 88].

31 UNIDO's International Centre for Genetic Engineering and Biotechnology has a Director and is making appointments, but has few funds – and receives little information about private BT.

32 It used to be claimed that breeding for erect leaves, by reducing mutual shading, improved the plant's use of sunlight. Plant scientists inform us that this is no longer widely believed; does not seem to have worked in practice; and may be ruled out if, as appears likely, a 'source-sink model' of the plant is valid.

33 It is widely argued that one reason for the apparent 'yield potential plateau' in rice and wheat is that the third option, increased partitioning efficiency, has been largely exhausted by the dwarfing programmes.

34 'Wrong' in the sense that the State, by correcting them and undertaking politically feasible redistribution of some of the benefits, could achieve gains for everybody.

35 Demand conditions in most export-crop markets – and the rather slow progress, and slower employment-creation, of most industrial production–severely restrict the role of non-food products as sources of technical or institutional change in poor people's lives in most LDCs.

36 These processes, near the point of production, lower weight/value ratios and render long-distance trade an economic proposition.

37 The absence of tropical roots and tubers as Western products – and the radical differences between cereal farming and root or tuber farming – has somewhat reduced the flow of innovation tending to cut costs of labour-displacing methods in such crops. It is, however, not clear what BT will achieve here. BT is closer to major breakthroughs at farmer level in roots and tubers than in most tropical cereals.

38 From 1950 to 1984, the price of cereals, deflated by the c.i.f. index of US $ prices of industrial countries' manufactured exports to LDCs, fell by 1.3 per cent per year compound [World Bank, 1986, p. 7]. The price *rises* after the 1988 US drought are probably just a blip.

39 US agricultural protection and support policies have similar effects to EEC's and Japan's for some crops – tobacco, cotton – but for cereals the USA tends to maintain farm incomes by methods that reward farmers for restraining production.

40 Since most LDCs are net cereals importers – while many are compensated for other EEC farm policies (on sugar and beef) by separate protocols for such exports – partial-equilibrium analysis suggests net benefit to LDCs as a whole [Matthews, 1985]; but general-equilibrium approaches reverse the conclusion, by allowing for the fact that if the EEC (and other rich countries) removed of pro-agricultural policy biases, this would shift 'Western' producers towards making manufactured goods, thus cutting their prices to LDC importers [*ibid.*; Burniaux and Walbroeck, 1985].

41 Themselves often largely declaratory and/or with little effect on total farm output [Lipton, 1987].

42 No doubt a case could also be made that they will fall faster!

43 Because EEC farmers do not get lower prices when world grain demand falls relative to supply; hence they supply as much as before (instead of cutting supplies), and worsen the excess supply [Koester, 1982; Matthews, 1985].

8. Conclusions: MV Research and the Poor

Table 1 (p. 2) reveals a huge MV spread [CIMMYT, 1987; Dalrymple, 1986, pp. 2, 82–8; 1986a, pp. 2, 105–11]. Dalrymple estimates that in 1982–3, over 50 million hectares of wheat and over 72 mn ha of rice were planted in LDCs to 'wheat varieties developed by CIMMYT, [rice varieties] developed by IRRI and CIAT and the offspring of those or similar varieties developed in national research programmes'. Almost all are semi-dwarf varieties, plus a few of intermediate height. They comprised 52 per cent of all wheat area, and 54 per cent of all rice area, in LDCs. A further 75–80 million LDC hectares were by the mid-1980s planted to maize MVs, defined as commercially purchased hybrids (or improved open-pollinated varieties) released in the past ten years. In all, about half the wheat, rice, and maize area in LDCs comprised MVs in 1985–6 – as against almost nothing two decades previously [ibid., pp. 107–11].

Even that understates the scale of MVs' spread:

- The definitions are narrow. A few widespread and substantially improved local varieties – without CIMMYT, IRRI, or CIAT parentage – are excluded.
- There are a few country exclusions for wheat and rice MVs, for want of data: North Korea, Taiwan, and a few areas under very short MVs in China [ibid., pp. 108–9; Dalrymple, 1986, pp. 85–6].

- Even on the narrow definitions and with the country exclusions, the LDCs' area in MVs has increased since 1982–3. By 1984–5, a *further* 5.2 million hectares had been planted to MV rice, 1.7 mn ha to MV wheat, and 10.3 mn ha to MV maize in India alone [FAI, 1986, p. II-107].
- The proportion of *area* in LDCs under MVs of rice, wheat, and maize (about 50 per cent) is considerably less than the proportion of *output* (some 60–70 per cent), because the MVs justify more inputs of fertilizer, water management, etc. and hence give higher yield.
- MVs' 'share' in impact on the poor also exceeds their share of area. They contribute more to *employment* from wheat, rice, and maize in LDCs than to area, but less than to output – probably some 55–60 per cent. Since MVs produce somewhat coarser grain, at a price discount in retail markets, their share in LDCs of *wheat and rice consumption by poor people* (say, the poorest one-fifth of LDC populations) is probably even larger than in output – perhaps 75 per cent.

Finally, these numbers exclude MVs of many 'poor people's foods' – millets, sorghum, cassava, etc. LDC-wide data are not available, but India alone estimated 9.1 million hectares under high-yielding sorghum and pearl millet in 1982–3, and 10.3 mn ha in 1984–5 [*ibid.*, p. II-107] – about 40 per cent of area in these crops (table 2, p. 6)

*   *   *

We now draw together a few conclusions, and suggest implications for international and national agricultural research planners seeking to help the poor. We have tried to avoid hectoring criticisms of MV researchers. The above numbers tell us much about their achievements. The persistence of mass poverty in some LDCs – either despite adopting MVs, or whilst neglecting them – should be blamed on socio-economic structures and resultant policy biases, rather than on the features of MVs themselves.

Indeed, if social scientists had in 1950 designed a blueprint for pro-poor agricultural innovation, they would have wanted something very like the MVs: labour-intensive, risk-reducing, and productive of cheaper, coarser varieties of food staples.

(Even better might have been a wider range of MVs, concentrating on less-favoured 'rainparched' areas, and on cassava and millet. But natural scientists could reasonably retort that, given the genetic potentials, such emphases could not in the 1960s have produced enough extra food to avoid disaster.)

However, it is not quite good enough – although it is *fair* enough – to blame socio-political distortions for the inadequate 'translation' of MVs' large spread into gains for the poor. MVs are an evolutionary technique (Ch. 6, section l-n), not one that requires (or stems from) a transformation of the structure of rural power. An evolutionary technique – especially if used first by richer, less risk-averse farmers, with better access to information and inputs – tends, when introduced into an entrenched power structure, to be used so as to benefit the powerful. Even labour-saving and 'consumption-cheapening' MVs may not, in highly stratified societies, bring gains mainly to poor people.

Planners of agricultural research cannot fine-tune its results to each recipient village. They should, however, allow for the general features of the societies into which research results are introudced. Moreover, those features are changing. Increasingly, the near-landless, not 'small farmers', comprise most of the poor.

Our understanding of how MV impacts work themselves out, too, is changing. We have learned that the employment gains per hectare, created by MVs, fall off as better-off farmers seek labour-replacing ways to weed and thresh. We have also learned that price restraint (due to extra MV-based food supply), while helpful to poor food buyers in the short run, is soon reflected in wage restraints as the supply of labour increases (Chapter 5, b).

Finally, our scientific understanding – both of poverty and of MVs' impact on it – is changing too. Social scientists now see most threats to poor people's food access, whether from population growth or from technical change, as operating to reduce entitlements to food, whatever the impact on its availability. Natural scientists now realise that the best MVs – apparently helping the poor (who are least able to bear risks) by strongly resisting crop pests and diseases – can endanger the poor by their very success. A successful variety tends to replace other crops and varieties, thereby reducing diversity in the field, stimulating new pathotypes, and thus creating new risks, against which the poorest are the most defenceless (Chapter 2, k).

How can national and international research centres improve the effect of agro-biological discoveries on the world's poor? At first glance, the scope is severely limited by national policy. In most poor countries, political pressures and preferences induce most parts of the State apparatus to discriminate against rural people (including farmers), except to the extent that such people provide food, savings, skills, or export products to the articulate urban élites. If very few resources flow to rural infrastructure, to locating and (where economic) managing groundwater, or to the health or education of farm families; if farm price policy is used mainly to extract surpluses for urban use; or if agricultural research itself receives scanty or fluctuating current funding – then how can such research do much, for poor people or anybody else?

While research success cannot overcome policy failure (or policy vacuum), it can improve agricultural policy. For example, it was not lectures on price policy that persuaded the Indian Government to offer farmers more attractive wheat prices in 1965–70; it was evidence that MVs would enable many farmers to respond to such price incentives with markedly increased outputs. Another example is that of Sri Lanka, where the repeated successes of plant breeders ever since the late 1950s in developing robust intermediate and modern rice varieties, suitable for widespread use, was partly cause and partly effect of the major involvement of leading rice researchers with agricultural policy decisions on many matters, from trade through land tenure to agricultural extension. If researchers show that they can deliver the goods, politicians become readier to listen to them on the agricultural policy issues that so greatly affect the impact of research.

So the scope of agricultural researchers for improving their impact on policy, and in particular on poor people, is quite large. In seeking to redesign their research to use that scope, they need to respond to seven main issues, identified in this book.

- The setting of pro-poor research priorities requires clear, published, formal decision rules. Such rules would encourage LDC and IARC researchers, managers and funders to be objective in face of pressures and fashions, scientific and political. Carefully selected rules can encourage, not reduce, nonconformist and interdisciplinary challenges to current research paradigms.

- Such priorities are needed, in part, to select – in the light of new scientific prospects and risks – ways to consolidate, extend and redirect past improvements in the efficiency of food plants in handling water, nutrients, light and pests.

- But 'efficiency' needs environmentally careful definition. The poor will suffer most if nutrients and water are used efficiently now in ways that prove unsustainable later; or if plants, efficient now, lose diversity and stimulate virulent new pests later.

- The weak impact of MVs in some circumstances – regions, crops, agroclimates, administrative regimes – endangers many poor people, and requires new actions from IARCs and national crop researchers.

- MV researchers need to identify more precisely the groups, in each environment, most needing help ('small farmers', large poor farming households, urban workers, or rural landless; men, women, or children). Large and growing proportions of the poorest depend on income from rural labour, and are threatened as their increasing population reduces their bargaining power and hence food *entitlements*. Yet the IARC system was designed to increase food *availability*, seen as threatened for small-farm and urban populations as they increased.

- MV research has produced extra food and income which have saved many lives. Yet its approach to human nutrition requires a more appropriate research menu. Today's menu is too optimistic that – even without specific decisions on policy priorities or research directions – MV research will reduce the prices of poor people's food (relative to their incomes). Also, protein requires much less emphasis from researchers; the energy needs and absorption of vulnerable groups, much more.

- MV research, in both natural and social sciences, now concentrates too heavily on farmers, including 'small farmers', in areas where MVs have spread. To remedy this, IARCs (and several national research centres in LDCs) will require to develop new approaches in natural and social sciences; greater readiness to analyse the interactions of MVs with total systems (of power, of ecology, of economic transactions); and more awareness of history.

*   *   *

Earlier we sketched how the choice between research into one or other region – or into one or other crop, varietal type, or characteristic – could formally take account of five things: a scientific estimate of (i) the probability, extent and timing of 'success'; a sociological estimate of (ii) the likely rate of adoption by various types of farmers; and socio-economic evaluations of (iii) the size of gains from adoption, (iv) the impact on real income of main income-groups of farmers, rural labourers, and townspeople, (v) the downside risks around the above four components. Such an explicit system of choices – while merely making more open and rational what is done already – will be opposed by those who now claim to make these choices by 'common sense'. Partly, these are people defending their freedoms or preferences. They also include, however, research directors and planners who fear that formal criteria may discourage nonconformist researchers.

We have stressed that agricultural research needs more mavericks, more people who question the paradigms. The past successes of IARCs with irrigated rice and wheat encourage people who want 'more of the same' research. Yet many things – BT, the shift to Africa, the need for field diversity, the need to improve crops grown in environments generally more diverse than rice and wheat, new knowledge on population, the effects of MVs on 'backward' regions – may require shifts of the classical paradigm, that of transfer of one robust technology (and MV group) to many places. The complex problem of developing MVs and farming methods that absorb labour, yet do not stimulate poor people to produce even larger populations of potential workers (Ch. 7), will not readily yield to conformist research. Nor will the need for research more responsive to the farming (and labouring!) systems and experiments of the rural poor themselves.

However, formal and quantitative published criteria for research funding will *help* the nonconformists. They can argue their case – say, for weed research, or for Leontief-type models, or for understanding agricultural progress in eleventh-century China – in the light of its expected returns: faster success in varietal innovation, more or faster adoption, more GNP gains from adoption, or a larger share of those gains for the poor. Today, a research nonconformist must often struggle against the undeclared criteria of a superior. These, while partly sensible, can also partly reflect the accidents of a research director's personal experience, the pressures of powerful political interests and scientific establishments, or on occasion even family or sectional concerns.

How could formal criteria, in choosing among research strategies, help directly? One example is the emphasis, in much African research, on farm practices or MVs that save labour where it is now scarce, at least seasonally. Formal research pre-assessment would compel intending researchers to estimate the pace of likely research success and adoption – and to assess whether, when the innovation come onstream, the growing workforce would have made the labour-saving innovation inappropriate. If hand-weeding were judged inadequate, researchers in such circumstances would be guided, by this quantitative pre-assessment, to seek MVs that diminished the main weed threat (or re-timed their growth, so that weeding requirements missed the peak times of labour demand) – rather than to test herbicides, or to seek MVs compatible with them.

Formalising the choice, among plant breeding goals or strategies, will not remove the need for judgement. Is the breeder to aim at productivity, robustness and sustainability in respect of food plants, of plant populations, or of dietary energy for the poor (p. 68)? At neglected regions containing mainly poor farmers, or at advanced regions whose surplus feeds mainly poor workers (sec. 2, i)? Should varieties be conserved *in situ* or *ex situ*? The answers are not obvious, nor

universal. But the five-component, formalised choice procedure at least helps to get the answering procedures clearer,
and more open.

It also, we admit, makes research administration harder. If
the decision procedure means confronting a scientific establishment – e.g. one that wants research into high-lysine maize,
when the probable gains to undernourished people are small –
the normal processes of compromise, by which the world
works, may suffer. Researchers' worries about their place in
their sub-disciplines – which is already sometimes endangered
by engagement in the not-quite-pure, multidisciplinary, at
times a bit political, atmosphere of an applied food-crop
research centre – are sharpened by mangement procedures
that may well induce research less publishable in learned
journals, especially if the most respected journals (and the best
promotion prospects) are founded in outstanding work within
the established paradigms of a single discipline. New incentives and career structures are needed, to insert young and
able researchers happily into the more 'target-orientated'
forms of research suggested here.

*　　　*　　　*

Our second conclusion is that those emphases in research
(both in IARCs and nationally) which have led to past successes
with MVs need to be protected, and modified, in the light of
experience and of new scientific approaches. For example,
MVs make better *use of plant nutrients* than older varieties; but
researchers need to ask what MVs and practices would
encourage (i) more use of organic nutrient sources (as complements, not substitutes, for commercial fertilizers), and (ii)
'substitution of labour for fertilizer' by better timing, placing
or combining of nutrient sources. Both the need to create
productive employment, and the secular (albeit interrupted)
rise in feedstock prices, render this important (pp. 49, 193).
The effects of MV-fertilizer options on outputs and employment in subsequent seasons and years, too, needs review
(Chapter 2, e). So does the right balance of research among
biotechnology, other routes to nitrogen fixation, and nitrogen
supplementation (Chapter 7, h). Answers will vary by crop and
agro-climate, and the question urgently demands the sort of

formal decision procedures advocated above. Poor farmers' special need to avoid costly inputs and higher risks – and even poorer labourers' need for innovations that do not destroy employment – will be critically affected by these nutrient-related research choices.

Management of *light response*, mostly by breeding for photo-period-insensitivity, is another area of successful MV research. However, it is increasingly recognised that farmers need plants that are insensitive to photoperiod at some seasons but not at others; especially for poor farmers, this is due to the risks of delayed access to fertilizer, and to the need to avoid harvesting when crops must be stored wet (pp. 53–5). MV research is only beginning to respond to these complex needs.

Resistance to moisture stress in some MVs, such as IR-20 rice, is greater than in the competing varieties they displace. However, maize MVS have displaced more robust (although lower-yielding) sorghums and millets, with special risks to the poor. MS is a complex matter; the effects on different crops, even varieties, depend critically – and differently – on when water is short. More work is needed to link agricultural researchers to irrigation researchers (p. 71); accepting the new International Institute for the Management of Irrigation into the IARC system would be a good start. Much more response by formal researchers to farmers' own methods of water risk management is also required (p. 71). The approach of formal research to MS, even with the help of BT, is limited by the fact that a food plant's response to MS partly depends on its surroundings – terrain, soil structure, and organisms (includ-ing other food plants, weeds, worms, etc.); most of this interaction is not heritable. Even the heritable part of a plant's MS response is polygenic and therefore hard to understand or manipulate (p. 59). Such complexity is likely to be summarised in rules of thumb: in the wisdom of experience. Hence, in handling MS, agricultural researchers need to give close attention to farmers' and irrigators' perceptions and practices – and to the possibility that, in many cases of alleged MS, it is in fact nitrogen shortage that does the damage (p. 63).

Despite earlier allegations that MV researchers neglected *pest control* in their search for higher yields, it is, increasingly, robustness in face of pests that gives new MVs their extra attractiveness. But, from the standpoint of the poor, is it the right sort of robustness, in face of the right sort of pests? More knowledge is needed about the scale of losses caused by different sorts of pests, and about whether the victims of such losses (workers and consumers as well as farmers) are poor and vulnerable. However, the quantity – and sometimes, compared to research into other pests, the specificity – of work on MVs' interaction with 'unfashionable' pests (weeds, rats and birds) is plainly far too low, relative to their economic importance (Chapter 2, i). Herbicides play a huge role in weed research, yet are costly for poor farmers, and may displace poorer weeding labourers. Bird-resistant varieties – though carefully selected by farmers, as with the awned rice varieties of West Africa – remain largely unresearched, as does rat resistance.

As for the triumphs of pest research in MV selection, they have indeed greatly reduced losses to many insects, fungi, and bacterial and viral diseases. However, this very success carries new dangers, especially to the poor. Such dangers imply new research strategies – as may the MV scientists' great successes in developing plants for improved nutrient management. New definitions of 'efficiency', for plant populations and for single plants, may be required.

*   *   *

Successful breeding, for yield and vertical resistance, has led to the expansion of MV rices and wheats which – despite the diverse origins of many a variety – are bred for extreme purity, drive out less successful varieties, and produce large areas with very little genetic diversity. These MVs challenge pests to develop virulent new pathotypes or die. Thus safe, successful single varieties in the short run often mean less diverse, riskier plant populations in the long run (Ch. 2, k). The 'efficiency formula' (p. 388) needs to allow for such risks. That apart, what more can MV researchers do about them?

The shift from vertical to horizontal resistance – and sometimes to tolerance – is widely agreed, verbally, to be desirable. Yet many researchers clearly doubt whether it is feasible; for their practice is mainly to seek new, better vertical resistance, to keep one step ahead of the pathotype (Chapter 2, j). In large part, this is a scientific judgement that outsiders, such as ourselves, should respect, even if we question it. We do, however, note that the benefits of success in such research last no longer than the resistance itself, while the risks if vertical resistance breaks down are borne wholly by farmers and consumers, and are severest for the poorest. Poor farmers, for example, cannot often afford emergency prices for back-up chemical pest control.

Crop diversity, within and among seasons and fields, is as important as varietal diversity in securng plant populations against dangers of eventual new pathotypes. This is a very important addition to the case for shifting MV research towards neglected, less successful crops, which are driven out of many areas by the very success of wheat, rice and maize MVs. The need for crop diversity as insurance also argues for more attention by researchers to mixed-cropping systems (pp. 00, 00–00) than an attempt to maximise *expected* GNP gains – even if weighted to emphasise gains to the poorest most – would indicate. Diversity further requres to be preserved by well administered, fully catalogued, and (at least) duplicated seed libraries, whether *ex situ*, *in situ*, or both (pp. 93–5).

Diversity, for long-run pest protection, is one of two main 'environmental' issues facing MV researchers. The other is the need to avoid creating varietal sources of, or incentives to, soil exhaustion. Given the degree of sustainable robustness, MVs need to be selected for better nutrient conversion efficiency always; for better partitioning efficiency sometimes (provided the crop parts 'selected against' are not vital to the poorest: p. 31); but for better extraction efficiency only rarely, and then consciously and overtly (p. 47). Experiments need to distinguish – as they seldom now do – between extra food yields due to four sources: each of the three above types of improved plant efficiency, and extra intakes of nutrients. Experiments should also assess the impact of the higher-yield package in later seasons, and upon mixed crops, on the affected area.

*   *   *

As MVs spread to more regions, crops and seasons, the threats to diversity and sustainability may become more worrying. Right now, however, this is for most poor countries a less immediate problem than the fact that large numbers of growers rely upon regions, crops, and agroclimates where MVs have made little impact. Such growers can well lose, as extra output from MVs elsewhere depresses their product prices. (The very poorest – landless labourers – may on balance gain from the process, but in most 'MV-unaffected' areas, even in Asia, most poor people are still farmers, not labourers, most of the time.)

There is a true 'regional dilemma' (Ch. 2, i): research on food crops in promising, advanced, well-watered areas omits many of the poorest *producers*; research on neglected areas may do less to raise food output, and hence may not help the poorest food *consumers*. However, the expertise and past success of researchers into MV lead regions, crops, and topics gives these specialists status and power, so that the possibly diminishing returns to their efforts tend to be overlooked, as do the threats to diversity (e.g. among North Indian wheats) and the limited gains to the poorest. The lower prestige of forms of research that have in the past been less tightly organised conceptually, and also less successful – work on mixed crops, on upland areas, or on weeds – is also self-confirming; it leads to missed opportunities for big gains by assigning more resources, and more of the ablest and least conformist researchers, to such matters (Ch. 2, i). The proportion of IARC efforts devoted to 'neglected' crops and agroclimates almost certainly needs to increase further, in view of these built-in biases to 'self-confirm' earlier, successful lines of enquiry.

However, the problem of 'neglected areas' need to be better defined. It is not simply a question of assigning cash to semi-arid uplands, humid valleys, etc. Researchers need to explore the reasons – unusually good water management (p. 160), or good infrastructure for input delivery [Ahmed and Hossain, 1987]? – for 'spots of success' with MVs in otherwise unsuccessful areas. Also, these areas need to be classified better, so as to allow for (i) areas that can diversify into crops abandoned in MV lead areas; (ii) areas that can take up 'second-generation MV crops', such as the improved finger-millets; (iii) areas where MVs do raise output, but too slowly to offset the cost-price squeeze; and (iv) areas where poverty can be reduced through migration to nearby MV lead areas – leading to remittances, smoother flows of income, and more land-per-person for family members who remain (Chapters 3, i, and 4, g).

Distinct research guidelines for each type of area are needed, because poor farmers are, in each area, differently affected by MVs elsewhere. Such guidelines become increasingly important as MVs of maize, sorghum and perhaps cassava spread in Africa. Many areas (perhaps some entire countries) in Africa will remain untouched by any MVs in the early 1990s. Yet poverty will in most countries still mean, for the most part, small-farm poverty, with few poor households mainly dependent upon hired work, and with non-farm incomes still largely dependent on farm outputs (e.g. via processing, or the manufacture of hoes or harnesses).

Most MV crops and regions have so far been linked to growing urban demand – for wheat, rice, even hybrid maize, from better-watered areas near big cities. As this changes, so should research priorities. In particular, attention to 'remote' areas or crops implies a new approach to farmers who produce mainly for 'subsistence'. Most researchers, agricultural or economic, assume that they can do little for such farmers, and can help poor farmers and their employees only with 'commercial' production. 'Near-subsistence' farmers are allegedly more or less outside the cash economy, and have little or no money to buy even highly cost-effective new inputs – seeds, fertilizer, tools, micro-irrigation. Yet we know that, even in remote places, 'near-subsistence' farmers typically spend a quarter of their time, and earn a third of their income, in non-farm occupations [Chuta and Liedholm, 1979]. These tasks bring in resources which can be used to buy farm inputs, if this is judged safe and profitable. Then, even food deficit farmers will use cash to buy MV seeds and fertilizers – rather than to buy food so as to make up for low levels of production from TVs.

This new perception of near-subsistence farming as a major client for MVs is a necessary part of any research shift to 'neglected' crops and areas. 'Neglected' crops are usually consumed by (or very near) the producers, partly because the high weight/value ratio precludes distant transport. 'Neglected' areas usually have costly and bad transport, partly because populations are dispersed over large areas of bad land (with few urban agglomerations, for want of rural food surpluses). So the shift to such crops and areas involves analysis of the timings, transport and delivery, and credit systems appropriate to local near-subsistence. Maize and sorghum research into composites rather than hybrids, for example, might be indicated, especially to reach the poorest farmers (pp. 41–2).

The basic reappraisal of 'near-subsistence' farming implies a new view of rural infrastructure. Often, this should be seen as a route to diversity and exchange *within* a region – not to national economic integration via long-distance trade *among* specialised regions. In particular, poor people's crops often have high weight/value ratios, or (like cassava) short shelf-life. MVs of such crops, in remote areas, can form the basis of dynamic modernisation only as tools for *intra*-regional trade, probably involving a nearby small town as 'growth pole' [Perroux, 1962].

Such a shift to neglected crops and areas, as a tool for poverty reduction through MVs, will also need specific changes in agricultural research priorities. Two examples suffice. (i) Upland crops (not only rice) – as compared with classic MV rice and wheat areas (p. 77) – require research to cope with denser, different, and differently timed weeds. (ii) Sorghum and millet are much likelier to be grown in mixed stands than wheat and rice; the need to shift research towards such mixtures (pp. 67–8) is well understood at ICRISAT, but meets severe resistance in practice among single-crop specialists in national research systems.

* * *

It is notable that research into the impact of MVs upon nutrition largely neglects areas and crops *not* affected by MVs (Ch. 3, i). Yet such areas and crops represent the best prospect for overcoming the 'entitlements problem': that – even where MVs raise food supply – not much extra demand builds up in the hands of the hungry, so that the extra output raises stocks (or cuts imports) but does little for nutrition (Ch. 5, b). If extra

MV-based food is mainly eaten by its hungry growers and their children – or else is paid to hire extra farmworkers, who eat it – the entitlements problem does not arise: extra MV food goes straight into better nutrition. If MV research is to perform better in alleviating poverty, a shift to 'near-subsistence' crops and regions is strongly indicated, especially if power-structures will otherwise steer benefits even of labour-using and food-producing research (pp. 322–4) to the better-off. (This is not a defence of restrictions upon trade in food: if local 'near-subsistence' is to be made less uncomfortable, it should be through production rather than protection.)

$$*\qquad\qquad *\qquad\qquad *$$

In the process leading to the establishment of the IARCs, the entitlements problem was largely disregarded. The poor were seen as (i) 'small farmers' in areas where MVs could be widely spread, who would gain (as yields rose) much more than enough to offset the cost-price squeeze; and (ii) food con-sumers, mainly urban, who would gain as MVs restrained food prices. These perceptions of 'who are the poor' since the early 1960s, have changed dramatically. Yet the consequential changes in MV research priorities are, at best, just beginning.

Even among farmers in MV areas, it is those with low household income (from all sources) per consumer-unit who face serious poverty. This at-risk group overlaps rather badly with 'small farmers', i.e. those with few acres (of whatever quality) per household (Ch. 3, a). Both agricultural scientists and socio-economists need to look more closely at the effect of MV options upon the capacity of poor farm households – i.e., for the most part, those with many members, high child/adult ratios, and limited income sources – to earn income, from non-MV and non-farm activities as well as from the MVs. Timing of labour requirements of MVs, to avoid clashes with options for casual unskilled wage-employment, could be crucial here.

However, neither agricultural nor socio-economic researchers need spend much more research time on offtake of, or benefit from, MVs among poorer farm households in MV lead areas. The facts are well known. Such households adopt MVs, and gain thereby. However, unless helped (by well understood methods, both agro-technical and socio-economic), they adopt later and gain less than the better-off, risk-taking innovators.

More important, poverty – and MVs' impact upon it – is not a problem mainly for farm households in MV lead areas, but for farm households outside them (Ch. 3, i) – and in most areas for households that depend for income mostly on labour. Already, these include most poor rural households in Bangladesh, Eastern India, Java, Mexico, North-East Brazil, and parts of the Philippines, Kenya and Rwanda. By 1995, rural labour poverty will be more prevalent than small-farm poverty in most of Asia and Latin America, and in large parts of Africa. Research planning decisions, taken now, will if successful, lead to widespread innovations in 1995–2005. Yet these decisions, where they emphasise poverty, still look almost entirely at the impact of MV options upon 'small farmers', not upon rural labourers. *The main needed shift in poverty-oriented MV research priorities is to move the focus from 'small farmers' to rural labourers* (p. 176).

The impact of MVs on demand for labour, while still clearly favourable, has been deteriorating sharply. Around 1965–70, if a hectare of land was shifted to MVs and thereby doubled its yield, demand for labour would rise by about 40 per cent. Today the rise would be about 10–15 per cent: still a gain, but only 3 to 6 years' worth of workforce increase. Yet this is a favourable instance, even for MVs, the most promising of labour-using innovations.

Part of this setback has been due to aid-supported research into ways to accompany MVs with labour-displacing innovations: herbicides, threshers, direct planting methods (pp. 184, 193, 200). Partly, it is due to failure to screen MVs for possible perverse effects on employment, e.g. of post-harvest labour (p. 194). Some MV-linked research, e.g. into methods of fertilizer placement (p. 193), has special prospects for improving employment, but does not appear to receive higher priority on such grounds. Commercial research is bound to respond to the bulk of effective demand by Western consumers and producers – and to some extent by big farmers in LDCs – to save labour-costs (Ch. 4, h); it is therefore of crucial importance that publicly-financed research, in IARCs and nationally, redress the balance. Given the expected rates of return, research directors should favour MVs, crops, regions, and characteristics where research outcomes will raise the demand for hired, unskilled labour – or reduce its supply, by generating innovations that make it more profitable or safer for deficit farmers to redirect family labour from the job markets to their own farms. The 5–15-year time-lag of research, together with steady growth in rural workforces, means that such labour-using priorities are usually correct even where, as in much of rural Africa, 'labour scarcity' is the theme of *today's* complaints and pressures (anyway mainly from not-so-poor employers).

Three problems arise. Should not MV research seek to moderate the peak-season demand for labour? Can it ensure that small, poor farmers and urban consumers not lose from a shift in research priorities towards labour-intensity? What will impede such a shift from inducing parents to opt for more children – tomorrow's workers?

Machines such as tractors, or inputs such as weedicides, can in principle merely add their forces to peak-season labour, thereby raising output, and hence jobs in other seasons. In practice, however, such equipment and inputs (i) seldom raise output, (ii) once acquired, are also used to displace workers in slack seasons. Alternative ways of spreading the labour peak, such as MVs with different maturity patterns (pp. 196–7) – or of dealing with it, e.g. by developing complementary farming systems (and migratory patterns) for nearby regions – should almost always be preferred to subsidised imports, or research, to displace even 'peak-season' labour (Ch. 4, f–g).

Although maximum welfare for poor *farmers*, as an aim, requires different research priorities from maximum welfare for poor *labourers*, there is seldom violent conflict. Farms with less land-per-person are likely to support poorer households and to make more use of labour. Steering resources to such farms thus helps not only 'small farmers' but also labourers, since such resources will be used with more extra labour – either from hired labourers, or from family workers who thus compete less than before against hired labourers for wage-work – than if they were assigned to larger farmers.

Could poor urban *consumers* be harmed by a switch to labour-oriented priorities in MV research? Food staples for consumers, like work for labourers, should be increased by small-farm emphasis (p. 137). Smaller farms usually produce more food and more employment per hectare – and work new inputs, including MVs, more labour-intensively also. The three anti-poverty aims – that poor groups should obtain more food, more work, and more small-farm income – may however, together constrain policy options quite seriously. If poor *farmers* are to gain from MVs, the productivity of farm labour must usually rise. It must rise faster than the demand for food; otherwise, poor *consumers* could lose, from rising food prices or growing import dependence. But it must rise more slowly than output per hectare; otherwise, once the expansion of arable farmland slows to a trickle, poor *workers* lose as farm employment falls off.

Of course, this does not mean that all MV-linked innovations always have to walk this tightrope! Some LDCs have enough foreign exchange for extra food imports *and* development imports; some have very few poor farm labourers, or can absorb almost all extra workers outside farming; some have spare land. Most have none of these advantages; but even then it is only farm growth as a whole – not every single MV innovation – that needs to have just the right critical impact on growth of labour productivity, large enough to relieve poor consumers, not so large as to 'unemploy' poor workers. All the same, the existence of the 'tightrope' underlines the critical importance of planning MV developments as a whole. Competitive, imported, and private-sector research – all on balance desirable – complicate, and sometimes invalidate, both predictions and requirements for an innovation's effect on labour markets. However, at very least, an 'onus of proof' rests upon any proposed innovation in food farming that raises labour productivity more than yield, or less than the demand for food. If alternative innovations can be sought that have a good chance of successfully 'walking the tightrope', there is an *a priori* case for this.

Another important way out for MV research, seeking innovations that absorb labourers yet enrich small farmers, is to make use of the double meaning of 'labour productivity'. More food output *per worker*, obtained by innovations that substantially raise demand for person-hours, could be consistent with only a tiny rise, or even a fall, in food output *per hour*.

Such innovations could help food output-per-worker to grow faster than food demand (thus assisting poor food consumers), while keeping the growth of food output-per-hour below the growth of yields (thus providing more employment per hectare, assisting poor workers). To achieve this, MV researchers will need to know how different MVs, timings, and cropping patterns fit into 'farm labour household systems' – not just farming systems (pp. 344).

Thus it is possible for MV research, if carefully planned, to serve poor farmers and consumers, while shifting the priority – as it must – to providing livelihoods for the labourers who by 1995 will form a growing majority of the world's poor. But would not such a shift of priority encourage poor parents to produce more children, in the hope that such children will soon bring back labour incomes? Policymakers should not ignore this risk. There is evidence, however, that widespread distribution of income gains, awareness of the spread of health benefits, and female education – separately or jointly – bring about fertility transition much faster than was once believed. Thus a more labour-oriented MV research programme, preferably but not necessarily linked with improved planning for health and educational provision, could well help transition.

*   *   *

The original thrust of MV research might suggest that these concerns about labourers are misplaced. Labourers, even if partly paid in kind, usually buy much more food than they sell. If MVs raise food supply and bring prices down, surely labourers must gain, as net buyers of food?

Unfortunately, this does not work well (Chapter 5, b). (i) In many countries, domestic food prices are determined largely by world prices, not by domestic food supply. (ii) Even if workers gain from food price restraint, there is an offsetting loss, because such restraint reduces the employer's incentive to hire them. (iii) As workers move into each MV area – and as population and workforce increase – their competition means that, within a year or two of the shift to MVs, employers can hire labourers at much the same real wage as before; food-price restraint (via MVs) permits money-wage restraint, and again the poor gain little.

There *has* been some price effect – and, in the lead areas, a small real wage effect – from MVs. Food supply has become more local (with lower transport cost), more smoothly available, and more concentrated on inexpensive (coarse) forms of rice and wheat, standing at a price discount. The risks attached to income from food production, and to availability of food, have (on sensible definitions of risk: pp. 148–9) been reduced. Nevertheless, the incidence and severity of 'food poverty' – income-per-person too low to afford enough food – and of frank undernutrition have, even in some leading MV regions, declined little if at all (p. 258).

The menu of MV-linked nutrition research may have made matters worse. It has diverted resources – from increasing, cheapening and stabilising sources of dietary energy, towards building in nutritional goals of secondary importance or less. The search for MVs rich in protein, or in specific amino-acids – while it was wisely aborted at ICRISAT – continues elsewhere, despite evidence that success would have scanty or no nutritional benefits, even if, as is not the case, it were clearly obtainable, heritable, and costless in terms of yield and robustness. The search for aesthetic or cooking qualities, unless clearly time-saving for poor working women, is likely to be harmful to the nutrition of poor people (Chapter 5, h).

Of nutritional goals of current MV research, only improved absorption – via breastmilk and via weaning foods – properly addresses primary needs among persons at risk. Even there, attention to low cost and to seasonal availability should replace some of the current concern for 'energy density', which could become as dubious a slogan as the 'protein gap' used to be. Low-cost, ample, labour-intensive, robust calories, with some attention to the special needs of vulnerable groups (especially infants and weanlings), remain the best main focus for MV research. That focus is blurred by superficial add-ons, requiring breeders to incorporate this or that nutrient or characteristic because rats die if they eat only a cereal lacking it.

Nutritionists do, however, have major (and largely undischarged) functions in food crop breeding institutions. They should investigate the main nutritional deficiencies in actual human diets, especially of poor infants, based on each main staple. They should establish, with breeders and economists, the prospects, costs and benefits of developing MVs that correct those deficiencies (and compare those costs and benefits with other approaches, e.g. fortification). Also, nutritionists should help to set MV research into the context of 'agriculture-health linkages'; infants' capacity to absorb different foods in infections, and measures to limit those infections alongside measures to improve access to food, may – given real incomes – largely determine the nutritional effectiveness for a household of an MV innovation. Finally, nutritionists should investigate the health-nutrition impact of labour requirements and income sources, especially for pregnant and lactating women, associated with alternative MV innovations.

*　　　　*　　　　*

Many of these specific suggestions would require agricultural researchers to communicate across disciplines, or to engage new specialists. Even within a discipline, the type of economist or agronomist who analyses how MVs affect labour is often different from the type who looks mainly at 'small farmers'. The experience of IARCs with anthropologists has not been uniformly happy. In national crop research centres, posts for economists are often either unfilled, or occupied by persons without the seniority, esteem, colleagues, or career structures to communicate effectively with crop scientists.

How, then, is the need to place MVs into total contexts to be met, given the discipline-centred nature of careers in crop science, and the effort (and strain) already involved in the unusually high degree of interdisciplinary task-orientation that characterises many IARCs and such leading national institutions as the Indian Agricultural Research Institute? These total contexts – 'systems', in the unpleasant jargon now current – are required for most of the anti-poverty work that we have outlined, from the analysis of MVs' effects on non-MV regions, through the appraisal of MV-related options relatively unlikely to be filtered to the well-off through local structures of power, to the general equilibrium approaches of sections 6, b to 6, f.

Four issues, raised in Ch. 7, c–j, increase the urgency of a new direction in the IARCs. *Population pressure* requires them to incorporate demographic skills; to examine the impact of MV options upon poor people's nutritional entitlements, especially in large households; to use, but change, their farming systems research capacity for this purpose; and to reinterpret the whole IARC brief, which was written in the early 1960s to embody a now outdated view of the relations between population, food and poverty (p. 346). *The shift to Africa* requires that IARCs develop clear views of their proper relationships to water research, to national research systems, and probably to 'slow and clean' versions of farming/labouring systems research (p. 302). As just shown, the reduction of *labour poverty* increasingly requires quite new research priorities.

Finally, and tugging the other way, *biotechnology* requires that the public sector's 'traditional plant breeders' widen and deepen their disciplinary knowledge. At present, public-sector research into poor people's food plants in LDCs is of the 'fifth order of smalls' for BT research (p. 365–6). Today's BT research (tissue culture development), however, is inherently pro-poor (pp. 368–70) unless non-competitively privatised. The risks and hopes for poor people from tomorrow's BT research are enormous. Herbicide bonding, for instance, could harm them (p. 372) – yet, in the long run, BT could well greatly increase the comparative advantage of tropical and sub-tropical labour-intensive farmers in producing food staples (p. 383). If the hopes are to be realised and the fears dispersed, LDCs need access to the highest quality of independent scientific BT research, closely linked to traditional plant-breeding skills. Only Brazil, China, India and Mexico – and perhaps Indonesia and Pakistan – could conceivably achieve this on their own. Other LDCs, and the tropics as a whole, certainly need international support. Yet IARCs' own comparative advantage does not clearly lie in BT, nor indeed in basic science (p. 381).

It would be dishonest to present a pat solution to this dilemma. Disciplinary deepening is needed by agricultural researchers seeking to serve the poor of the LDCs in face of a huge, still vague set of hopes and threats from BT. Inter-disciplinary and contextual work, also incorporating new subjects from history to demography, is needed if agricultural researchers are to transcend more of their past, agro-economistic, partial-equilibrium oversimplifications.

It is no solution to cry for more of everything. African countries need help to set up small, simple, indigenised research systems, with a few clear priorities. Within any research system, including that of the IARCs, communication across disciplines – or within a discipline, if its scope is allowed to become increasingly deep or complex – suffers from very rapidly rising marginal cost.

We shall not, however, close this book with a hedge. The IARCs have shown great comparative advantage in the poverty-oriented and applied linkage of 'traditional' plant breeding to other sciences, biological and socio-economic. They can afford *poverty-orientation* better than many otherwise excellent national crop research centres, which are pressed to meet the needs of powerful groups: urban labour and capital élites; big farmers in MV lead areas who supply them with food and exportables. IARCs are also better able to encourage *applied* and *interdisciplinary* work, because their high quality and past achievement reduce the reluctance of young researchers to risk deviations from the purer and more single-discipline approaches that usually advance a career in 'normal science'.

A twin, massive, and inherently interdisciplinary challenge is posed to the IARCs' technology-transfer paradigm: by the success of MVs without major poverty alleviation in much of South Asia; and by the very limited spread of MVs to the main crops and agroclimates of Africa. To meet that challenge will require more interdisciplinary work (often jointly with farmers and labourers), and more risk-taking with new disciplines – demography, hydrology, history – and new paradigms. Probably, only the IARCs can do these things on the requisite scale, and with the requisite capacity to concentrate on poor people rather than pure science.

IARCs have been right to move into BT, and also right to do so cautiously and cheaply. Their main role here is to know, say and do just enough to pre-identify and avert specific major risks (and to identify, and to seize upon, specific major potential gains) from BT to the poorest, by main crops, regions, and types of poverty group. IARCs should, in our view, not undertake major basic, or applied BT research. It would divert them from their main tasks and their main comparative advantages. The first is in traditional plant breeding (and its allied agricultural and socio-economic sciences) orientated towards: yield and robustness in smallholder conditions; preserving and increasing diversity; avoiding soil or water mining; and embracing the newly enhanced needs for labour-intensity, and for a spread to new regions, crops and farming/labouring systems. The second is in applied interdisciplinary work to identify which MVs and associated farm practices can best home in on major poverty groups – given that, as we now know, an innovation's capacity to cut risks, raise output of cheap food, and employ labour need not *suffice* to enable poor farmers, workers and consumers to gain from that innovation, because such gain must be 'sucked through the filter' of a power-structure that favours the strong.

Except with unusually dramatic innovations, IARCs can reliably and substantially raise farm output – let alone help the poor – only in countries with functioning national research systems, able to screen and adapt exotic germ plasm for local conditions, and to deal with new pathotypes in time. Yet some African and a few Asian governments, under financial pressure, have repeatedly slashed research funds – not only preventing proper research, but not seldom forcing the research station to use its fields to grow food for its unpaid workers! More frequently, domestic agricultural research in LDCs make a real contribution, but is overburdened, over-diffused, and neglectful of main food crops, especially those consumed mainly in rural areas. In very few countries is BT work a sensible goal for national agricultural research systems.

This raises (at least) two main problems for the future of poverty-reducing research into MVs. First, can or should LDCs other than Brazil, China, India and Mexico – and perhaps Pakistan and Indonesia – divert, or muster, cash or researchers to provide, at least, early information on specific major 'hopes and fears' for poor people from BT for their main crops? Neither national centres nor IARCs are obviously suitable. UNIDO's new institute has few funds, and is apparently not concentrating on food crops. Perhaps some LDCs could use aid funds to purchase BT information from private and public sources; but problems of what to buy – and of how to interpret it, apply it, and integrate it into national research and extension – will remain. A low-cost solution, that does not divert IARC or national personnel or cash from their prime tasks, is urgently needed.

An even bigger problem is how to develop effective, poverty-reducing agricultural research in countries now deficient in this. ISNAR, one of the IARCs, is the 'Institute for the Support of National Agricultural Research', and had developed – and helped governments to implement – recommendations in many LDCs. In few cases, however, has ISNAR found explicit commitment in national agricultural research to poverty alleviation; in Kenya it found that planning priorities for this were unreflected in research priorities [Lipton, 1985].

Such a situation is perhaps defensible. Some 'standard economists' might say: produce efficiently first, then redistribute to the poor. Even the radical economist Joan Robinson upbraided Sri Lanka for its welfarism, for seeking 'to eat the fruit before it has planted the tree'. Yet all this, in the context of research priorities, is abstraction of the wrong sort. Focusing national research upon poor people's crops, regions or techniques is one of the few hopeful ways to achieve lasting gains, without dependency, for poor rural people in most LDCs. Such a focus – involving more use of labour, and more research attention to long-neglected crops and areas – is not normally in conflict with growth.

In such countries as Kenya, Botswana or the Ivory Coast – which have a considerable infrastructure for national research, though one in need of concentration on fewer key issues – donors could reasonably ask that their aid be used, among other things, to help steer that research more firmly towards poverty reduction. A much more serious problem is posed by the many countries, in Africa probably a majority, whose governments have not shown much commitment towards research into main food crops. Why should they? Dispersed smallholders, with little or no experience of gains from such research, seldom press for it effectively. Money and experts are scarce. Governments, lacking even approximate knowledge of areas and outputs under main food crops, cannot assign research wisely among crops or regions; for main export crops, far more is usually known (and past results are known to have helped to finance governments and to pay off debts). Finally, most research takes 5 to 15 years from inception to widespread adoption; many governments are hard pressed to survive financially, even sometimes physically, in the next 5 to 15 *months*.

How is a constituency for national-level food-crop research to be built up under such circumstances? It is tempting to say: pick winners; don't complicate the issue by seeking poverty-orientation. The temptation should be resisted. *Socially*, experience in many countries shows that 'winners', having obtained a pattern of research to help them get further ahead, gain strength to prevent this pattern from shifting to help the poor afterwards. *Economically*, 'winners' tend to supply the extra food so capital-intensively as to create very little extra demand for it; for example, Botswana's grain needs could be supplied in most years by irrigating big areas of maize for a handful of tractor/combine farms along the Limpopo River, but almost all the incomes would go to rich farmers, input suppliers, etc. who would not demand much of the extra grain (and who do not require it nutritionally). *Politically*, a broad base of support for agricultural research requires that many farmers, or many consumers, expect gains; a few growers of food staples, even if each is big, are unlikely to have lasting power in countries as dominated by urban interests (and food imports) as most of those that currently lack basic agricultural research systems. *Internationally*, a counterweight to the excellent but 'biased' research embodied in imported machinery, inputs, and techniques needs to be constructed. Hence, even in countries starved of national food-crop research, it is mistaken to delay poverty-orientation until such research is well developed.

*   *   *

MV research in and for LDCs is, in part, a prisoner of its own success. This naturally leads practitioners and funders to ask for 'more of the same'. Yet change is urgently required: by diminishing returns to 'the same', by radical changes in our understanding of poverty, and by the exposure of big gaps in MVs' impact upon it. Luckily, some of the necessary changes can build upon the proven comparative advantages of the IARCs. Others can be developed only through successful national agricultural research systems. Many LDCs have, or are constructing, such systems. Even for LDCs that are not, however, broad constituencies for poverty-reducing MV research have to be developed, alongside that for research itself. For poverty-orientated, MV-based food agriculture is the only chance to provide the growing poor populations of Asia and Africa with livelihoods during the many decades before their widespread industrialisation.

References

Abeyratne, E., 'The extension of HYVs', Seminar paper No. 26, in [ARTI-IDS, 1973].

Abrigo, W. *et al.*, 'Somatic cell culture at IRRI', in IRRI, 1985a.

Acharya, S.S., 'Green revolution and farm employment', *Indian J. Agric. Econ.*, July-Sept. 1973.

——, 'Comparative efficiency of HYVP: a case study of Udaipur District', Economic and Political Weekly, IV, 44, Nov. 1, 1969, pp. 1755–7.

Adams, D., Graham, D. and Pischke, J. von, *Undermining Rural Development with Cheap Credit*, Westview, Boulder, 1984.

ADB (Asian Development Bank), Rural Asia: *Challenge and Opportunity*, Federal, Singapore, 1977.

——, *Asian Agricultural Survey*, University of Tokyo Press, Tokyo, 1969.

Adelman, I., and Robinson, S., *Income Distribution Policies in Developing Countries: a Case Study of Korea*, World Bank, Oxford University Press, 1978.

Advanced Technology Alert System (ATAS), 'Tissue culture, technology and development', ATAS Bulletin, No. 1, November 1984.

Affan, K., 'Effect on Aggregate Labour Supply of Rural-Rural Migration to Mechanized Farming: a Case Study in Southern Kordofan, Sudan', D.Phil., Sussex University, 1981.

Agarwal, B., 'Tractorization, productivity and employment: a reassessment', *J. Development Studies*, 16, 3, April 1980, pp. 375–86.

——, 'Rural women and HYV rice technology', *Economic and Political Weekly, 19*, 13, March 31, 1984, pp. A39–52.

———, *Rural Women and the HYV Rice Technology in India*, mimeo, Inst. of Econ. Growth, New Delhi, 1984a.

———, 'Tractors, tubewells and cropping intensity in the Indian Punjab', *J. Development Studies, 20*, 4, July 1984b, pp. 290–302.

Aggarwal, P.C., *The Green Revolution and Rural Labour: A Study in Ludhiana*, New India Press, Delhi, for Sri Ram Centre, 1973.

Ahluwalia, M., 'Rural poverty and agricultural performance in India', *J. Development Studies, 14*, 3, Apr. 1978, pp. 298–323.

———, 'Rural poverty, agricultural production, and prices: a re-examination', in Mellor and Desai, 1985.

Ahmad, Q.K. and Hossain, M., *Rural Poverty Alleviation in Bangladesh: Experience and Policies*, BIDS/FAO (WCCARD follow-up program), M/Q 7485, Feb. 1984.

Ahmed, C.S. and Herdt, R.W., 'An intersectoral analysis of the effects of rice mechanization in the Philippines', Paper No. 81–24, Dept. of Ag. Econ., IRRI, 1981.

Ahmed, I., 'Technology and rural women in the Third World', *International Labour Review, 122*, 4, Jul.–Aug. 1983, pp. 493–505.

———, 'The bio-revolution in agriculture: key to poverty alleviation in the Third World?', *International Labour Review, 127*, 1, 1988.

Ahmed, R., 'Agricultural price policies under complex socioeconomic and natural constraints: the case of Bangladesh', *Research Report No. 27*, IFPRI, Washington, DC, 1981.

Ahmed, R. and Hosain, M., *Infrastructure and Development of a Rural Economy*, mimeo, IFPRI/BIDS, Washington, DC, 1987.

Aiber, M. *et al.* (eds.), *Genetic Manipulation: Impact on Man and Society*, Cambridge, 1984.

Akerlof, G., 'A theory of social custom, of which unemployment may be one consequence', *Quarterly. J. Economics, 94*, June 1980, as reprinted in his *An Economic Theorist's Book of Tales*, Cambridge, Mass., 1984.

Alauddin, M. and Tisdell, C., 'Market analysis, technical change and income distribution in semi-subsistence agriculture', *Agricultural Economics, 1*, 1, Dec. 1986, pp. 1–18.

Ammerman, A. and Cavalli-Sforza, L., 'Measuring the rate of spread of early farming in Europe', Man, 9, 6, 1971.

Anand, S., *Inequality and Poverty in Malaysia: Measurement and Decomposition*, World Bank, Oxford University Press, 1984.

Anderson, J. and Pandey, S., 'Assessing impact of agricultural research centres on efficiency and equity objectives', mimeo, Australian Ag. Econ. Soc., Annual Conf., Armidale, Feb. 1985.

Anderson, P., *Passages from Antiquity to Feudalism*, New Left Books, London, 1974.

Anderson, R.G. (ed.), *Proceedings of the Wheat–Triticale–Barley Seminar*, CIMMYT, Mexico, 1973.

Ando, A., Fisher, F.M. and Simon, H., *Essays on the Structure of Social Science Models*, MIT Press, Cambridge, Mass., 1963.

Anthony, K.R. *et al.*, *Agricultural Change in Tropical Africa*, Cornell, Ithaca, 1979.

Arnold, M. and Innes N., 'Plant breeding for crop improvement with special reference to Africa', in D. Hawksworth (ed.), *Advancing Agricultural Revolution in Africa*, Commonwealth Agricultural Bureau, Farnham Royal, 1984.

Arrow, K., *Collected Economic Papers: Vol. 2, General Equilibrium*, Blackwell, Oxford, 1983.

——, 'Economic equilibrium', in D. Sills, (ed.), *International Encyclopedia of the Social Sciences*, Vol. 4, Macmillan, New York, 1968, reprinted in Arrow, 1983.

—— and Debreu, G., 'Existence of an equilibrium for a competitive economy', *Econometrica*, 1954, reprinted in Arrow 1983.

ARTI-IDS: Agrarian Research and Training Institute (Colombo) and Institute of Development Studies (Brighton), 1973 Kandy Seminar on economic and social implications of HYVs: papers and proceedings (mimeo).

Asaduzzaman, M., 'An analysis of adoption of HYV paddy in Bangladesh, D.Phil., Univ. of Sussex, Brighton, 1980.

Bardhan, P., 'Types of labour attachment in agriculture: results of a survey in West Bengal 1979', *Economic and Political Weekly*, XV, 35, Aug. 30, 1980, pp. 1477–84.

——, Peasant labour supply: a statistical analysis', in *Land, Labour and Rural Poverty*, Oxford University Press, Delhi, 1984.

—— and Rudra, A., 'Terms and conditions of labour contracts in agriculture: results of a survey in West Bengal, 1979', *Bulletin of the Oxford Inst. of Econ. and Stat., 43*, (Feb.) 1981.

Barker, R., 'The status of agricultural development in S. and S.E. Asia as a result of the new food grain technology', Conference on Appraisal of Agricultural Development in Less Developed Countries, Annual Meeting of U.S. University Directors of International Agricultural Development Programs, Univ. Hawaii, 1971.

——, 'Yield and fertilizer input', in IRRI 1978a.

——, 'Impact of prospective new technologies on crop productivity: implications for domestic and world agriculture', mimeo, National Academy of Sciences, Washington, D.C., 1987.

—— and Herdt, R. 'Equity implications of technology changes', in IRRI 1978, pp. 83–108.

——, 'Who benefits from the new technology?', ch.10 in their *The Rice Economy of Asia*, Resources for the Future, Washington, DC, 1984.

Barlow, C., Jayasuriya, S. and Price, E., *Evaluating Technology for New Farming Systems: Case Studies from Philippine Rice Farms*, IRRI, Los Banos, 1983.

Barnum, H. and Squire, L., A Model of an Agricultural Household: Theory and Evidence, *World Bank Staff Occasional Papers No. 27*, Washington, DC, 1979.

Barrett, J.A., 'The evolutionary consequences of monoculture', in Bishop and Cook [1981].

Barry, P. and Baker, C. 'Reservation prices and credit use: a measure of response to uncertainty', *Amer. Econ. Rev., 53*, 2, 1971.

Bartsch, W.J., *Employment Effects of Alternative Technologies in Asian Crop Production*, ILO, Geneva, 1973.

Beachell, H.M., Khush, G.S. and Aquino, R.C., 'IRRI's international breeding program', in IRRI 1972a.

Becker, G., *A Treatise on the Family*, Harvard University Press, Cambridge, Mass., 1981.

Beckerman, W., *In Defence of Economic Growth*, Cape, London, 1974.

Beek, M. 'Breeding for disease resistance in wheat: the Brazilian experience', in Lamberti *et al.*, 1981.

Behrman, J. and B. Wolfe, 'More evidence on nutrition demand: income seems over-rated and women's schooling under-emphasized', *J. Development Econ., 14*, 1984.

Bell, C., 'Alternative theories of sharecropping – some tests using evidence from North India', *J. Development Studies, 13*, 4, July 1977, pp. 317–46.

——, Hazell, P. and Slade, R., *Project Evaluation in Regional Perspective*, Johns Hopkins, Baltimore, 1982.

Bell, M. and Richards, P., 'Exploiting the potential of technology', mimeo, TOES/AMPS Summit, Libreville, 1986.

Berg, A., *Malnutrition: What Can Be Done?*, Johns Hopkins, Baltimore, 1987.

Berg, E., 'Discussant's comments', in ICRISAT, 1980, pp. 289–92.

Berry, A. and Cline, W., *Agrarian Structure and Productivity in Developing Countries*, Johns Hopkins, Baltimore, 1979.

Berry, A. and Sabot, R.H., 'Labour market performance in developing countries: a survey', in P. Streeten and R. Jolly, *Recent Issues in World Development*, Pergamon Press, Oxford, 1981.

Bersten, R.H., Siwi, B.H. and Beachell, H.M., 'Development and diffusion of rice varieties in Indonesia', Research Paper No. 71, IRRI, Jan. 1982.

Bhaduri, A., 'A study in agricultural backwardness under semi-feudalism', *Economic Journal, 83*, 329, March 1973.

Bhalla, G.S. and Chadha, G.K., *Green Revolution and the Small Peasant*, Concept, New Delhi, 1983.

—— and Alagh, Y.K., *Performance of Indian Agriculture: a Districtwise Study*, Sterling, New Delhi, 1979.

——, and Alagh, Y. K., Thind, S.S. and Sharma, R.K., *Foodgrains Growth – A Districtwise Survey*, JNU/Planning Comm., Centre for the Study of Regional Devel., J. Nehru University, New Delhi, 1983.

Bhalla, Sheila, 'Real wage rates of agricultural labourers in the Punjab, 1961–77: a preliminary analysis', *Economic and Political Weekly*, XIV, 26, June 30, 1979, pp. A57–A68.

Bhalla, Surjit, Appendix, in [Berry and Cline, 1979].

Bharadwaj, K., *Production Conditions in Indian Agriculture*, Dept. of Applied Economics, Occ. Paper No. 33, Cambridge University Press, 1974.

Biggs, S., 'Generating agricultural technology: triticale for the Himalayan hills', *Food Policy*, 7, 1, pp. 69–82, Feb. 1982.

—— and Clay, E., 'Sources of innovation in agricultural technology', *World Development*, 9, 4, April 1981, pp. 321–36.

Binswanger, H., *The Economics of Tractorization in South Asia*, ADC, Washington DC, 1978.

——, 'Income distribution effects of technical change – some analytical issues', *S.E. Asian Econ. Rev.*, 1, 3, Dec. 1980, pp. 179–218.

——, 'Attitudes toward risk: theoretical implications of an experiment in South India', *Economic Journal*, 91, 364, Dec. 1981, pp. 867–90.

——, and Rosenzweig, M., *Contractual Arrangements, Employment and Wages in Rural Labour Markets: A Critical Review*, ADC, New York, 1981.

——, and Rosenzweig, M. 'Behavioural and material determinants of production relations in agriculture', *J. Development Studies*, 22, 3, April 1986.

—— and Pingali, P., 'Resource endowments and technology priorities for sub-Saharan Africa', mimeo, Agricultural Research Unit, World Bank, 1986.

—— and Ruttan, V. (eds.), *Induced Innovation: Technology, Institutions and Development*, Johns Hopkins, Baltimore, 1977.

——, and Ryan, J. 'Efficiency and equity issues in the *ex ante* allocation of research resources', *Indian J. Agric. Econ.*, XXXII, 3, July–Sept., 1977, pp. 217–31.

Bishop, J.A. and Cook, L.M. (eds.), *Genetic Consequences of Man-made Change*, Academic Press, London, 1981.

Bliss, C.J. and Stern, N.H., *Palanpur: the Economy of an Indian Village*, Clarendon, Oxford, 1982.

Bloch, M., 'The rise of dependent cultivation and seignorial institutions', in Postan, 1966.

Blyn, G., 'The green revolution revisited', *Econ. Development and Cultural Change*, 31, 4, 1983, pp. 705–25.

Bolton, F.R. and Zandstra, H.G., *Evaluation of Double Cropped Rainfed Wetland Rice*, Research Paper No. 63, IRRI, June 1981.

Bond, C., *Women's Involvement in Agriculture in Botswana*, Overseas Development Ministry, London, 1974.

Borgstrom, G., 'The green revolution', *Seminar*, 183, Nov. 1974.

Borlaug, N., 'Breeding wheat for high yield, wide adaptation and disease resistance', in IRRI, 1972a.

Boserup, E., *Conditions of Agricultural Progress*, Asia, Bombay, 1965.

BOSTID (Board on Science and Technology for International Development), *Priorities in BT Research for International Development*, National Research Council, Washington DC, 1982.

Boxall, R., Greeley, M. and Tyagi, D. (*et al.*), *The Prevention of Farm-level Foodgrain Storage Losses in India: A Social Cost-Benefit Analysis*, IDS Research Report, Brighton, 1977.

Brass, P., 'International rice research and the problems of rice growing in Uttar Pradesh and Bihar', in Robb, 1984.

Braudel, F., *The Structures of Everyday Life: the Limits of the Possible*: Vol. 1, *Civilization and Capitalism*, Collins, London, 1981.

Bray, F., *The Rice Economies: Technology and Development in Asian Societies*, Blackwell, Oxford, 1986.

Brown, D., *Agricultural Development in India's Districts*, Harvard University Press, Cambridge, Mass., 1971.

Brown, L., *Seeds of Change*, Praeger, New York, 1970.

Browning, J. and Frey, K., 'The biology of using multilines to buffer pathogen population and prevent disease loss', *Indian J. Genetics and Plant Breeding*, 39, 1979, pp. 3–9, 105–6.

Buddenhagen, I.W., 'Disease resistance in rice', in Lamberti *et al.* eds., 1981.

—— and de Ponti, O., 'Crop improvement to minimize future losses to diseases and pests in the tropics', *FAO Plant Protection Bull.*, *31*, 1, 1983.

Bull, D., *A Growing Problem: Pesticides and the Third World Poor*, Oxfam, Oxford, 1982.

Burch, D., 'Overseas aid and the transfer of technology: a case study of agricultural mechanization in Sri Lanka', D.Phil., Sussex University, Brighton, 1980.

Burgess, T., 'The revolution that failed', New Scientist, Nov 1, 1984.

Burke, R.V., 'Green revolution technology and farm class in Mexico', *Econ. Development and Cultural Change*, 28, 1, 1979, pp. 135–54.

Burniaux, J.-M. and Walbroeck, J., 'Agricultural protection in Europe: its impact on developing countries', in C. Stevens, (ed.), *Pressure Groups, Politics and Development*, Hodder and Stoughton, London, 1985.

Buttel, F., Kenny, M. and Kloppenburg, J., 'From green revolution to biorevolution', *Econ. Development and Cultural Change, 34*, 1, 1985.

Byerlee, D. and Harrington, L., 'New wheat varieties and the small farmer', paper to International Agric. Econ. Assoc. Conference, Djakarta, Aug 24 – Sept 19, 1982, mimeo, CIMMYT Economics Prog., 1982.

——, Harrington, L. and Winkelmann, D.L., 'Farming systems research', *American J. Agric. Econ., 64*, 5, Dec. 1982a, pp. 405–54.

Byres, T.J., 'The dialectic of India's green revolution', *S. Asian Review, 5*, 2, 1972.

——, 'The new technology, class formation and class action in the Indian countryside', *J. Peasant Studies, 8*, July 1981.

Cain, M., 'Risk and fertility in India and Bangladesh', *Population and Development Rev., 7*, 3, 1981, pp. 435–74.

Campbell, B., 'Agricultural progress in medieval England: some evidence from East Norfolk', *Econ. Hist. Rev., 36*, 1, Jan. 1983, pp. 26–46.

Carloni, A.S., 'Sex disparities in the distribution of food within rural households', *Food and Nutrition, 7*, 1, 1981, pp. 3–12.

Carr, C. and Myers, R.H., 'The agricultural transformation in Taiwan: the case of *ponlai* rice, 1922–42', in Shand, 1973.

Case, H., 'Neolithic explanations', *Antiquity, XLIII*, 1969, pp. 176–86.

Cassen, R., *India: Population, Economy, Society*, Macmillan, London, 1976.

CGIAR (Consultative Group for International Agriculture Research), *1986/87 Annual Report*, Washington DC, September 1987.

Chadha, G.K., *Dynamics of Rural Transformation: a Study of Punjab 1950–80*, Centre for Regional Development, J. Nehru University, New Delhi, 1983.

Chambers, R. (ed.), 'Rural development: whose knowledge counts?', *IDS Bull., 10*, 2, 1979.

——, Rural Development: Putting the Last First, Longmans, London, 1983.

——, Longhurst, R. and Pacey A. (eds.), *Seasonal Dimensions to Rural Poverty*, Pinter, London, 1981.

Chang, T.T., 'Germplasm of rice and its utilization by plant breeders', in Singh and Chonchalow (eds.), 1982.

——, 'Conservation of rice genetic resources: luxury or necessity?', Science, 224, Apr. 20, 1984.

Chattopadhyay, M. and Rudra, A., 'Size-productivity revisited', *EPW*, Sep. 1976, pp. A104–16.

Chaudhry, M.G., 'Green revolution and redistribution of rural incomes: Pakistan's experience', *Pakistan Development Rev., 21,* Autumn 1982, pp. 173–205.

Chayanov, A.V., *Theory of Peasant Economy* (tr. B. Kerblay, ed. D. Thorner), Irwin, Homewood, Illinois, 1966.

Chen, L., Haq, E. and D'Souza, S., 'Sex bias in the family allocation of food and health care in rural Bangladesh', *Population and Development Rev., 7, 1,* 1981.

Chuta, E. and Liedholm, C., *Rural Non-farm Employment: A Review of the State of the Art,* Michigan State Univ., 1979.

Chuta, E. and Liedholm, C., *Rural Non-farm Employment: A Review of the State of the Art,* Michigan State University, 1979.

CIAT, Annual Reports, Cali, as cited.

——, *A Summary of Major Achievements in the Period 1977–83,* Cali, 1984.

CIMMYT, *Reviews* (annual), as cited.

——, *Report on Maize Improvement 1978–9,* El Batan, 1981.

——, *Report on Wheat Improvement 1980,* El Batan, 1983.

——, *Report on Wheat Improvement 1981,* El Batan, 1984.

——, *Report on Wheat Improvement 1982,* El Batan, 1984a.

——, *Report on Wheat Improvement 1983,* El Batan, 1985.

——, *1986 CIMMYT World Maize Facts and Trends,* Mexico D.F., 1987.

——, *1986 World Maize Facts and Trends,* Mexico D.F., 1987

Clark, C. and Haswell, M., *The Economics of Subsistence Agriculture,* Macmillan, Basingstoke, 1970.

Clark, N. and E. Clay, The Dryland Research Project at Indore 1974–80 – an Institutional Innovation in Rural Technology Transfer, *IDS Discussion Paper No. 222,* Brighton, Sussex, 1986.

Cleaver, H.M., 'The contradictions of the green revolution', *American Econ. R.,* LXII, 2, May 1972, pp. 177–88.

Clower, R.W., 'A reconsideration of the microfoundations of economic theory', *Western Economic Journal,* 6, 1968.

Cohen, J.M., 'Effects of green revolution strategies on tenants and small-scale landowners in the Chilalo region of Ethiopia', *J. Developing Areas, 9,* April 1975, pp. 335–58.

Collinson, M.P., 'Farming systems research in East Africa', *Int. Devel. Papers, 3,* Dept. Ag. Econ., Michigan State Univ., East Lansing, 1982.

——, 'A low-cost approach to understanding small farmers', *Agricultural Administration, 8,* 6, Nov. 1981.

Colmenares, J.H., *Adoption of Hybrid Seeds and Fertilizers among Colombian Corn Growers,* CIMMYT, 1975.

Connell, J., Dasgupta, B., Laishley, R. and Lipton, M., *Migration from Rural Areas: The Evidence from Village Studies*, Oxford Univ. Press, Delhi, 1976.

Cordova, V. *et. al.*, 'Changes in rice production technology and their impact on rice earnings in Central Luzon, Philippines, 1966–1979', Dept. of Agric. Econ., Paper No. 81–19, IRRI, 1981.

Cornia, A.E., 'Farm size, land yields and the agricultural production function: an analysis for fifteen developing countries', *World Development, 13*, 4, pp. 513–34, 1987.

Craswell, E. and Vlek, P., 'Fate of fertilizer nitrogen applied to wetland rice', in IRRI, 1979a.

Crill, P., Ikehashi, H. and Beachell, H.M., 'Rice blast control strategies', in IRRI, 1982, pp. 129–46.

Crisostomo, C.M. *et al.*, 'New rice technology and labour absorption in Philippine agriculture', mimeo, Conf. on Manpower Problems in East and S.E. Asia, Singapore, May 1971.

CSO (India), *Statistical Abstract*, 1984.

Cutie, Jesus T., *Diffusion of Hybrid Corn Technology: the Case of El Salvador*, CIMMYT, El Batan, 1975.

Dalrymple, D., *Development and Spread of HYVs of Wheat and Rice in the Less Developed Nations*, Foreign Agric. Econ. Report No. 95, Econ. Research Service., USDA, Washington, DC, 1976.

——, 'The adoption of high yielding grain varieties in developing nations', *Agric. History, 53*, 4, Oct. 1979, pp. 704–26.

——, *Development and Spread of HYVs of Wheat and Rice in the Less Developed Nations*, 7th ed., CGIAR/USAID, 1985.

——, *Development and Spread of High-yielding Wheat Varieties in Developing Countries*, USAID, Washington, DC, 1986.

——, *Development and Spread of High-yielding Rice Varieties in Developing Countries*, USAID, Washington, DC, 1986a.

Dandekar, K. and Sathe, V. 'Employment guarantee scheme and food-for-work programme', *Economic and Political Weekly, XV*, 15, April 12, 1980, pp. 707–13.

Dasgupta, B. *The New Agrarian Technology and India*, Macmillan, New Delhi, 1977.

——, *Village Society and Labour Use*, Oxford University Press, New Delhi, 1977a.

Datta, G. and Meerman, J., *Household Income and Household Income per Capita in Welfare Comparisons*, Staff Working Paper No. 378, World Bank, Washington, DC, 1980.

David, C., 'Government policies and farm mechanization in the Philippines', mimeo, Hangzhon (China) Conf. on Small Farm Mechanization, June 1982.

David, C. and Barker, R., 'Modern rice varieties and fertilizer consumption', in IRRI, 1978, pp. 175–220.

Davis, K., *The Population of India and Pakistan*, Princeton, 1951.

Day, P.R., 'Genetic vulnerability of crops', *Amer. Review of Phytopathology, 11*, 1973.

—— (ed.), *The Genetic Basis of Epidemics in Agriculture*, Annals of the New York Academy of Science, 1977.

Deane, P. and Cole, W., *British Economic Growth, 1688–1959*, 2nd ed., Cambridge, 1967.

Delgado, C. and Ranade, C., 'Technological change and agricultural labour use', in Mellor *et al.*, 1987.

——, Mellor, J. and Blackie, M., 'Strategic issues in accelerating food production in sub-Saharan Africa, in Mellor *et. al.* 1987

Dernburg, T. and McDougall, D., Macroeconomics: *The Measurement, Analysis and Control of Aggregate Economic Activity 6th ed.*, McGraw Hill, Kogakusha, Tokyo, 1980.

Deuster, C., 'The green revolution in a village of West Sumatra', *Bull. Indonesian Econ. Studies.*, XVII, 1, Mar. 1982.

Dixon, J., 'Consumption', in Falcon *et al.*, 1984.

Dodgshon, R., *The Origin of British Field Systems*, Academic, London, 1980.

Doggett, H., 'A look back at the 1970s', in ICRISAT, 1982, vol.2.

Doyle, J., *Altered Harvest: Agriculture, Genetics and the Fate of the World's Food Supply*, Viking Penguin: New York, 1985.

Drèze, J., Leruth, L. and Mukerjee, A., 'Rural labour markets in India: theories and evidence', mimeo, paper presented at 8th World Congress of International Economics Association, New Delhi, Dec. 1986.

Duff, B., 'Mechanization and modern rice varieties', in IRRI, 1978, pp. 145–72.

Duvick, D., 'Genetic diversity in major farm crops on the farm and in reserves', *Econ. Botany*, 38, 2, Apr.-June 1984.

Edirisinghe, N., The Food Stamp Scheme in Sri Lanka: Costs, Benefits, and Options for Modification, *Research Report*, No. 58, IFPRI, Washington, DC, 1987.

Edwards, E., 'American agriculture: the first three hundred years', in USDA, 1941.

Eicher, C. and Baker, D. *Research on Agricultural Development in sub-Saharan Africa*, International Development Paper No. 6, Dept. of Agric. Econ., Michigan State Univ., East Lansing, 1982.

—— and Staatz, J.M., *Agricultural Development in the Third World*, Johns Hopkins, Baltimore, 1984.

Engels, F., 'The Peasant Question in France and Germany' (1894), in K. Marx and F. Engels, *Selected Works in Two Volumes*, Foreign Languages Publishing House, Moscow, 1951, vol. 2.

Epstein, T.S., *South India: Yesterday, Today, Tomorrow*, Manchester, 1973.

Ercelawn, A., 'Income inequality in rural Pakistan: a study of sample villages', *Pakistan J. Applied Econ., III*, 1, Summer 1984, pp. 1–28.

Evenson, R. and Flores, P., 'Social returns to rice research', in IRRI, 1978.

—— and Kislev, Y., *Agricultural Research and Productivity*, Yale, 1976.

FAI (The Fertilizer Association of India), *Fertilizer Statistics 1985–86*, New Delhi, October 1986.

Falcon, W.P., 'The green revolution: generations of problems', *American J. Agric. Econ., 52*, 3, 1970.

——, Jones, W.O., Pearson, S.R. *et al.*, *The Cassava Economy of Java*, Stanford, 1984.

FAO, *Introduction and Effects of HYVs of Rice in the Philippines*, Rome, 1971.

——, *Report of the FAO/UNEP/IBPGR International Conference on Crop Genetic Resources*, Rome, 1981.

——, *Integrating Nutrition into Agriculture and Rural Development: Six Case Studies*, Nutrition in Agriculture Series, No. 2, Rome, 1984.

——, 'Breeding for durable resistance in perennial crops', *Plant Production and Protection Paper 70*, Rome 1986.

Farmer, B. (ed.), *Green Revolution?*, Macmillan, London, 1977.

Farrington, J. and Abeyratne, F., *Farm Power and Water Use in the Dry Zone of Sri Lanka*, Development Study No. 22, Univ. of Reading, 1982.

Faulkner, W., Linkage between industrial and agricultural research: BT in pharmaceuticals', D. Phil., Sussex Univ., 1986.

Feder, G. and O'Mara, G.T., 'Farm size and the diffusion of Green Revolution technology', *Econ. Development and Cultural Change, 30*, 1, 1981, pp. 59–76.

Fernando, H.E., 'Resistance to insect pests in rice varieties', Seminar Paper No. 22, in ARTI-IDS.

Fischer, K.S., Johnson, E.C. and Edmeades, G.O., 'Breeding and selection for drought resistance in tropical maize', mimeo. CIMMYT, 1983.

Flinn, N.C. and O'Brien, D.T., 'Economic considerations in the evaluation of urea fertilizers in dry land wet farming', Dept. of Agric. Econ., Paper 82–06, IRRI, 1982.

—— and Unnevehr, L.J., 'Contributions of modern rice varieties to nutrition in Asia – an IRRI perspective', in Pinstrup-Andersen *et al.*, 1984.

Flores, P., Evenson, R. and Hayami, Y., 'Social returns to rice research in the Philippines: domestic benefits and foreign spillover', *Econ. Development and Cultural Change*, 26, 3, 1978, pp. 591–607.

Fogel, R.W. and Engerman, S.L., *Time on the Cross*, Harvard University Press, Cambridge, Mass., 1974.

Ford, E.B., *Genetic Polymorphism*, All Souls Studies, V, Oxford, 1965.

Ford Foundation, *India's Foodgrain Crisis and Steps to Meet It*, Ministry of Food and Agriculture and Ministry of Community Development, New Delhi, 1959.

Fowler, P.J., 'Later prehistory', in Renfrew, 1973.

Franke, R.W., The green revolution in a Javanese village, D. Phil., Harvard University, 1972.

Frankel, F.R., *India's Green Revolution*, Princeton, 1971.

Frankel, H., 'Genetic perspectives of germplasm conservation', in Aiber *et al.* (eds.), 1984.

Gafsi, S. and Roe, T., 'Adoption of unlike high-yielding wheat varieties in Tunisia', *Econ. Development and Cultural Change*, 28, 1, 1979, pp. 119–31.

George, P.S., 'The changing patterns of consumer demand for foodgrains in India', *Indian J. Agric. Econ.*, 35, 1, 1980, pp. 53–6.

Gerhart, J., *The Diffusion of Hybrid Maize in Kenya*, CIMMYT, 1975.

Ghai, D. and Radwan, S. (eds.), *Agrarian Policies and Rural Poverty in Africa*, ILO, Geneva, 1983.

—— and Smith, L., 'Food price policy and equity', in Mellor *et al.* 1987.

Ghodake, R., 'Economic evaluation of traditional and improved technologies in dryland agriculture', mimeo, ICRISAT, Aug. 1983.

Gibbon, D., de Koninck, R. and Hasan, I., 'The green revolution – its distributional impact: a study in regions of Malaysia and Indonesia', mimeo, n.d.

Gill, G.J., *Farm Power in Bangladesh*, vol. 1, Development Study No. 19, University of Reading, 1981.

——, 'An *ex ante* evaluation of impact of improved dryland technology on factor shares', mimeo, ICRISAT Economics Program, Hyderabad, August 1983a.

—— and Kshirsagar, K., 'Employment generation potential of deep vertisol technology in semi-arid tropical India', mimeo, ICRISAT, Hyderabad, December 1983.

Godfrey, M., 'Trade and exchange-rate policies in sub-Saharan Africa', *IDS Bull.*, 15, 3, July 1985.

Goldsmith, E. and Hildyard, N., *The Social and Environmental Effects of Large Dams: Vol. 1, Overview*, Wadebridge Ecological Centre, Camelford (UK), 1984.

Gonzales, L.A. and Regaldo, B.M., 'The distributional impact of food policies on nutritional intake: a first approximation for the Philippines', IRRI, Los Banos, 1983.

Goodman, M., review in *Qtly. Rev. Biol., 59*, Sept. 3 1984.

Goodwin, R., 'The multiplier as matrix', *Economic Journal, LIX*, 1949, pp. 537–55.

Goody, J., 'Rice-burning and the green revolution in Northern Ghana', *J. Development Studies, 16*, 2, Jan. 1980, pp. 136–55.

Gough, K., 'The green revolution in South India and North Vietnam', *Social Scientist, 6*, 1, August 1977.

Govt. of Bangladesh and USAID, *Joint Evaluation of the Fertilizer Distribution Improvement Project*, Dhaka, Nov. 1982.

Grabowski, R., 'The implications of an induced innovation model', *Econ. Development and Cultural Change, 30*, 1, 1981, pp. 177–81.

Greeley, M., 'Farm-level post-harvest food losses: the myth of the soft third option', *IDS Bulletin, 13*, 3, 1982.

——,'Food technology and employment: the farm-level post-harvest system in developing countries', *J. Agric. Econ., 37*, 3, 1986.

——, *Post-harvest Losses, Technology and Unemployment: the Case of Bangladesh*, Westview, Boulder, 1987.

—— and Begum, S., 'Women's employment and agriculture: extracts from a case study' *Women in Bangladesh: Some Socio-economic Issues*, Women for Women, Dhaka, 1983.

Grewal, P.S. and Bhuller, B.S., 'Impact of Green Revolution on the cultivation of pulses in Punjab', *Indian J. Agric. Econ., 37*, 3, 1982.

Griffin, K., The Political Economy of Agrarian Change, Macmillan, 1975.

—— and Khan, A.R. (eds.), *Poverty and Landlessness in Rural South Asia*, ILO, Geneva, 1977.

Grist, D., Rice, 5th ed., Longmans, London, 1975.

Hamid, M. A., *et. al., Irrigation Technologies in Bangladesh*, Rajshahi University, 1978.

Han, H., 'Use of haploids in crop improvement', in IRRI, *Biotechnology in Agricultural Research*, Los Banos, 1985.

Hansen, M. *et al.*, 'Plant breeding and biotechnology', *Bioscience, 36*, 1, 1986.

Harbert, L. and Scandizzo, P., *Food Distribution and Nutritional Intervention: the Case of Chile*, Staff Working Paper No. 512, World Bank, Washington DC, 1982.

Hargrove, T.R., Cabanilla, V. and Coffman, W., *Changes in Rice Breeding in Ten Asian Countries: 1965–84*, Research Paper Series No. 111, IRRI, Manila, June 1985.

Harrison, P., *The Greening of Africa*, Penguin, New York, 1987.

Harriss, B., 'Appraisal of rice processing projects in Bangladesh', *Bangladesh J. Agric. Econ.*, *1*, 2, Dec. 1978.

——, 'The intra-family distribution of hunger in South Asia', paper for WIDER Project on Hunger and Poverty: Seminar on Food Strategies, 1986.

——, 'Post-harvest rice processing systems in rural Bangladesh', *Bangladesh J. Agric. Econ.*, 2, 1, June 1979.

——, 'Regional growth linkages from agriculture: discussion', *J. Development Studies*, *23*, 2, pp. 275–89, 1987.

—— et al., 'Exchange Relations and Poverty in Dryland Agriculture', ESCOR 3326/3683, ODA, London, mimeo, 1982.

Harriss, J., 'The limitations of HYV technology in North Arcot: the view from a village', in Farmer, 1977.

——, 'Bias in perception of agrarian change in India', in Farmer, 1977a.

——, *Capitalism and Peasant Farming*, Oxford University Press, Bombay, 1982.

Hart, G., 'Agrarian Change in Java: Issues and Controversies', mimeo, n.d.

——, 'Exclusionary labour arrangements: interpreting evidence on employment trends in rural Java', *J. Development Studies*, *22*, 4, 1986.

Hartmann, B. and Boyce, J., *A Quiet Violence: View from a Bangladesh Village*, Zed, London, 1983.

Hartmans, E., 'Increasing the pace of development in sub-Saharan Africa: the role of IITA', mimeo, Ibadan, 1985.

Hawkes, J.G., *Plant Genetic Resources: the Impact of the IARCs*, CGIAR Study Paper No. 3, World Bank, Washington, DC, 1985.

Hayami, Y., 'Induced innovation, green revolution and income distribution: comment', *Economic Development and Cultural Change*, *30*, 1, 1981, pp.169–76, reprinted in Eicher and Staatz, 1984.

—— and Hafid, A., 'Rice harvesting and welfare in rural Java', *Bull. Indonesian Econ. Studies*, *XV*, 2, July 1979.

—— and Herdt, R., 'Market price effects of technical change on income distribution in semi-subsistence agriculture', *American J. Agric. Econ.*, *59*, 2, May 1977, pp. 245–56.

—— and Kikuchi, M., *Asian Villages at the Crossroads*, Univ. of Tokyo, 1981.

—— and Ruttan, V., *Agricultural Development: an International Perspective*, Johns Hopkins, Baltimore, 1971.

Hazell, P., *Instability in Indian Foodgrain Production*, Research Report No. 30, IFPRI, May 1982.

——, 'Sources of increased instability in Indian and US cereal production', *American J. Agric. Econ.*, *60*, 3, August 1984, pp. 302–11.

——, 'Changing Patterns of variability in cereals prices and production', in J. Mellor and Ahmed Rais ud-Din (eds.), *Agricultural Price Policy for Developing Countries*, Johns Hopkins, Baltimore, for IFPRI, 1987.

—— (ed.), *Summary proceedings of a Workshop on Cereal Yield Variability*, IFPRI and DSE, Washington, DC, 1987a.

—— and Roell, A., *Rural Growth Linkages: Household Expenditure Patterns in Malaysia and Nigeria*, Research Report No. 40, IFPRI, Washington, DC, September 1983.

Herath, H.M.G., 'Production efficiency, return to scale and farm size in rice production: evidence from Sri Lanka', *Agric. Admininstration*, *12*, March 1983, pp. 141–53.

Herdt, R.W., 'Focusing research on future constraints to rice production', Dept. of Agric. Econ. Paper No. 81–06, IRRI, 1981.

—— and Capule, C., *Adoption, Spread and Production Impact of Modern Rice Varieties in Asia*, IRRI, 1983.

—— and Garcia, L., 'Adoption of modern rice technology: the impact of size and tenure in Bangladesh', mimeo, IRRI, Dec. 1982.

—— and Mandac, A.M., 'Modern technology and economic efficiency of Philippine rice farmers', *Econ. Development and Cultural Change*, *29*, 2, 1981, pp. 375–99.

—— and Wickham, T.M., 'Exploring the gap between actual and potential yields: the Philippine case', in IRRI, 1978, pp. 3–24.

—— and Garcia, L., 'Adoption of modern rice technology: the impact of size and tenure in Bangladesh', mimeo, IRRI, Dec. 1982.

Hewitt, C. de Alcantàra, *La Modernizacion de la Agricultura Mexicana*, 1940–1970, Editores Siglo Veintiuno, Mexico, 1978.

Hibino, H., 'Rice virus disease problems and need for MAB technology', in IRRI, 1985a.

Hicks, J.R., *Value and Capital*, 2nd ed., Oxford, 1946.

Hill, P., *Population, Prosperity and Poverty: Rural Kano, 1900–1970*, Cambridge University Press, Cambridge, 1977.

——, *Dry Grain Farming Families*, Cambridge University Press, Cambridge, 1982.

Hobsbawm, E., 'The British standard of living, 1790–1850', *Econ. Hist. Review, X*, 1, 1957.

Holden, J.H., 'The second ten years', in Holden and Williams, 1984.

—— and Williams, J.T., *Crop Genetic Resources: Conservation and Evaluation*, IPBGR, Allen and Unwin, London, 1984.

Howes, M., 'The creation and appropriation of value in irrigated agriculture: a comparison of the deep tubewell and the handpump in rural Bangladesh', in M. Howes and M. Greeley, *Rural Technology, Rural Institutions and the Rural Poorest*, CIRDAP/IDS, Dhaka, 1982.

Hunt, D., *The Impending Crisis in Kenya: the Case for Land Reform*, Gower, Aldershot, 1984.

ICRISAT, *Annual Reports*, as cited.

——, *Socioeconomic Constraints to Development of Semi-Arid Tropical Agriculture*, Patancheru, 1980.

——, *Sorghum in the Eighties* (2 vols.), Patancheru, 1982.

——, *Second International Workshop on Striga*, 5–8 Oct. 1981, Patancheru, 1983.

IITA, *Annual Report and Research Highlights 1985*, Ibadan, 1986.

ILO, *Matching Employment Opportunities and Expectations: A Programme of Action for Ceylon*, Geneva, 1971.

Ingrams, C.R. and Williams, J.T., '*In situ* conservation of wild relatives of crops', in Holden and Williams, 1984.

Iqbal, F., 'The demand for funds by agricultural households: evidence from rural India', *J. Development Studies*, 20, 1, 1983.

IRRI (International Rice Research Institute, Los Banos, Philippines), *Reporter* and *Annual Reports*, as cited.

——, *Rice, Science and Man*, Los Banos, 1972.

——, *Rice Breeding*, Los Banos, 1972a.

——, *Major Research in Upland Rice*, Los Banos, 1975.

——, *Economic Consequences of the New Rice Technology*, Los Banos, 1978.

——, *Changes in Rice Farming in Selected Areas of Asia*, Los Banos, 1978a.

——, *Long Range Planning Committee Report*, Los Banos, 1979.

——, *Nitrogen and Rice*, Los Banos, 1979a.

——, *Symposium on Drought Resistance in Crops with Emphasis on Rice*, Los Banos, 1981.

——, *Rice Research Strategies for the Future*, Los Banos, 1982.

——, *Weed Control in Rice*, Los Banos, 1983.

——, *Proceedings of the Second Upland Rice Conference*, Los Banos, 1985.

——, *Biotechnology in International Agricultural Research*, Los Banos, 1985a.

Jairath, J., 'Technical and institutional factors in utilization of irrigation', *Econ. and Political Weekly*, XX, 13, March 30, 1985, pp. A2–10.

Jamison, D. and Lau, L., *Farmer Education and Farm Efficiency*, Johns Hopkins, Baltimore, 1982.

Janick, J. et al., *Plant Science: an Introduction to World Crops*, Freeman, San Francisco, 1981.

Jansen, H., Adoption of Modern Cereal Cultivars in India: determinants and implications of regional variation in speed and ceiling of diffusion, D.Phil. (unpub.), Cornell, 1988.

Janvry, A. de and Sadoulet, E., 'Agricultural price policy in general equilibrium models: results and comparisons', *American J. Agric. Econ.*, 69, 2, 1987, pp. 230–46.

—— and Subbarao, K., 'On the relevance of modelling for the analysis of food price policy', *Econ. and Political Weekly, XXII*, 25, June 20, 1987.

Jayasuriya, S. K. and Shand, R.T., 'Technical change and labour absorption in Asian agriculture: some emerging trends', *World Development*, 14, 3, 1986.

——, Barlow, C., Garcia, L., Castillo, G. and Kikuchi, M., 'Impact of technology adoption on labour utilization in Iloilo', Paper No. 8, 1–14, Dept. of Agric. Econ., IRRI, 1981.

——, Te, A. and Herdt, R.W., 'Mechanization and cropping intensification: power tillers in the Philippines', Paper No. 82–14, Dept. of Agric. Econ., IRRI, 1982.

Jennings, P.R., Coffman, W. and Kaufman, H.E., *Rice Improvement*, IRRI, 1979.

Jiggins, J., *Gender-related Impacts and the Work of the International Agricultural Research Centers*, CGIAR Study Paper No. 17, World Bank, Washington, DC, 1986.

Jodha, N.S., 'Intercropping in Traditional Farming Systems', *J. Development Studies, 16*, 4, 1980, pp. 427–42.

——, 'Decline of common property resources and its implications for livestock farming in Rajasthan', mimeo, ICRISAT, Patancheru, 1983.

—— and Singh, R.P., 'Factors constraining growth of coarse grain crops in semi-arid tropical India', *Indian J. Agric. Econ., 37*, 3, pp. 346–54, 1982.

Joffe, S., *Biotechnology and Third World Agriculture: A Literature Review and Report*, Institute of Development Studies, Brighton, 1986.

Johnson, A.A., *Indian Agriculture in the 1970s*, Ford Foundation, New Delhi, 1970.

Johnston, B.F. and Clark, W.C., *Redesigning Rural Development: A Strategic Perspective*, Johns Hopkins, 1982.

Jones, D.B., 'Discussion', in ICRISAT, 1982, vol. 2, pp. 717–22.

Jones, E., *Agriculture and the Industrial Revolution*, Blackwell, Oxford, 1974.

Jose, A.V., 'Trends in real wage rates of agricultural labourers', *Econ. and Political Weekly*, IX, 13, Mar. 30 1974, pp. A25–30.

Joshi, P.K., Bahl, D.K. and Jha, D., 'Direct employment effect of technical change in UP agriculture', *Indian J. Agric. Econ., 36*, 4, pp. 1–6, 1981.

Judd, A., Boyce, J. and Evenson, R., *Investing in Agricultural Supply*, Discussion Paper No. 44, Econ. Growth Centre, Yale Univ., 1983.

Junankar, P.N., 'Has the green revolution increased inequality?', *Development Research Digest*, IDS, 1978, pp.16–17.

Kahlon, A.S., 'High-yielding varieties', *Seminar*, 183, Nov. 1974.

Kahn, R.F., 'The relation of home investment to unemployment', *Economic Journal, XLI*, 162, June 1931.

Kaldor, N., 'The equilibrium of the firm', *Economic Journal, XLIV*, 173, March 1934.

Kaneda, H., 'Mechanization, industrialization and technical change in rural Punjab', in Shand, 1973.

Kautsky, K., *Die Agrarfrage*, Dietz, Stuttgart, 1899.

Kennedy, L., 'The first agricultural revolution: property rights in their place', *Agric. History, 56*, 2, April 1982.

Kenney, M. and Buttel, F., 'Biotechnology: prospects and dilemmas for Third World development', *Econ. Development and Cultural Change, 34*, 1, 1985.

Ketkar, S.L., 'The long-run impact of the Green Revolution on the distribution of rural incomes in India', *Indian J. Agric. Econ., 35*, 3, 1980, pp. 51–8.

Keynes, J.M., *The General Theory of Employment, Interest and Money*, Macmillan, London, 1936.

——, 'Relative movements of real wages and unemployment', *Economic Journal, XLIX*, 193, Mar. 1939.

Khan, A.H., *Reflections on the Comilla Rural Development Project*, OLC Paper No. 3, American Council on Education, Overseas Liaison Committee, Washington, 1974.

Khan, M.H. and Maki, D.R., 'Relative efficiency by farm size and the green revolution in Pakistan', *Pakistan Development Rev., 19*, Spring 1980, pp. 51–64.

Khush, G.S., 'Breeding for resistance in rice', in Day, 1977.

Khusro, A., *The Economics of Land Reform and Farm Size in India*, Macmillan (India), 1973.

Kikuchi, M., and Hayami, Y., 'New rice technology, intra-rural migration and institutional innovation in the Philippines', Research paper No. 86, IRRI, Jan. 1983.

——, Huysman, A. and Res, L., 'New rice technology and labour absorption: comparative histories of two Philippine rice villages', mimeo, Paper 82–17, Dept. of Agric. Econ., IRRI, 1982.

Kiyosawa, S., 'Dynamics of blast resistance', in IRRI, 1975.

Koester, U., *Policy Options in the Grain Economy of the EC: Implications for Developing Countries*, Research Report No. 35, IFPRI, Washington DC, 1982.

Kohli, S.P., *Wheat Varieties in India*, Indian Council for Agricultural Research, New Delhi, 1968.

Kosuge, T., Meredith, C. and Hollaender, A. (eds.), *Genetic Engineering of Plants: an Agricultural Perspective*, Plenum, New York, 1983.

Krishnaji, N., 'Inter-regional disparities in per capita production and productivity of foodgrains', *Economic and Political Weekly, 10*, 33–5, Aug. 1975, pp. 1377–86.

Krueger, A.O., 'The political economy of the rent-seeking society', *American Econ. Rev.*, June 1974.

Kuhn, T.S., *The Structure of Scientific Revolutions*, Chicago, 1973.

Kuile, C.H., 'The humid and sub-humid tropics', in Mellor *et al.* 1987.

Kumar, P. and Sharma, B.M., 'Growth rates of agriculture in Haryana – a case study', *Indian J. Agric. Econ., 38*, 2, April-June 1983, pp. 202–7.

Kumar, S., 'Role of the household economy in determining child nutrition at low income levels: a case study of Kerala', Occasional Paper No. 95, Div. of Nutritional Sciences, Cornell, 1977.

Lal, D., 'Agricultural growth, real wages, and the rural poor in India', *Economic and Political Weekly, XI*, 26, June 1976, pp. A47–61.

Lamberti, F., Walter, N.H. and van der Graaff, N.A., *Durable Resistance in Crops*, Plenum, New York, 1981.

Leaf, M.J., 'The green revolution and cultural change in a Punjab village', *Economic Development and Cultural Change, 31*, 1983, pp. 227–70.

Leijonhufvud, A., *On Keynesian Economics and the Economics of Keynes: a Study of Monetary Theory*, Oxford Univ. Press, New York, 1968.

Lenin, V.I., *The Development of Capitalism in Russia (1899 and 1904)*, Progress, Moscow, 1964.

Levi-Strauss, C., *The Savage Mind*, Weidenfeld, London, 1966.

Levinson, F.J., *Morinda: An Economic Analysis of Malnutrition among Young Children in Rural India*, Cornell/MIT, International Nutrition Policy Series, Cambridge, Mass., 1974.

Lingard, J. and Baygo, A.S., 'The impact of agricultural mechanization on production and employment in rice areas of West Java', *Bull. Indonesian Econ. Studies, XIX*, 1, Apr. 1983.

——, *Why Poor People Stay Poor: Urban Bias in World Development*, Temple Smith, London and Harvard University Press, 1977.

——, 'Inter-farm, inter-regional and farm-non-farm income distribution: the impact of the new cereal varieties', *World Development, 6*, 3, March 1978, pp. 319–37.

——, 'The technology, the system and the poor: the case of the new cereal varieties', in Inst. of Social Studies, *Development of Societies: the Next 25 Years*, Nijhoff, The Hague, 1979.

——, 'Agricultural risk, rural credit and the inefficiency of inequality', in J.-M. Boussard, J. Roumasset and I. Singh (eds.), *Risk, Uncertainty and Agricultural Development*, SEARCA, Laguna and ADC, New York, 1979a.

Lipton, M., 'Towards a theory of land reform', in D. Lehmann (ed.), Agrarian Reform and Agrarian Reformism, Faber, London, 1974.

——, 'Agricultural finance and rural credit in poor countries', *World Development, 4, 7*, 1976, reprinted in Streeten, P. and Jolly, R. (eds.), *Recent Issues in World Development*, Pergamon, Oxford, 1981.

——, 'Migration from rural areas of poor countries: the impact on rural productivity and income distribution', in R. Sabot (ed.), *Migration and the Labor Market in Developing Countries*, Westview, Boulder, 1982.

——, *Poverty, Undernutrition and Hunger*, World Bank Staff Working Paper No. 597, Washington, DC, 1983.

——, *Demography and Poverty*, World Bank Staff Working Paper No. 623, Washington DC, 1983a.

——, 'Conditions of poverty groups and impact on economic development and cultural change: the role of labour', *Development and Change, 15*, 4, Oct. 1984.

——, Labour and Poverty, World Bank Staff Working Paper No. 616, Washington, DC, 1984a.

——, 'Urban bias revisited', in Harriss, J. and Moore, M.P. (eds.), *Development and the Rural-Urban Divide*, Cass, 1984b.

——, *The Place of Agricultural Research in the Development of sub-Saharan Africa*, Discussion Paper No. 202, Institute of Development Studies, Brighton, March 1985.

——, 'India's agricultural development and African food strategies: a role for the EEC', in W.M. Callewaert and R. Kumar (eds.), *EEC India: Towards a Common Perspective*, Peeters, Leuven, 1985a.

——, *Land Assets and Rural Poverty*, World Bank Staff Working Paper No. 744, Washington, DC, 1985b.

——, 'Coase's Theorem versus Prisoner's Dilemma: a case for democracy in less developed countries', in R.C.O. Matthews (ed.), *Economics and Democracy*, Macmillan, 1985c.

——, 'The limits of agricultural price policy: which way at the World Bank?', Development Policy Review, 5, June 1987, pp. 197–215.

——, and de Kadt, E., *Agriculture-Health Linkages*, WHO, Geneva, 1987.

—— and Longhurst, R. (eds.), *Food Problems in South Asia, 1975–90*, IDS Discussion paper No. 78, Brighton, 1975.

Lluch, C., and Mazumdar, D., 'Wages and employment in Indonesia', World Bank, mimeo, 1981.

Lockwood, B., Mukherjee, P.K. and Shand, R.T., *The HYV Program in India: Pt. I*, Planning Commission (India) with Australian National University, New Delhi, 1971.

Locy, R., 'Notes on Principles and Applications', in ATAS, 1984.

Longhurst, R., *The Energy Trap: Work, Nutrition and Child Malnutrition in Northern Nigeria*, Cornell International Nutrition Monograph Series No. 13, 1984.

—— 'Cash Crops, Household Food Security and Nutrition', *IDS Bulletin* Vol. 19, 2, Apr. 1988, pp. 28–36.

—— and Lipton, M., 'Secondary food crops and the reduction of seasonal food insecurity: the role of agricultural research', in D. Sahn (ed.), *Causes and Implications of Seasonal Variability in Household Food Security*, Washington DC: International Food Policy Research Institute, 1988 (in progress).

—— and Payne, P., *Seasonal Aspects of Nutrition*, IDS Discussion paper No. 145, University of Sussex, Brighton, 1979.

Low, A., *Agricultural Development in Southern Africa: Farm-Household Economics and the Food Crisis*, Chorley, London, 1986.

Lowdermilk, M.K., 'Diffusion of dwarf wheat production technology in Pakistan's Punjab, Ph.D., Cornell University, 1972.

Lynam, J.K., 'On the design of commodity research programmes in the international centers', IIMI–Rockefeller–CIAT, Lahore, 1986.

—— and Pachico, D., 'Cassava in Latin America: current status and future prospects', draft, CIAT, Cali, 1982.

McIntire, J. *et al.*, 'Evaluating sorghum cultivars for grain yield and for feed', mimeo, ILCA, Addis Ababa, 1985.

Malaos, A.A., 'The management approach to rural development planning in Cyprus', M.A., University of East Anglia, 1975.

Malla, P.B., 'Logit analysis of technology adoption by rice farmers in Dhanusha district, Nepal', Research Paper No. 22, Agric. Projects Services Centre (Nepal), June, 1983.

Malone, C., *Indian Agriculture: Progress in Production and Equity*, Ford Foundation, New Delhi, 1974.

Malthus, T.R., *A Summary View of the Principle of Population*, 1826.

Mandac, A.M. and Flinn, J.C., 'Farm level management and N response in rainfed rice, Bicol, Philippines', Agric. Econ. Dept., Paper No. 83–10, IRRI, 1983.

Marshall, A., *Principles of Economics (1890)*, 8th ed., Macmillan, London, 1961.

Matlon, P., 'The size distribution, structure and determinants of personal income among farmers in the North of Nigeria', D. Phil. thesis, Cornell University, 1977.

Matthews, A., *The Common Agricultural Policy and the Less Developed Countries*, Trocaire (Gill and MacMillan), Dublin, 1985.

Mauss, M. Cunnison, I., trans. I. *The Gift: Forms and Functions of Exchange in Archaic Societies*, Cohen, London, 1970.

Maxwell, S., 'The role of case studies in farming systems research' and 'Health, nutrition, agriculture: linkages', *IDS Discussion Paper No. 198*, University of Sussex, 1985.

Mehra, S., *Instability in Indian Agriculture in the Context of the New Technology*, Research Report No. 25, IRRI, July 1981.

Mellor, J. and Desai, G. (eds.), *Agricultural Change and Rural Poverty*, Johns Hopkins, Baltimore, 1986.

——, Delgado, C. and Blackie, M. (eds.), *Accelerating Food Production in sub-Saharan Africa*, Johns Hopkins, Baltimore, 1987.

Mellor, J.W., *Instability in Indian Agriculture in the Context of the New Technology*, Research Report No. 25, IFPRI, Washington DC, 1981.

Milliano, W. de, 'Breeding for disease resistance in wheat: the Zambian experience', in Lamberti *et al.*, 1981.

Minchinton, W. (ed.), *Essays in Agrarian History*, David and Charles, Newton Abbott, 1968.

Mingay, G., 'The agricultural revolution in English history: a reconsideration', in Minchinton, 1968, vol. 2.

Minhas, B.S., Jain, L.R., Kansal, S.M. and Saluja, M.R., 'On the choice of appropriate consumer price indices and data sets for estimating the incidence of poverty in India', mimeo, Indian Statistical Institute, Delhi, 1987.

Mohan, R., 'Contribution of research and extension to productivity change in Indian agriculture', *Econ. and Political Weekly*, IX, 39, Sept. 28, 1974, pp, A97–104.

Monyo, J.H. *et al.* (eds.), *Intercropping in Semi-arid Areas*, IDRC-076e, Ottawa, 1976.

Mooney, P., *Seeds of the Earth: A Private or Public Resource?*, International Coalition for Development Action, 1979.

——, 'The law of the seed revisited', *Development Dialogue*, 1985.

Moran, P.B., 'Investment appraisal of the IRRI mechanical reaper', Ag. Eng. Dept., IRRI, 1982.

Morineau, M., *Les Faux-semblants d'un Démarrage Economique*, Cahiers des Annales, 30: Librarie Armand Colin, Paris, 1971.

Moris, J., 'Options for science-based interventions in African agriculture', mimeo, ODI, London, 1988.

Mueller, R.A. and von Oppen, M., 'Women and Agricultural Technology', for Bellagio Inter-Centre Seminar on Developing Agricultural Technologies for Diverse Target Groups in the SAT, mimeo, ICRISAT, Patancheru, 1985.

Mujeeb-Kazi, A. and Jewell, D.C., 'CIMMYT's wide cross programme for wheat and maize improvement', in IRRI, 1985a.

Munchik de Rubinstein, E., 'The impact of technical change in rice on calorie intake by low-income households in Cali, Colombia', project report, mimeo, IARC Impact Study, Nov. 1984.

Mundle, S., *Backwardness and Bondage: Agrarian Relations in a South Bihar District*, Indian Inst. of Public Administration, New Delhi, 1979.

Murty, K., 'Consumption and nutritional patterns of ICRISAT mandate crops in India', Economics Program Progress Report No. 53, ICRISAT, 1983.

Myers, N., *A Wealth of Wild Species: Storehouse for Human Welfare*, Westview, Boulder, 1983.

Myrdal, G., *Economic Theory and Underdeveloped Regions*, Vora, Bombay, 1958.

Nabors, M. and Dykes, T., 'Tissue culture of cereal cultivars with increased salt, drought, and acid tolerance', in IRRI, 1985a.

Nadkarni, M., ' "Backward" crops in Indian agriculture', *Econ. and Political Weekly*, *XXI*, 38 and 39, Sept. 20–27, 1986, pp. A113–18.

Nair, K., *In Defense of the Irrational Peasant*, Chicago, 1979.

Nakamura, J.I., *Agricultural Production and the Economic Development of Japan, 1873–1922*, Princeton University Press, 1966.

Namboodiri, N.V. and Choksi, S.N., 'Green revolution – changes in the supply, sources and per capita availability of proteins and calories', *Indian J. Agric. Econ.*, *XXXII*, 3, pp. 27–33, July-Sept. 1977.

Narain, D. and Roy, S., *Impact of Irrigation and Labour Availability on Multiple Cropping: a Case Study of India*, Research Report No. 20, IFPRI, Washington, DC, Nov. 1980.

Narvaez, I., 'Technical aspects of cereal production change', in Anderson, 1973.

Ng, S. and Hahn, K., 'Applications of tissue culture to tuber crops at IITA', in IRRI, 1985a.

Norman, D., 'Rationalising mixed cropping under indigenous conditions: the example of Northern Nigeria', *J. Development Studies*, 11, pp. 3–21, 1974.

——, Hays, H. and Simmonds, E., 'Farming Systems in The Nigerian Savannah', *Research Strategies for Development*, Westview Press, Colorado, 1982.

North, D. and Thomas, R.P., 'The first economic revolution', *Econ. Hist. Review*, 2nd series, *30*, 1977, pp. 229–41.

Oberai, A.S. and Ahmed, I., 'Labour use in dynamic agriculture: evidence from Punjab', *Econ. and Political Weekly*, *XVI*, 13, Mar. 28, 1981, pp. A2–4.

——, and Manmohan Singh, H.K., 'Migration flows in Punjab's Green Revolution belt', *Econ. and Political Weekly*, *XV*, 13, March 21, 1980, pp. A2–12.

Okhawa, K., Johnston, B.F. and Kaneda, H., *Agriculture and Economic Growth: Japan's Experience*, Princeton and Tokyo University Press, 1970.

Okigbo, B. 'Nutritional implications of projects giving high priority to the production of staples of low nutritive quality: the case for cassava in the humid tropics of West Africa', *Food and Nutrition Bulletin*, 2, 4, 1980, pp. 1–10.

Okigbo, B.N. and Ay, P., 'Nutrition in the research and training activities of IITA', in Pinstrup-Andersen *et al.*, 1984.

Olson, M., *The Rise and Decline of Nations*, Yale, New Haven, 1982.

Olson, R.A., Sander, D.H. and Dreier, A.F., 'Soil analyses – are they needed for data interpretation?', in USDA *et al.*, 1941.

Oppen, M. von and Rao, P.P., 'Sorghum Marketing in India', in ICRISAT, 1982, vol. 2, pp. 659–74.

O'Toole, J.C., Chang, T.T. and Somrith, B., 'Research strategies for improvement of drought resistance in rainfed rices', in IRRI, 1982, pp. 201–22.

Ou, S.H., 'Genetic defence of rice against disease', in Day, 1977.

Overton, M., 'Estimating crop yields from probate inventories: an example from East Anglia', *J. Econ. Hist.*, *39*, 1979.

Pachico, D.H., 'Nutritional objectives in agricultural research – the case of CIAT', in Pinstrup-Andersen *et al.*, 1984.

Pal, B.P., 'Modern rice research in India', in IRRI, 1972.

Pal, I.K., 'The problem of fragmentation: Orissa, India', in IRRI, 1978a, pp. 141–6.

Palmer, I., *Science and Agricultural Production*, UNRISD *Report No.72. 8*, Geneva, 1972.

Panda, N. and Heinrichs, E.A., 'Tolerance to brown planthopper in rice', Saturday Seminar, July 26 1982 (mimeo), IRRI.

Papanek, G., 'Poverty in India', mimeo, Boston University, Oct. 1986.

Parain, C., 'The evolution of agricultural technique', in Postan, 1966.

Parikh, A. and Mosley, S., 'Fertiliser response in Haryana', *Economic and Political Weekly*, XVIII, 13, Mar. 26, 1983, pp. A24–31.

Parthasarathy, G., *Summary* of discussions on HYVs and farm labour, *Indian J. Agric. Econ.*, July–Sept., 1974.

—— 'Employment, wages, and poverty of hired labour', in Hirashima, S. (ed.), *Hired Labour in Rural Asia*, Inst. of Developing Economies, Tokyo, 1977.

Patnaik, U., 'Capitalist development in agriculture', *Econ. and Political Weekly*, *VI*, *39*, 1971, pp. 123–30.

Peiris, J.W.L., 'Recent advances in the development of HYVs', in ARTI-IDS, 1973, Paper No. 28.

Perrin, R. and Winkelmann, D., 'Impediments to technical progress on small versus large farms', *American J. Agric. Econ.*, *58*, 5, Dec. 1976 (proceedings).

Perroux, F., *L'Economie des Jeunes Nations*, Columbia, New York, 1962.

Piggott, S., 'Early pre-history', in Thirsk, 1981.

Pingali, P., Bigot, Y. and Binswanger, H., 'Agricultural Mechanization and the Evolution of Farming Systems in sub-Saharan Africa', mimeo, ARD, World Bank, Washington, DC, May 1985.

Pinstrup-Andersen, P., 'Decision-making on food and agricultural research policy: the distribution of benefits from new agricultural technology among Colombian income strata', *Agric. Administration*, *4*, 1, Jan. 1977, pp. 13–28.

——, 'Modern agricultural technology and income distribution: the market price effect', *European Rev. of Agric. Econ.*, *6*, 1979, pp. 17–46.

——, *Agricultural Research and Technology in Economic Development*, Longmans, Harlow, 1982.

—— and Hazell, P., 'The impact of the green revolution and prospects for the future', *Food Review International*, *1*, 1984, pp. 1–18.

——, de Londono, N.R. and Hoover, E., 'The impact of increasing food supply on human nutrition', *American J. Agric. Econ.*, *58*, pp. 131–42, May 1976.

——, Berg, A. and Forman, M. (eds.), *International Agricultural Research and Human Nutrition*, IFPRI/ACC-SCN, Washington, DC, 1984.

Podoler, H. and Appelbaum, S., 'Basic nutritional requirements of larva of bruchid beetle', *Stored Products Research*, 7, 1971, pp. 187–93.

Pollak, S., 'Rice genetics, production techniques and social relations: the Javanese experience', M.A. research paper, Sussex University, 1986.

Popkin, B., 'Time allocation of the mother and child and nutrition', *Ecology of Food and Nutrition*, *9*, 1978, pp. 1–14.

Postan, M.M. (ed.), *Economic History of Europe*: Vol 1, *The Agrarian Life of the Middle Ages*, 2nd ed., Cambridge University Press, 1966.

Prahladachar, M., 'Income distribution effects of the green revolution in India: a review of empirical evidence', *World Development, 11*, 11, Nov. 1983, pp. 927–44.

Prescott, J.M., Bekele, G. and Burnett, P.A., 'Pathology and disease surveillance', in CIMMYT, 1985.

Quizon, J. and Binswanger, H., 'Income distribution in agriculture: a unified approach', *American J. Agric. Econ.*, *65*, 13, Aug. 1983, pp. 526–38.

Raghavan, V., 'The applications of embryo rescue in agriculture', in IRRI, 1985a.

Rahman, K., 'Insect pests of bread grain in Punjab and their control', *Ind. J. Agricci.*, *12*, 1984, pp. 564–7.

Rajagopalan, V. and Varadarajan, S., 'Nature of new farm technology and its implications for factor shares – a case study in Tamil Nadu', *Indian J. Agric. Econ.*, *38*, 4, Oct.-Dec. 1983.

Rajaram, S. *et al.*, 'Bread wheat', in CIMMYT, 1984.

Rajaraman, I., 'Poverty, inequality and economic growth: rural Punjab, 1960/61–1970/71', *J. Development Studies, 11*, 4, 1975 pp. 278–90.

Rajpurohit, A. *et al.*, 'Recent trends in agricultural growth rates in Karnataka', *Indian J. Agric. Econ.*, *38*, 4, Oct.–Dec. 1983.

Ramachandran, I., 'Food consumption in rural Indian households: has it improved in recent years?', NFI Bulletin, 8, 1, 1987.

Ramaiah, K., '*Striga* research at ICRISAT's Upper Volta Center', in ICRISAT, 1983.

Randhawa, N.S., *Green Revolution: Case Study of Punjab*, Vikas, Bombay, 1974.

Ransome, J., 'Agronomy', in CIMMYT, 1985.

Rao, C.H. Hanumantha, *Technical Change and Distribution of Gains in Indian Agriculture*, Macmillan, Delhi, 1975.

——, 'Changes in rural poverty in India: implications for agricultural growth', Rajendra Prasad Memorial Lecture, 30th Annual Conference of Indian Society of Agricultural Statistics, Akola, December 1985.

Rao, G.H., Caste and poverty: a case study of a scheduled caste in a delta village, D.Phil. thesis, Andhra Univ., 1975.

Rao, N., 'Transforming traditional sorghums in India', in ICRISAT, 1982, vol. 1.

Ravallion, M., *Markets and Famines*, Clarendon Press, Oxford, 1987.

Rawls, J., *A Theory of Justice*, Cambridge, 1972.

Ray, S.K., 'An expirical investigation of the nature and causes for growth and instability in Indian agriculture: 1950–80', *Indian J. Agric. Econ.*, *XXXVIII*, 4, Oct.–Dec. 1983, pp. 459–75.

Renfrew, C.C., 'Monuments, mobilization and social organisation in neolithic Wessex', in C.C. Renfrew (ed.), *The Explanation of Culture Change*, London, 1963.

Repetto, R., *Economic Inequality and Fertility in Developing Countries*, Johns Hopkins, Baltimore, 1979.

Reutlinger, S., *Poverty and Hunger*, World Bank, Washington, DC, 1986.

Richards, P., *Indigenous Agricultural Revolution*, Hutchinson, London, 1985.

Riches, N., *Agricultural Revolution in Norfolk* (1937), Cass, London, 1967.

Robb, P. (ed.), *Rural South Asia: Linkages, Change and Development*, Curzon, London, 1983.

Robbins, L., 'On the elasticity of demand for income in terms of effort', *Economica, X*, 1930, pp. 123–9.

Robinson, J., 'Keynes and Kalecki', *Collected Economic Papers*, vol. 3, Blackwell, Oxford, 1965.

Roche, F., 'Production systems', in Falcon *et al.*, 1984.

Rochin, R.I., A microeconomic analysis of smallholder response to HYVs of wheat in Pakistan, D. Phil thesis, Michigan State University, as extracted in USAID, 1973.

Roger, Z. *et al.*, 'Screening of pearl millet cultivars for resistance to *Striga hermonthica*', in ICRISAT, 1983.

Rogers, E., *Diffusion of Innovations*, Free Press, Glencoe, New York, 1962.

——, and L. Svenning, *Modernization among Peasants: The Impact of Communication*, Holt, Rinehart and Winston, New York, 1969.

Rosenzweig, M. and T. Schultz, 'Market opportunity, genetic endowments and the intra-family distribution of resources: child survival in rural India', Center Discussion Paper No. 347, Economic Growth Center, Yale University, 1980.

Roumasset, J., *Rice and Risk*, North Holland, 1976.

——, Boussard, J.M. and Singh, I.J. (eds.), *Risk, Uncertainty and Agricultural Development*, Agricultural Development Council, New York, 1979.

Roumassett, J. and Smith, J., 'Population, technical change and landless workers', *Population and Development Review*, 7, 3, 1981, pp. 401–20.

Roy, P., 'Transition in agriculture: empirical studies and results', *J. Peasant Studies, 8*, 2, 1981.

Rudra, A., 'Class relations in Indian agriculture', *Economic and Political Weekly, XII*, 22, pp. 916–23; 23, pp. 963–8; and 24, pp. 998–1003, 1978.

——, 'Extra-economic constraints on labour in agriculture', mimeo, Working Paper, ILO (ARTEP), Bangkok, 1982.

Ruthenberg, H., *Farming Systems in the Tropics*, (2nd edn), Oxford University Press, 1974.

Ruttan, V., 'The green revolution: seven generalizations', *Internat. Development Rev., XIX*, 4, 1977, pp. 16–23.

Ryan, J.G., 'Effects of IARCs on human nutrition', in Pinstrup-Andersen *et al*, 1984.

—— and Asokan, M., 'The effects of the green revolution in wheat on the production of pulses and nutrients in india', *Indian J. Agric. Econ., 32*, pp. 8–15, July-Sept. 1977.

—— and Subrahmanyam, K.V., *An Appraisal of the Package of Practices Approach in Adoption of Modern Varieties*, ICRISAT Economics Dept., Occasional paper No. 11, May 1975.

Saari, E.E., 'South and South-east Asian region', in CIMMYT, 1985.

—— and Wilcoxson, R., 'Plant disease situation of high-yielding dwarf wheats in Asia and Africa', *Annl. Rev. of Phytopathology, 12*, 1974, pp. 49–68.

Saunders, D. and Hobbs, R., 'Agronomy', in CIMMYT, 1984a.

Sawant, A., 'Investigation of the hypothesis of deceleration in Indian agriculture', *Indian J. Agric. Econ., XXXVIII*, 4, Oct.–Dec. 1983.

Sawyer, W., 'Applications for animals', *ATAS Bulletin, 1*, 1984.

Saxena, R. and Jadawa, M.G., 'Study on the trends and yields of rice and wheat in India during the first three five-year plans', *J. Ind. Soc. Agric Stats*, Dec. 1973.

Schaffer, B. and G. Lamb, *Can Equity Be Organized? Equity, Development Analysis and Planning*, Gower, 1981.

Schendel, W. van, *Peasant Mobility: The Odds of Life in Rural Bangladesh*, van Gorcum, Assen, 1981.

Schluter, M.G.G., *Differential Rates of Adoption of the New Seed Varieties in India: The Problem of Small Farms*, Occ. Paper No. 47, USAID/ Dept. of Ag. Econ., Cornell Univ., Ithaca, New York, 1971.

——, *The Interaction of Credit and Uncertainty in Determining Resource Allocation and Incomes on Small Farms, Surat District, India*, Occ. paper No. 68, USAID/Dept. of Agric. Econ., Cornell University, Ithaca, New York, 1974.

Schofield, S., 'Seasonal factors affecting nutrition in different age groups', *J. Development Studies, 11*, 2, Oct. 1974, pp. 24–47.

——, *Development and the Problems of Village Nutrition*, Croom Helm, London, 1979.

Schultz, T.W., *Transforming Traditional Agriculture*, Yale, 1964.

—— (ed.), *Distortions of Agricultural Incentives*, Indiana University, Bloomington, 1978.

Scobie, G.M. and Posada R., 'The impact of technical change on income distribution: the case of rice in Colombia', *American J. Agricultural Economics, 60*, 1, 1978, reprinted in Eicher and Staatz, 1984.

Sen, A.K., *Poverty and Famines*, Clarendon Press, Oxford, 1981.

——, 'Peasants and dualism with and without surplus labour' (1966), as reprinted in his *Resources, Values and Development*, Blackwell, Oxford, 1984.

——, 'Food, economics and entitlements', *Lloyds Bank Review*, April 1986.

Sen, S., *A Richer Harvest*, Tata-Hill McGraw, Bombay, 1974.

Shackle, G.F., *Expectations in Economics*, 2nd ed. Cambridge University Press, 1952.

Shah, C.H., 'Some aspects of agricultural economics: a trend report', in Indian Council for Social Science Research, *A Survey of Research in Economics*: vol. III, *Agriculture (1)*, Allied, Bombay, 1975.

Shahid, A. and Herdt, R., 'Land tenure and rice production in four villages of Bangladesh', Agric. Econ. Dept. mimeo 82–25, IRRI, Los Banos, 1982.

Shand, R.T. (ed.), *Technical Change in Asian Agriculture*, Australian National University Press, Canberra, 1973.

Sharma, U., 'Contribution of HYVs to cereal output, yield and area in Gujarat', *Indian J. Agric. Econ.*, *36*, 1, 1981, pp. 79–81.

Shingi, P.K., Fliegel, F.C. and Kivlin, J.E., 'Agricultural technology and unequal distribution: an Indian case study', *Rural Sociology, 46*, pp. 430–45, Fall 1981.

Siamwalla, A. and Haykin, S., *The World Rice Market: Structure, Conduct and Performance*, Research Report No. 39, IFPRI, Washington, DC, 1983.

Sidhu, S., 'Economics of technical change in wheat production in the Indian Punjab', *American J. Agric. Econ.*, *56*, May 1974, pp. 217–26.

Simmonds, N.W., 'Variability in crop plants, its use and conservation', *Biol. Review, 37*, 1962, pp. 422–65.

——, *Principles of Crop Improvement*, Longmans, Harlow 1979 reprinted with corrections 1981.

——, 'Genotype (G), environment (E) and GE components of crop yields', *Experimental Agriculture, 17*, 1981a, pp. 355–62.

——, 'Plant breeding: the state of the art', in Kosuge, 1983.

——, 'A plant breeder's perspective of durable resistance', *FAO Plant Protection Bulletin, 33*, 1, 1985.

——, *Farming Systems Research: A Review*, Technical Paper No. 43, World Bank, Washington, DC, 1985a.

——, 'The practical use of crop plant collections with special reference to the tropics', Plant Breeding Group, Association of Applied Biologists, Churchill College, April 1987.

Sin, H., 'Engineering crops that resist weed-killers', *Science, 231*, July 21, 1986.

Singh, B., *Agrarian Structure, Technical Change and Poverty*, Agricole, New Delhi, 1985.

Singh, D., V.K., and R., 'Changing patterns of labour absorption on agricultural farms in Eastern U.P.', *Indian J. Agric. Econ.*, *36*, 4, Oct.–Dec. 1981, pp. 39–44.

Singh, I.J. and Day, R., *Economic Development as an Adaptive Process: the Green Revolution in the Indian Punjab*, Cambridge University Press, 1977.

Singh, I.J., Squire, L. and Strauss, J. (eds.), *Agricultural Household Models, Extensions, Applications of Policy*, World Bank, Washington, DC, 1986.

Singh, K., 'The impact of new technology on farm income distribution in the Aligarh district of U.P.', *Indian J. Agric. Econ.*, Apr.–June 1978.

Singh, R.B., and Chonchalow, N. (eds.), *Genetic Resources and the Plant Breeder*, IBPGR, Bangkok, 1982.

Singh, R.I. and Kanwar, R., 'Impact of new agricultural technology on labour earnings in District Kanpur, U.P.', *Indian J. Agric. Econ.*, Conference Number 1, July–Sept. 1974.

Smith, J., Umali, G., Rosegrant, M. and Mandac, A., 'Risk and fertilizer use on rainfed rice in Bicol, Philippines', mimeo, IRRI Agric. Econ. Dept., Saturday Seminar, 15 Oct. 1983.

—— and Gascon, F., 'The effect of the new rice technology on family labour utilization in Laguna', Research Paper Series No. 42, IRRI, 1979.

——, Cordova, V. and Herdt, R., 'Trends in labour absorption and earnings: the case of rice production in the Philippines', Dept. of Agric. Econ. paper No. 81–13, IRRI, Los Banos, 1981.

Smith, M. in FAO, 1981.

Somel, R., 'Nutritional dimensions of agricultural research at ICARDA', in Pinstrup-Andersen *et al.*, 1984.

Sondahl, M.R. *et al.*, 'Applications for agriculture: the potential for the Third World', *ATAS Bulletin*, 1, 1984.

Spate, O.H.K. and Learmonth, T.A., *India and Pakistan*, Methuen, London, 1967.

Sriramulu, M., 'Studies in the varietal resistance of paddy to lesser grain borer', M.Sc. (Ag.), Dept. of Agric. Entomology, Agricultural College of Bapatla, 1973.

Subbarao, K., 'Institutional credit, uncertainty and adoption of HYV technology: a comparison of East U.P. with West U.P.', *Indian J. Agric. Econ.*, 35, 1, 1980, pp. 69–90.

——, 'Interventions to fill nutrition gaps at the household level: a review of India's experience', Conference on Poverty in India, mimeo, Queen Elizabeth House, Oxford, October 1987.

Swaminathan, M.S., 'The next phase', *Seminar*, Nov. 1974.

——, 'Rice', *Scientific American*, 250, 1974a.

——, 'Sustainable Nutrition Security for Africa: Lessons from India', Project Paper No. 5, The Hunger Project, New York, 1986.

Swenson, C.G., *The Effects of Increases in Rice Production on Employment and Income Distribution in Thanjavur District, S. India*, D.Phil. thesis, Michigan State Univ., 1973.

——, 'The distribution of benefits from increased rice production in Thanjavur, S. India', *Indian J. Agric. Econ.*, XXXI, 1, Jan.-March 1976, pp. 1–12.

Taylor, L., *Structuralist Macroeconomics: Applicable Models for the Third World*, Basic Books, New York, 1983.

Taylor, T. *et al.*, *Kenya's National Agricultural Research System*, ISNAR, The Hague, 1981.

Thirsk, J. (ed.), *Agrarian History of England and Wales*: Vol. I.i, *Prehistory*, Cambridge University Press, 1981.

Thompson, W., *Population and Progress in Asia and the Far East*, Univ. of Chicago Press, 1959.

Thorner, A., 'Semi-feudalism or capitalism? Contemporary debate on classes and modes of production in India', *Economic and Political Weekly*, 17, 49, pp. 1961–8; 50, pp. 1993–9; and 51, pp. 2061–6, Dec. 1982.

Torrey, J., 'The development of plant BT', *American Scientist*, 73, July-Aug. 1985.

Tripp, R., 'Farmers and traders: some economic determinants of nutritional status in Northern Ghana', *J. Tropical Pediatrics*, 27, 15, 1981.

——, 'Nutrition in agricultural research at CIMMYT', in Pinstrup-Andersen *et al.*, 1984.

Tuckman, B., 'The green revolution and the distribution of agricultural income in Mexico', *World Development*, 4, 1, Jan. 1976, pp. 17–24.

Turner, M., 'Agricultural productivity in eighteenth-century England: further strains of speculation', *Economic History Review*, 37, 2, May 1984.

UN, Dept. of International Economic and Social Affairs, *Estimates and Projections of Urban and City Populations, 1950–2025: the 1982 Assessment*, UN, New York, 1985.

USAID, *Spring Review of Small Farm Credit* (SR 114), rd. XIV, Washington, DC, Feb. 1973.

U.S. Dept. of Agriculture, *Yearbook of American Agriculture*, Washington, DC, 1941.

Valle-Riestra, J., 'Nutrition in agricultural research at CIMMYT', in Pinstrup-Andersen *et al.*, 1984.

Valverde, V. *et al.*, *The Patulul Project: Production, Storage, Acceptance and Nutritional Impact of Opaque 2 Corn in Guatemala*, INCAP, 1983.

Vergara, B.S. and Dikshit, N.N., 'Research strategies for rice areas with excess water', in IRRI, 1982, pp. 187–99.

Visaria, P., 'Rapporteur's report on rural unemployment (employment effects of the "green revolution")', *Indian J. Agric. Econ.*, 27, 4, Conf. No., pp. 179–89, Oct.-Dec. 1972.

Vyas, V.S., 'Some aspects of structural change in Indian agriculture', Presidential Address, *Indian J. Agric. Econ.*, *34*, 1, 1979.

Wade. R., 'Management of Irrigation Systems: Finding a Cooperative Solution', *World Development*, Apr. 1988 (forthcoming).

Walker, T.S., 'Trends in HYV adoption, production instability, and sorghum growing area in Peninsular India: implications for sorghum improvement strategies', All India Sorghum Improvement Workshop, AICSIP, Dhanwad campus, 1984.

—— and Singh, R., 'Improved seeds of change in dryland agriculture', mimeo, ISAE/ICRISAT/AICRIPDA Seminar, Patancheru, Aug. 22–24, 1983.

——, Singh, R., Asokan, M. and Binswanger, H., 'Fluctuations in income in three villages of India's SAT', Economics Program Progress Report No. 57, ICRISAT, Patancheru, Dec. 1983.

Walras, L. (1902, 4th ed.), *Principles of Pure Economics* (tr. Jaffé), Allen and Unwin, 1954.

Whitcombe, E., 'The new agricultural strategy in Uttar Pradesh, India, 1968–70: technical problems', in Shand, 1973.

White, L., *Medieval Technology and Social Change*, Oxford University Press, 1962.

Whitehead, R., Lawrence, M. and Prentice A., 'Incremental dietary needs to support pregnancy', in T. Taylor and N. Jenkins (eds.), *Proceedings of the XIII International Congress of Nutrition*, Libbey, London.

Wickham, T.H., Barker, R. and Rosegrant, M.V., 'Complementarities among irrigation, fertilizer and modern rice varieties', in IRRI, 1978, pp. 221–40.

Wilkes, G., 'Overview from an Economic Botanist' in ATAS, 1984.

Withers, L. and Williams, S., 'Research on long-term storage and exchange of in-vitro plant germ plasm', in IRRI, 1985a.

Wittfogel, K., *Oriental Despotism*, Yale, New Haven, 1957.

Wolfe, B.L. and Behrmann, J.R., 'Determinants of child mortality, health and nutrition in a developing country', *J. Development Economics*, *11*, pp. 165–93, 1982.

Woodburn, J., 'Egalitarian societies', *Man, 17*, 3, pp. 431–51, Sept. 1982.

World Bank, *World Development Reports 1981, 1983, 1984, 1986, 1987*, Washington, DC, 1981, 1983, 1984, 1986, 1987.

Wright, W.C., 'Cultural practices to maximize yields under rainfed and irrigated conditions', in Anderson, 1973.

Yoxen, E., *The Gene Business: Who Should Control Biotechnology?*, Pan, London, 1983.

Zaag, D.E. van der and Horton, D., 'Potato production and utilization in world perspective with special reference to the tropics and sub-tropics', *Potato Research, 29*, CIP, Peru, 1983.

Author Index

Subject Index